ICTs for Agricultural Extension

ICTs
for Agricultural Extension

Global Experiments, Innovations and Experiences

Edited By

DR. R. SARAVANAN

Associate Professor and In-Charge
School of Social Sciences
College of Post Graduate Studies
Central Agricultural University (CAU)
Barapani- 793 103, Meghalaya, INDIA

NEW INDIA PUBLISHING AGENCY
PITAM PURA, NEW DELHI-110 034

Published by
Sumit Pal Jain *for*
New India Publishing Agency
101, Vikas Surya Plaza, CU Block, L.S.C. Mkt.,
Pitam Pura, New Delhi- 110 088, (India)
Phone: 011-27341717, Fax: 011-27341616
E-mail: newindiapublishingagency@gmail.com
Web: www.bookfactoryindia.com

ISBN : 978-93-80235-24-0

Composed and Designed by NIPA

List of Contributors

Anastasios Michailidis
Lecturer of Agricultural Extension
Department of Agricultural Economics
University Campus, School of Agriculture
Aristotle University of Thessaloniki
Thessaloniki, GREECE

Er. Asmahan Farred Hattar
Agricultural Engineer
Director, Department of Information and Farm Management, National Center for Agricultural Research and Extension (NCARE), Ministry of Agriculture
Postal Code: Baq'a (19381). P.O.Box 639
Amman, JORDAN

Ashraf Saber Alhawamdeh
National IPM/FFS Project Coordinator
Extension Department
National Centre for Agricultural Research and Extension (NCARE)
Ministry of Agriculture, JORDAN

Benjamin K Addom
School of Information Studies (iSchool)
Syracuse University
126 Remington Avenue, Apt. B, Syracuse
NY 13210, USA

B. Gal
Head of the Extension Service Farm Management and Economics Department
Ministry of Agriculture and Rural Development, ISRAEL

Charles Karemangingo
Senior Lecturer
Soil and Agricultural Engineering
Higher Institute of Agriculture and Animal Husbandry, BP 210, Musanze, RWANDA

Christopher Tafara Gadzirayi
Senior Lecturer
Agriculture Education and Extension
Bindura University of Science Education
P.Bag 1020 Bindura, ZIMBABWE

Claudette de Freitas
Head, Information Management
Technical Services Division
Caribbean Agricultural Research and Development Institute (CARDI)
TRINIDAD & TOBAGO

D. Wolfson
Senior Extension Specialist, Floriculture Division, Ministry of Agriculture and Rural Development, ISRAEL

E. Gelb
Center for Agricultural Economic Research
Samuel Neaman Institute for Advanced Studies in Science
and Technology, ISRAEL

Edward Mutandwa
Lecturer
Rural Development and Agri-buisness
Higher Institute of Agriculture and Animal Husbandry, BP 210, Musanze, RWANDA

Edwin Joseph
Associate Professor
Geography & Planning
Grand Valley State University
B-4-200 Macinac Hall, 1 Campus Drive
Allendale MI, USA

Eliada Gudza
Executive Director
Southern Africa Regional Office
World Links Zimbabwe
10 Windermere Lodge, Mazowe Street
Fife Avenue, Harare
Pox CY 2968 Cuseway Harare, ZIMBABWE

Esmat AlKaradsheh
Researcher
Water, Soil and Environment Research Directorate
National Center for Agricultural Research and Extension (NCARE)
Ministry of Agriculture, Amman, JORDAN

Frank Mugisha
Teaching Assistant, Rural Development and Agribusiness
Higher Institute of Agriculture and Animal Husbandry, BP 210, Musanze, RWANDA

Godfrey Sibanda Title
Technical Manager
Research Council of Zimbabwe
Science and Technology Ministry
45 Colin Avenue, Letombo Park, Amby
Harare, ZIMBABWE

Hossein Shabanali Fami
Associate Professor of Agricultural Extension
Department of Agricultural Extension & Education, Faculty of Agricultural Economics & Development, University of Tehran, IRAN

Ike Nwachukwu
Associate Professor
Director, Extension Services
Department of Agricultural Extension and Rural Sociology
Michael Okpara University of Agriculture, Umudike, NIGERIA

Javad Ghasemi
M.Sc. (Agricultural Education) Graduate
Department of Agricultural Extension & Education, Faculty of Agricultural Economics & Development, University of Tehran, IRAN

Justin Mupinda
Country Programs Coordinator
World Links Zimbabwe
P.O. Box MP 965, Mt Pleasant, Harare
ZIMBABWE

Kristin Elizabeth Davis
Knowledge, Capacity and Innovation Division, International Food Policy Research Institute, P.O. Box 5689, Addis Ababa
ETHIOPIA

M.F.M. Rizwan
Programme Assistant (Agriculture)
Audio Visual Centre
Department of Agriculture
Audio Visual Centre, Gannoruwa
Peradeniya
Postal code: 20400, SRI LANKA

Malihe Falaki
Ph.D. Scholar of Agricultural Education
Department of Agricultural Extension & Education, Faculty of Agricultural Economics & Development, University of Tehran, IRAN

Masami Yamada
Technical Executive Officer
Dept. of Agriculture, Forestry and Fisheries
Fukui Prefectural Government, JAPAN

Moses Okwusi
Head of Unit, Marketing information and ICT Services, Department of Agricultural Extension Services, National Root Crops Research Institute, Umudike, NIGERIA

Nathan Taremwa Kanuma
Lecturer
Rural Development and Agribuisness
Higher Institute of Agriculture and Animal Husbandry
BP 210, Musanze, RWANDA

Olufemi Martins Adesope
Lecturer
Agricultural Economics and Extension
University of Port-Harcourt
Port Harcourt, Rivers State, NIGERIA

Pádraig Wims
Lecturer in Rural Development
School of Biology and Environmental Science
Agriculture and Food Science Centre
University College Dublin
Belfield, Dublin 4, IRELAND

Rafaa Ashamallah Ghobrial
Information Research Professor Assistant
Head of Information Services and Systems
Documentation and Information Centre
National Centre for Research, Ministry of Science and Technology
Postal Code: P.O. Box 2404, SUDAN

Rohan Wijekoon
Assistant Director (Audio Visual)
Department of Agriculture
Audio Visual Centre, Gannoruwa
Peradeniya-20400, SRI LANKA

R. Saravanan
Associate Professor &
In-Charge, School of Social Sciences
College of Post Graduate Studies
Central Agricultural University (CAU)
Barapani- 793 103, Meghalaya, INDIA

Rubaiya Ahmad
Policy Associate
Access to Information (A2I) Programme
(A UNDP Funded Programme)
Prime Minister's Office, Bangladesh
Old Sangshad Bhaban, Tejgaon
Dhaka-1215,BANGLADESH

Samia Nadim Akroush
Director of Socio Economic Directorate
Socio-Economic Research Directorate
National Center for Agricultural Research and Extension (NCARE)
Ministry of Agriculture,Amman, JORDAN

Wayne G. Ganpat
Deputy Director
Extension Division
Ministry of Agriculture land and Marine Resources, WEST INDIES

General Outline of the Country Wise Chapter

1. Overview of National agriculture scenario, National extension systems-scope and importance
2. Need of ICTs in agricultural extension/ Rationale for experimenting/ proposing e-agriculture/ cyber extension
3. Brief account of IT scenario in the country during last two decades and IT penetration in rural areas
4. National policy on IT in agricultural extension
5. Case studies / review of information on *"Best (ICT) practices"* of e-agriculture/ cyber extension (or) *"Innovations"* in ICTs for agricultural extension
   - Information village projects/ Village knowledge centre/ Information Kiosk/ Touch screen kiosk/ Mobile telephony etc.
   - Mobile ICT units for agricultural extension
   - Web portals/ data base/ digital networks for agricultural extension services provision
   - Web based discussion forums/ video-conference/ net meeting etc.
   - Offline multimedia CDs, Decision support systems/ expert systems etc.
   - Innovative Radio – community radio/ TV programmes/ initiatives
   - Online farm magazines/ news papers/ farm advisory publications etc.
   - Open Distance Learning (ODL) projects for agricultural technology transfer
   - Any other ICT initiatives and innovations

Common Understanding of the Terms Used in the Chapters

Agricultural extension

Agricultural extension refers to the public, private, NGOs and civil society institutions that provide a broad range of services (advisory, technology transfer, training, promotional, information and knowledge) on wide variety of topics (such as; agriculture-includes horticulture, animal husbandry & veterinary sciences, agro-forestry, fisheries, soil and water conservation, marketing, environmental education, and also rural health) that farmers, other rural people and farm stakeholders require to better manage their agricultural systems and livelihoods[1].

ICTs

Information and Communication Technologies (ICTs) are diverse set of technological tools and resources used to communicate, and to create, disseminate, store, and manage information. These technologies include computers, internet, and network hardware and software, satellite systems, broadcasting technologies (radio and television), and telephony (land lines and cellular). As well as the various services and applications associated with them, such as web portals, email, SMS, video-conferencing, DSS, MIS and expert system, *etc*[2].

e-Agriculture

Refers to the use of ICTs by the farmers, other rural people and farm stakeholders for information exchange, communication and learning processes, which is useful for the better management of their agricultural systems and livelihoods[3].

e-Extension/ Cyber Extension

e-Extension/ Cyber Extension means using the power of online networks, computer communications and digital interactive multimedia to facilitate agricultural extension. Cyber Extension includes effective use of information and communication technology, national and international information networks, internet, expert systems, multimedia learning systems and computer based training systems to improve information access to the farmers, other rural people and farm stakeholders[4].

Mobile Telephony

A mobile phone (also known as a wireless phone, cell phone, or cellular telephone) is a long-range, electronic device used for mobile voice or data communication over a network. In addition to the standard voice function of a mobile phone, telephone, current mobile phones may support many additional services, and accessories, such as SMS for text messaging, email, packet switching for access to the Internet, gaming, Bluetooth, infrared, camera with video recorder and MMS for sending and receiving photos and video, MP3 player, radio and GPS[5].

Innovative Radio

Innovative radio broadcasting refers to the community radio, digital radio broadcasting, and mobile radio broadcasting handset initiatives for the agricultural extension services provision.

Decision Support System (DSS)

A Decision Support System is a computer based system designed to help decision-makers using data, knowledge and communications technology to identify problems and make decisions to solve those problems[6].

Management Information System (MIS)

Management information system is basically concerned with processing data into information and information is communicated for appropriate decision making. Data collection and processing involves the use of ICTs comprising: computers and telecommunications networks (e-mail, voice- mail, internet, telephone, *etc.*)[7].

Expert System

An expert system is an intelligent computer program that uses knowledge and inference procedures for solving specific problems requiring human expertise for their solution. An expert system is a software system that attempts to reproduce the performance of one or more human experts[8].

e-Learning

e-Learning encompasses learning at all levels, both formal and non-formal, that uses an information network—the Internet, an intranet (LAN) or extranet

(WAN) – whether wholly or in part, for course delivery, interaction, evaluation and or facilitation. Others prefer the term online learning. Web-based learning is a subset of e-learning and refers to learning using an Internet[9].

Open Distance Learning

Open distance learning refers to the approaches of learning that focus on freeing learners from constraints of time and place while offering flexible learning opportunities. ODL is delivered using a variety of print and electronic systems either through synchronous communication (learning in which all parties participate at the same time) or through asynchronous communication (learning in which parties participate at different times). The main systems are mediated either by correspondence, audiovisual means (television and radio), multimedia (audio and text files), and the Internet[10].

Common understanding of the terms are adopted (with slight modifications) from the following sources (Accessed on 11th June, 2009);

[1]Agricultural Investment Sourcebook, Module 3 - Investments in Agricultural Extension and Information Services, http://go.worldbank.org/HRRLDN4DD0

[2]http://en.wikibooks.org/wiki/ICT_in_Education/Definition_of_Terms

[3]www.e-agriculture.org

[4]Sharma, V.P., 2003. Cyber extension: Connecting farmers' in India - Some experience, http://www.gisdevelopment.net/proceedings/mapasia/2003/papers/i4d/i4d003.htm

[5]http://en.wikipedia.org/wiki/Mobile_phone

[6]www.bestpricecomputers.co.uk/.../decision-support-systems.htm

[7]http://www.scribd.com/doc/9643550/Management-Information-System

[8]www.it.iitb.ac.in/~palwencha/mtp_first/CBR(presentaion)(ver%201).ppt

[9]http://en.wikibooks.org/wiki/e-learning

[10]portal.unesco.org/education/ev.php?URL_ID

....

Acknowledgement: The above "ICTs for Agricultural Extension" related terms are synthesised from the different sources are gratefully acknowledged.

Contents

Chapter 1

BANGLADESH

Rubaiya Ahmad

1. INTRODUCTION

Bangladesh is a very small country of 1,47,570 square kilometer area having a population of 150 millions. The country has only 8.29 million hectare (Bangladesh Krishi Diary, 2003) cultivable land against the huge population. The land area is gradually decreasing because of population growth, industrialization and other infrastructure development. This results in a declining trend of per capita land availability from 0.13 hectare to 0.06 hectare during the last few decades (1960 to 2000) (Food and Agriculture Organisation, Bangladesh Country Profile).

In order to ensure food security of this vast population and with the emergence of "Green Revolution" in 1960s, priority was given to produce more food in terms of grain through intensification of land usage. New crop variety (HYV), chemical fertilizers, pesticides and irrigation were introduced. As a result, immediate objective of more grain (rice) production had increased and Bangladesh had temporarily achieved self sufficiency in food (rice). Bangladesh made steady progress in crop agriculture in the post-independent period. Cropping intensity increased from 148 to 179 per cent and food grain production almost doubled during the period from 1969/70 to 1992/93 (Hoque, 2007).

However, a lot of HYVs, hybrids, which were introduced, required increased amount of chemicals, and over time they took a toll on the fertility of soil, gradually decreasing acreage production despite usage of high doses of chemical fertilizer and pesticides. Moreover, the natural disasters such as floods, droughts, *etc.*, continued to result in fluctuation in production. Wide fluctuations in production led to large instability in food grain prices having serious implications for household food security and the welfare of the people. The attainment of food security and rural economic development, therefore, could no longer rely on the production of rice alone.

Bangladesh is endowed with favourable climate and soils for the production of a variety of crops throughout the year. The production of minor crops such as pulses, oilseeds, vegetables, fruits, spices, *etc.*, is currently inadequate to provide for the required nutrition in the daily diet of the people. Accelerated production of minor crops is complementary but not competitive with major food crops production as per the requirement for reducing pressure on cereals. This will improve the daily diet of the people, generate rural employment by intensive cropping with improved technologies, increase farm income and thereby increase the purchasing power of the rural masses. It will also improve the balance of payments by promoting import substitution and export oriented crops.

Fruits and vegetables are important sources of minerals and vitamins and they provide a part of the calorie requirement in the daily diet of the people. The nutritional status of the Bangladeshi diet is on a declining trend due to low intake of vegetables, fruits and spices. The increased production and intake of vegetables by the people will help compensate for debilitating nutritional deficiencies. Moreover, the commercial value of these crops, which if adequately harnessed, will contribute significantly to rural income and thus to overall economic prosperity (Overview of Agriculture Sector - http://countrystudies.us/bangladesh/67.htm).

The Poverty Reduction Strategy (PRS) of Bangladesh identifies agriculture and rural development as the topmost priority for rapid poverty reduction. Despite that, the agriculture of Bangladesh is currently passing through a critical time. Over the years, the inability of the human-based extension system to deliver relevant, timely and appropriate information to support the farmers has increasingly led to a decline in the overall performance of the sector. In order to meet challenges of the 21st century and to reach the full potential of the agricultural sector, it is now time for the extension system of Bangladesh to undergo a complete re-structuring and a makeover. The Government of Bangladesh recognizes ICT as an indispensable tool in the fight against poverty. Consequently, it is also understood that ICT can enhance the contribution to agriculture and rural development tremendously. Use of ICTs in extension can lead to a multi-stakeholder, people-centric, cross-sectoral system that will bring together all stakeholders, especially farmers and enable them to access timely and relevant information, exchange opinions, experiences, good practices and resources related to agriculture. With the aim of improving communication and learning processes between various actors, an ICT based extension system can also ensure integration of technology with multimedia, knowledge and culture.

2. AGRICULTURAL TRENDS IN BANGLADESH

2.1. Overall Growth

Over the past decade, the agricultural sector has increased productivity. The sector started modest diversification into value-added products, such as fruits, vegetables, poultry, dairy and fish. This improvement was achieved despite a large population and high population density, scarce cultivable land and regular occurrence of natural calamities.

In recent years, the fisheries and livestock sector have been playing an increasingly important role in the economic uplift of Bangladesh. It is a labour-intensive and quick-yielding sector which augments growth and alleviates poverty. Around 1.3 million people are directly employed in the fishery sector alone (Bangladesh Fisheries Research Institute - http://www.fri.gov.bd).

With an annual growth rate of over 8 per cent since 1993, the contribution of the livestock sub-sector to GDP and the agriculture sector as a whole is currently 3.2 per cent and 10.11 per cent respectively. It has much potential to develop as a commercial sector with employment and income generating opportunities both in the rural and urban areas (Bangladesh Livestock Research Institute- http://www.blri.gov.bd/). A large number of enterprises-cattle, poultry and dairy farms have grown in the private sector in recent years.

2.2. Agricultural Diversification

Income growth and urbanization have shifted consumption toward higher value agricultural products such as livestock and fishery products, fresh fruits and vegetables. Rising incomes in rural and urban areas in Bangladesh have also spurred demands for processed food products, a trend that will continue to increase. Organic agriculture also constitutes a large part of the agri-business industry *via* increased use of locally produced organic fertilizer. Given that there is bound to be an increasing global demand for organic food products, locally produced and processed organic food products could constitute a sizeable portion of our exports. These changes have induced rapid growth in supermarkets and have influenced the structure of agricultural production and demand for agro-processing (Comparative Advantage and Crop Diversification).

2.3. Involvement of Private Sector

Until the early 1990s, state-owned enterprises in sub-sectors such as sugar, jute and fertilizers carried out most of the agribusinesses. Private sector involvement in agribusiness was relatively modest and limited to sub-sectors such as tea, rice and wheat milling, and marketing of agricultural products. In the 1990s, private agribusinesses, as well as trade and industry associations, began to grow. This accelerated in the second part of the decade and continued in the early part of the 2000s. Most growth took place in sub-sectors such as seeds, poultry, shrimp, potato and cold storage, fruit processing, and supermarket chains. Several factors were responsible for the emergence of private agribusiness during the 1990s (National Agriculture Policy). Market reforms in the 1980s and 1990s provided a policy framework that was increasingly supportive of private sector involvement in agribusiness. Sustained growth of the economy has increased purchasing power, particularly in urban areas, with the consequent increase in demand for non-staple foods and more processed items. Also, infrastructural developments and improved opportunities for development of enterprises (MoA, 2004). Improvements in road and telephone (*e.g.*, the mobile revolution and internet) communication, electrification and energy infrastructure (*e.g.*, gas pipelines), while far from being adequate, have stimulated greater trade volumes and investment outside the two major urban centers of Dhaka and Chittagong.

2.4. Emergence of Agri-Business

While agricultural production grew by about 37 per cent during the 1990s, agribusiness grew by about 66 per cent or roughly 5.2 per cent annually over a decade. The growth of agribusiness, though faster than the 3.2 per cent annual expansion in agriculture in that period, was still insufficient to generate employment and income growth required to reduce poverty significantly and accelerate the overall growth of the rural economy. Demand-driven agribusiness activities are critically important to Bangladesh economy. Agribusiness accounts for about 10 per cent of the country's Gross Domestic Product (GDP), according to estimates (ADB-Report, 2004). Because the market demand is income-elastic, the growth of agribusiness is much higher than the growth of agricultural production. With proper support, however, agribusiness growth could accelerate to an estimated 7 per cent a year or more, which would speed progress toward poverty alleviation (Bangladesh Agriculture Facts). In Bangladesh, agribusinesses comprise about 3 million enterprises, of which 90 per cent have less than 10 workers or employees and 36 per cent have only one worker. Nonetheless, they are estimated to account for half of household income for about 15 million people (ADB-Report, 2004).

3. AGRICULTURAL EXTENSION IN BANGLADESH

3.1. Evolution of Extension in Bangladesh

Bangladesh follows a linear model of technology transfer, which assumes a relationship between research, extension and farmer with organised publicly funded science as the source of innovation. This kind of extension models are usually top-down structures, often located within the Ministry of Agriculture. One of the examples is the Training and Visit (T&V) System promoted by the World Bank in 1970s (ADB-Report, 2004). This system had been established as public sector extension services and became a major model for providing and managing extension in many developing countries including Bangladesh. The T&V system had experienced apparent success in at least for a period of time. However, the supply-driven and top-down system, promoting agricultural messages that had been designed and developed by research scientists, with limited input from the technology users (farmers), of the system was finally abandoned in late 1990s for the following reasons (Yuan Zhou, (n.d.)):

i. The system was rigid and top-down and farmer's needs of different resource bases were not taken into consideration,

ii. Farmers' participation in decision making in farm management extension activities was a missing link,

iii. Field demonstration was neglected in favour of carrying messages to farmers. Demonstrations were based on the partial focus of the entire farming systems,

iv. The needs of resource poor farmers were not given attention to cater primary and basic needs and resource endowment,

v. The linkages between research and extension did not improve as was expected, and

vi. The technology transfer to 'contact farmers' did not have the desired impact on other farmers.

Once T&V era was over, the public funding for extension had declined and led to reduction in staff and inadequate operational budgets. Distant and remote farmers started to suffer the most due to poor service by the public sector and weak integration into the market. Also with the diversification of agriculture grew the demands for technological information related to not only rice, but also other grains, vegetables, fisheries, poultry *etc.*, Moreover, the mandate of extension was no longer limited to dissemination of technology.

Addressing new challenges required extension to play an expanded role with a diversity of objectives, which included: linking farmers more effectively and responsively to domestic and international markets; enhancing crop diversification; coupling technology transfer with other services relating to input and output markets; poverty reduction and environmental conservation; viewing agriculture as part of a wider set of rural development process that includes enterprise development and non-farm employment; and capacity development in terms of strengthening innovation process, building linkages between farmers and other agencies, and institutional development to support the bargaining position of farmers. Accordingly, the extension services in Bangladesh that has been seen as a public good and run by the state, passed through a transition. The institutional context and setting of the agricultural extension and advisory services in Bangladesh had undergone several organisational metamorphoses (ASIRP, 2003a).

3.2. Major Activities

Extension services to farmers are largely done by the Department of Agricultural Extension. The DAE's mission is to provide efficient and effective need based extension services to all categories of farmers, to enable them to optimise their use of resources, in order to promote sustainable agricultural and socioeconomic development. Apart from its own direct interventions, there are various bilateral programs/projects which complement DAE in providing extension services to the farmers and other client groups. The farm management and agriculture extension services in 'broad agriculture', as far as the farmers of Bangladesh are concerned, are also provided mostly by the public sector organisations. The major activities of the public sector agricultural and farm management in extension are summarised below (Agriculture Extension Manual, 1999):

1. The Department of Agricultural Extension (DAE)
    - Executing Farmers Information Need Assessment (FINA)
    - Bottom-up extension planning
    - Providing demand-led extension
    - Working with all categories of farmers
    - Use of a range of extension methods including group extension method
    - Use of innovative extension approaches (Integrated extension, Farmer-led extension *etc.*)
    - Partnership with related organisations (GOs and NGOs, *etc.*) for providing integrated extension and for sharing resources.
2. The Department of Livestock (DoL)
    - Animal disease diagnosis and disease control
    - Livestock and poultry breed improvement
    - Artificial insemination
    - Training on beef fattening, dairy farming, calf rearing, broiler farming, layer farming and hatchery management
    - Establishment of nursery for fodder production and distribution of planting materials of fodder crops, *etc.*
3. The Department of Fisheries (DoF)
    - Motivation of fish and shrimp farmers through direct communication (personal contact), group method (result demonstration) and mass communication method (radio and television programmes, booklets and posters, *etc.*).

- Technical inputs for bank credits to farmers
- Training of fish farmers and fishermen
- Fish seed production

Table 1 : Public sector organisations involved in farm management extension and advisory services.

Extension organisation	Ministry	Major responsibilities
The Department of Agricultural Extension (DAE)	Ministry of Agriculture	Crops sector extension services
The Bangladesh Agricultural Development Corporation (BADC)	Ministry of Agriculture	Crops sector extension services
The Department of Livestock (DoL)	Ministry of Livestock and Fisheries	Livestock and poultry extension services
The Department of Fisheries (DoF)	Ministry of Livestock and Fisheries	Fisheries extension services
The Forest Department (FD)	Ministry of Environment and Forst	Forestry/agro-forestry extension services
The Bangladesh Water Development Board (BWDB)	Ministry of Water Resources and Irrigation	Irrigation and drainage *vis a vis* agricultural extension
Local Government Engineering Department (LGED)	Ministry of Local Government, Rural Development and Cooperatives	Small-scale irrigation and rural infrastructure development *vis a vis* agricultural extension
Bangladesh Rural Development Board (BRDB)	Ministry of Local Government, Rural Development and Cooperatives	Farmers' cooperatives and 'One Household, One Farm' programme of extension services
Bangladesh Academy of Rural Development (BARD)	Ministry of Local Government, Rural Development and Cooperatives	Rural development *vis a vis* agricultural extension
Rural Development Academy (RDA)	Ministry of Local Government, Rural Development and Cooperatives	Rural development *vis a vis* agricultural extension
Bangladesh Agricultural University (BAU)	Ministry of Education	Farming systems *vis a vis* agricultural extension

Source: Ministry of Agriculture (Bangladesh) website-wwww.bangladeshgov.org/moa

Additionally, the NGO Sector is also increasingly making its mark in agricultural extension in Bangladesh. Initially NGOs were associated with the agricultural extension and small farm management with especially small scale vegetable crops, poultry, fisheries and agro-forestry /social forestry. Subsequently, NGOs are moving towards extension and farm advisory activities on some of the major crops like hybrid rice, maize, *etc.*, More than one hundred local (coverage wise), national and international NGOs are involved in agricultural extension and advisory activities of some sorts. The following are worth mentioning:

BRAC, PROSHIKA, CARE-Bangladesh, Mennonite Central Committee (MCC), Helen Keller International, Christian Commission for Development Bangladesh (CCDB), Rangpur-Dinajpur Rural Services (RDRS), Friends in Village Development Bangladesh (FIVDB), Gono Unnayan Prochestra, Save the Children Funds-USA (SCF-USA), Gono-Kalyan Trust (GKT), Voluntary Paribar Kalyan Association (VPKA), Karmajibi Kalyan Society (KKS), World View International Foundation (WIF), Development Service Centre of Bangladesh Mission, UPAKAR, Jogorani Chakra (JC), Thangamara Subuj Mohila Sangha(TMSS) *etc.*, (CARE: Rural Livelihoods Program).

NGOs form groups of target farmers, mostly of landless and marginal categories and in many cases they focus on women. Services provided by NGOs usually include credits, inputs supply and training. In some areas, NGOs target categories of farmers those who are practically fell out of the public sector extension services. The public sector extension and advisory agencies have relatively narrow focus (*i.e.* DAE focuses on crops, DoL focuses on livestock, DoF on fisheries *etc.*,); NGOs, on the other hand, have become attuned to providing multiple services to their beneficiaries. Most of the NGOs are, however, handicapped by the limited land area of their beneficiaries (landless and marginal farmers).

The private sector agencies related to agricultural extension and advisory services include the burgeoning seed companies, fertiliser dealers and pesticide dealers. Also included in the private sector are small plant nurseries, often owned by individual farmers, which have sprung up throughout the country during the decades of '80s and '90s. Many of these nurseries have, however, been established with NGO assistance.

These nurseries have been the major sources of planting materials of fruit and timber trees, vegetable crops, flower and other beautifying plants, *etc.*, These planting materials were hardly seen, if at all, in the market twenty years ago. The nursery owners are now making these available virtually in every corner of the country. The role of the public sector in the agricultural extension and advisory services of these planting materials, if any, has been minimal.

The public sector extension in Bangladesh, working on the tenet of progressive farmer, has indeed not been known for its effectiveness (Hossain, 2004). NGOs in Bangladesh did not have, until recently, a ready access to information and the technologies developed in the public sector research system. However, in recent years, the roles of NGOs in the national development arena have been widely recognised and the public sector is moving towards NGO collaboration, in most cases with stimulus from donors.

3.3. Challenges

Up to the time of T&V system, various direct supports from government organisations went largely to the large prosperous farmers, who had different possibilities for developing their farms. They had better access to resources and were usually better informed. NGOs, on the other hand, focused attention to poverty alleviation programmes where landless and marginal farmers are the targets. Small farmers, who constitute the bulk of the country's farming community, have largely escaped the attention of major extension efforts (Hossain, 2005).

Addressing the needs of small-scale farmers through information exchange between agricultural research and extension is still a challenge that the government faces the most. There is also a lack of felt need on the part of virtually all categories of farmers for information on new technologies and practices.

Rapid advances in agricultural technology have been noticed in the recent days. Bangladesh Agricultural Research Institute (BARI), Bangladesh Rice Research Institute (BRRI), Bangladesh Fisheries Research Institute (BFRI) and Bangladesh Livestock Research Institute (BLRI) have developed hundreds of technologies in the field of crops, horticulture, aquaculture, livestock and forestry. The challenge, however, has been the dissemination of these valuable technologies to farmers including those in less favourable and more remote areas. Although New Agricultural Extension Policy (NAEP) mentions use of mass media *i.e.* radio, television, newspaper, *etc.*, for rapid dissemination of agricultural technologies, the current situation calls for a much more comprehensive information and communication strategy. Various forces including the Department of Agricultural Extension (DAE) are at work to change agricultural extension from a process of technology transfer (research institution to farmer) to a process of facilitating a wide range of communication, information and advocacy services (demand-driven, pluralistic and decentralized extension). However, the current extension system still struggles to meet the timely and relevant information demand of its key stakeholders. Following are the main reasons for that.

3.3.1. Agriculture has Diversified, but Extension hasn't

The attainment of food security has always been the focus of the country's extension system. To feed the ever increasing population of Bangladesh, the Government launched "Green Revolution" and placed emphasis on the production of cereal crops, mainly rice. With the introduction of High Yielding Varieties (HYVs) of rice and wheat both from home and abroad the extension workers were specially trained to assist farmers in adopting the new technologies. As a result, production of major cereal crops *i.e.*, rice and wheat, increased tremendously. But the production of minor cereal crops like pulses, oilseeds, vegetables, fruits and spices declined gradually year after year and finally failed to keep pace with the rapid population growth. This involved the use of a huge amount of valuable foreign exchange in importing those minor cereal crops to improve the nutritional and nutrient status of the people and the soils.

Minor cereal crops are relatively cheap sources of protein and calorie and they also have a positive impact on the nutrient balance of the soil. The farmers can easily cultivate these crops after harvesting the major ones. With this realization and to reduce the huge drain of foreign exchange in importing those crops, the Government of Bangladesh launched a Crop Diversification Programme in the 1990's. After creation of mass awareness through demonstrations and other mass media, the farmers started adopting this programme. With land degradation and population overflow, today, crop diversification is considered a necessity for sustainable agricultural development and attainment of food security. However, the extension system is in many ways has not evolved to meet the demands of present time. More often than not, farmers rely not on the extension workers but on their own experience and self- gained knowledge to increase the production. As a result, the commercial crops are not receiving the attention they deserve and are hence failing to fetch the benefits for farmers.

3.3.2. SAAOs are Overburdened

The major function of an Sub Assistant Agriculture Officer (SAAO) is to disseminate appropriate information to farmers. Their role becomes crucial in the sense that farmers in Bangladesh are largely illiterate, unable to read printed material and information dissemination is done through personal contact with extension workers. Unfortunately, the ratio of an extension worker to farmers is very low (1: 1700+). Thus, the extension worker cannot be effective in disseminating agricultural information to farmers in Bangladesh through personal contacts. The farmer is therefore denied information when it is needed. Even when extension agents visit farmers they usually focus their activities on the male farmers, hardly reaching out to the women who constitute a substantial proportion of farmers in Bangladesh.

Distance from home and travel time to place of posting also have significant negative relationships with job performance. The SAAOs are also often burdened with other activities that are outside the mandate of agricultural extension. For example, extension agents have frequently been involved in organizing the supply of subsidized inputs and implementing credit schemes. They may also be asked to get involved in implementing public health programs and other government schemes unrelated to agriculture.

3.3.3. Extension is often not Demand Based

In Bangladesh, the Linear Model of Technology has been followed predominantly which works on the following principles:

- Researchers consider themselves as the sole source of technology.
- The model does not give emphasis on the synergy among different agencies involved in and there is clear division of labour: research generates technology, extension delivers it, and farmers use it.
- The technology transfer agencies and farmers are excluded from the technology generation process.
- The linkage between researchers and extensionists or between extensionists and farmers amounts practically to brief communication.

The concept of demand-driven extension emphasizes the need to provide services that meet the needs and priorities of farmers. The term *farmer-driven* or *farmer-led extension* might be used to address the goal of making these services meet the needs and priorities of farmers. The current linear model of extension system, however, is failing to do just that due to various reasons. First, in the absence of a market mechanism, public sector extension providers have trouble in determining the types of knowledge and advice farmers actually need. Making this determination is at the heart of making public sector extension demand-driven.

3.3.4. No Incentives for Extension Workers

Incentive problems also lead to state failures. Public sector agencies can use various instruments to create incentives, such as merit-based promotion.

However, when public budgets do not keep pace with inflation and general improvements in standard of living, as is often the case, the possibilities to use a differentiated system of promotion associated with salary incentives deteriorate. In addition, extension agents often have a lower social status than many other public sector employees and a lower rank in the 15 civil service system, which affect their morale. Another factor affecting agents' morale is that they do not have the operational funds to get to the field and work effectively. The incentive problems inherent in public sector extension agencies are closely linked with capacity problems. Public sector agencies often lack the incentives to invest in the capacity of their extension staff so as to keep their knowledge up-to-date.

3.3.5. Lack of Coordination between Public and Private Sector

The lack of coordination between the public and private extension service providers leads to extension failures. The public sector should assume the responsibility of providing authentic and relevant information and services to the farmers while private sector players and NGOs can best articulate the delivery mechanism and ensure timely and effective access to information and services at the community level. Lack of coordination among players leads to either low coverage of farmers or outdated and irrelevant delivery of information. NGOs in Bangladesh did not have, until recently, a ready access to information and technologies developed in the public sector research system. However, in recent years, the roles of NGOs in the national development arena have been widely recognised and the public sector is moving towards NGO collaboration, in most cases, with stimulus from donors.

3.3.6. Lack of Monitoring and Supervision

Information problem that extension managers often face is the inability to determine what extension agents actually do in the field and to supervise them. Considerable information asymmetries exist between extension agents and their managers because of the spatially dispersed nature of agriculture and because agricultural outcome, such as crop yields, is influenced by many factors, other than extension.

4. ICT IN ADDRESSING EXTENSION CHALLENGES

Although Bangladesh is a relatively late entrant in e-Agriculture, the foundation for leveraging ICTs for agricultural capacity building and marketing has existed for a long time. It is only recently that the Ministry of Agriculture of Government of Bangladesh (GoB) through the Department of Agricultural Extension and the Department of Agricultural Marketing has started harnessing ICTs more effectively to deliver information and services to the farmers. For the last twenty years, TV and radio programs have inspired agricultural innovations across the country. Government of Bangladesh and non-government players are now introducing interactive programs such as phone-in programs and talk shows to farmers and extension workers for interacting with agricultural policy makers and experts. The department of Agricultural Marketing has been disseminating web-based commodity price information to regional offices, which are then displayed on a notice-board of a local market. Agriculture focused telecentres are being planned around already existing farmers' clubs for more efficient, timely and cost effective dissemination of critical information on inputs, weather and marketing channels. These initiatives have not only created possibilities for public private-partnership in extension and marketing, but also have set the stage for exciting farmer-oriented policy reform. Agricultural extension, which depends to a large extent on information exchange between and among farmers on the one hand, and a broad range of other actors on the other hand, has been identified as one area in which ICTs can have a particularly significant impact. Extension agents (whether public servants, private input providers or NGO staff) as intermediaries between farmers and other actors in the agricultural knowledge and information system, are especially well-placed to make use of ICTs to access expert knowledge or other types of information. Some of the major benefits of using ICTs in extension are as mentioned below.

4.1. Greater Coverage of Extension

Till now, Extension, a department dedicated to expanding and enhancing the agricultural sector in Bangladesh was involved in guiding the farmers, and acting as a bridge between them and research institutions. This was done through seminars, personal visits and meeting between the farmers and extension workers, and sometimes through distribution of brochures. Various factors discussed in the "challenges" section leave a large group of farmers, especially women, out of the extension umbrella. ICTs have the power to reach thousands of farmers at a time through leveraging mass media, print media and other tools. Popular agriculture related TV programme "Mati-o-Manush" of terrestrial TV channel "Bangladesh Television" has been a trusted friend of rural farmers for the last 40 years. Other radio programmes and recently, some private channel run TV programmes have also been popularized by both rural and urban farmers.

In the context of decentralization of governmental structures and privatization of agricultural extension services, along with changing paradigms in agricultural extension itself (from top-down to bottom-up and more diversity in service providers), the number of actors has greatly increased to include the following:

- Farmers and their associations
- NGOs/CSOs
- The media
- Private companies (input suppliers; brokers/traders; telecentres)
- The formal (government) extension system
- National Agricultural Research System (NARS)
- Agricultural Universities
- Donors
- Think tanks and networks at various levels
- The political environment (local, national, regional, international governance structures, conventions, treaties, *etc.*,)

This multitude of actors requires effective and efficient information flows and communication between all levels. Linear information flows that dominated the traditional model of technology transfer from the formal research system to the farmers by way of the formal, government-owned extension system, are being replaced by pluralistic information flows between farmers as the demanders of services, and various providers of these services. These information flows can be enhanced through the use of ICTs but probably to a different extent at different levels in the system, as different actors have different kinds of information needs and communication problems, and different access to ICTs.

4.2. Capacity Building of Extension Workers

Extension workers, equipped with ICTs, can also add tremendous strength to already existing wide range of service. Some of those strengths include:

- A new range of additional media that can be part of the communication for developmental "mix" of traditional and/or appropriate media.
- Bottom up articulation and sharing of information on needs and local knowledge;
- Increased efficiency in use of development resources because information is more widely accessible;
- Less duplication of activities

- Rapid speed of communication - locally, nationally and globally
- Reduced communication costs in comparison to other available communication choices;
- Global access to information and human resources.

Department of Agricultural Extension (DAE) prepares Master Training Plan every year which has been formulated based on the Training Needs Analysis carried out by the DAE Training Wing. DAE's training strategy is to relate training to the new DAE policies and strategies, incorporate all training carried out by projects operating within DAE or that have a component that involves DAE staff, institutionalize training through the use of existing DAE facilities and other organizations in Bangladesh, strengthen the role of the DAE Training Wing in the management and implementation of DAE training programs, improve the skills and knowledge of all DAE staff, adopt a qualitative approach as opposed to a quantitative approach, and utilize training funds from all projects effectively. DAE training plans cover all DAE officers and staff including farmers and other client groups (ASRIP, 2003b).

DAE conducts a fortnightly/monthly training session for thana and district staff (Bangladesh is divided into six administrative divisions, each named after their respective divisional headquarters: Barisal, Chittagong, Dhaka, Khulna, Rajshahi and Sylhet. Divisions are subdivided into districts (*zila*). There are 64 districts in Bangladesh, each further subdivided into *upazila* (subdistricts) or *thana* ("police stations"). The upazilas are the lowest level of administrative government in Bangladesh. (Bangladesh, at present, has 482 upazilas and 599 administrative thanas). The course deals with issues such as fertilizer recommendations, soil plant nutrients, working with farmers and NGOs, preparedness for disaster/natural calamities, IPM in rice, vegetables and fruits, rodent control, plant disease management, safety in the use of chemicals, water management, gender awareness and technical upgrading for female block supervisors.

Efforts should be targeted at improving the quality of these trainings with the help of ICT, increasing ICT literacy level of extension personnel as well as ensuring their access to PCs. In the short-run, training and retraining of extension personnel for enhanced job performance should be vigorously pursued through distance learning or other methods using ICTs. However, in the long-run, a succession plan for extension personnel should be implemented in order to replace older generation of extension workers with younger and dynamic, computer literate, ICT-compliant ones who would be well empowered through training and re-training *via* refresher courses as well as being provided with the wherewithal to access up-to-date facilities in ICT.

4.3. Capacity Building of Farmers

Education in agriculture plays an important role in preparing farmers, researchers, educators, extension staff and all other agricultural and agri-business stakeholder to make productive contributions. New and emerging technologies such as those involving integrated crop management and improved nutrient balance are increasingly becoming complex and location specific. Precision agricultural technologies including GIS and other spatial software also require significant information and skill for successful adoption.

Information and Communication Technologies (ICTs) are becoming an increasingly important component of rural education. One of the largest and most effective uses of new technologies is distance education. Distance education is the delivery of learning or training to those who are separated mostly by time and space from those who are teaching or training. The teaching is done with a variety of "mediating processes" used to transmit content, to provide tuition and to conduct assessment or measure outcomes. In a country like Bangladesh, where affordability and geographic locations are barriers to access to education, distance education can play a vital role. Distance learning may largely be web-centric, but it may also use multimedia, which besides delivering content, would also enable a high level of interaction among farmers, service providers, consumers and administration both synchronously and asynchronously.

Literacy lessons for farmers can be combined with computer courses. When computer communication is an option, farmers can use one of the numerous distance education programs. The e-mail may provide them a lifeline to remote fellow students and teachers. The valuable knowledge, including indigenous and traditional knowledge of farmers can be shared through ICTs. For instance, farmers from one village can tell their stories and knowledge that they want to share on video. The video can be replicated for (electronic) library use, at community centers, in research institutes *etc.*,

If the right conditions are met, ICTs have great potentials also to support rural women to deploy their resources for development. Women's organizations, from agricultural community groups *via* health organizations to micro-financing cooperatives, may use ICTs to organize and exchange outside the local territory.

On the radio they may hear about the activities of remote organizations, working in the same field, and how they solve their problems. Women may run or work at small ICT enterprises such as telecenters or CD-ROM content design. Generally good in small enterprising, rural women can earn money by offering their information and communication services to fellow community members. Teleworking can be a source of income for women that need the flexibility to combine the family and household duties with money-generating activities. But it is also an opportunity for handicapped, immobile women and girls.

4.4. Coordination among Different Players

Extension organizations have a key role in brokering between communication technologies, providers of those technologies and services, and the client groups they serve. Local organizations and groups require a voice to highlight their judgment on ICT priorities and implementation of policies, programmes and projects so as to negotiate on equal footing with outside institutions in choices of communication for development approaches and media choices. Extension can play a role in helping catalyze and support this voice; In this role, they must be able to examine the appropriateness of various agriculture related ICTs, the accessibility of ICTs in rural and remote areas, how best to reconcile costs and benefits, and how to insure that ICT access includes a diversity of cultures, languages, social strata and age groups, and is gender sensitive. External institutions seeking to enable local organizations and groups to participate in ICT initiatives need to establish a rapport with the local groups to enable a trustful, learning relationship to emerge.

4.5. Promotion of Agribusiness

ICT can serve to bridge the digital divide and provide relevant business and market information to rural areas to reduce their isolation and foster new income-generating activities in the agribusiness and other non-farm activities by improving communication linkages between growers, processors and retailers for a better transfer of knowledge and technology, developing forecasting of market prices and disseminating prices to producers *etc.*,

It's hard to imagine the 21st century agribusiness environment without proper information and communication technologies. Worldwide, e-mail, sms, chats, video conferencing and internet are replacing old age technologies including faxes, telex machines *etc.* Old ICTs can no longer compete with high efficiency, quality and easiness of the new ones. Although ICT plays a vital role in knowledge sharing, it is also imperative that all agricultural and rural actors have the necessary tools to collect, provide, organize and disseminate accurate and well-adapted information to satisfy their own needs. In addition to having a well-developed knowledge management strategy, human resources must be developed, relevant indigenous content must be collected and be accessible and information services must be strengthened in order to guarantee transfer of knowledge and major use of ICT in facilitating and enhancing agribusinesses.

Individuals or groups in geographically based ICT enabled communities may find themselves connected to one or more different forms of online communities. Online participation can lead to greater face-to-face participation with the geographic community. Relationships therefore form in both directions, with one potentially increasing the livelihood of the other. Such interactions have implications for the evolution of nature of service delivery by the e-Governance and the growth of new agribusiness products to consumers and communities.

In no other area is the necessity of ICT so apparent than the area of agricultural marketing. Awareness of up-to-date market information on prices for commodities, inputs and consumer trends can improve farmers' livelihoods substantially and have a dramatic impact on their negotiating position.

Such information is instrumental in making decisions about future crops and commodities and about the best time and place to sell and buy goods. In many countries, initiatives have appeared that seek to address this issue. Simple websites to match offer and demand of agricultural produce are a start of more complex agricultural trade systems. These sites tend to evolve from local selling/buying websites and price-information systems, to systems offering marketing and trading functions. The Department of Agricultural Marketing website could potentially grow into such a system.

4.6. Farmers' Participation in Research and Policy Decisions

The Government should respond to the demands and unique needs of the regions and districts, considering their agro-ecological systems, state of economic development, poverty incidence, local synergies and economies of scale. Use of ICTs can enable inclusion of community based organizations, local NGOs, Private sector workers which also provide information or extension service to the rural community in the national policy dialogues that can help create and or shape positive policy change towards effective extension service. If they are not part of national policy dialogue, extension policies, programmes and regulatory reform initiatives risk neglecting the needs of the very people the initiatives are meant to serve. Call-in radio and TV talk shows, audio or video conferencing, SMS based polling systems may be tools to ensure communication of all stakeholders with the policy makers.

5. e-KRISHI VISION 2025

As part of its main activities, the "Access to Information" Programme of the Prime Minister's Office had developed separate vision documents for the five key sectors (Agriculture, Health, Education, Local Government, Establishment) of Bangladesh.

The e-Krishi vision 2025 is one of the 5 documents, which is an output of a comprehensive effort of the Ministry of Agriculture and the "Access to Information" Programme vision team.

The primary focus of this document is to explore the possibilities of leveraging ICT for the enhancement and development of the Agriculture sector of Bangladesh. In order to identify the key areas where ICT can play a vital role, this document gives an overview of the sector, paints a vision for 2025 and explores how ICT can help to implement the vision.

The entire vision setting exercise was built on a participatory process through consultations, interviews, workshops, FGDs with policy makers, opinion leaders, sector experts, civil society leaders and community opinion leaders.

The aim of e-Krishi Vision 2025 is to explore the use of ICTs in:

i. Fostering rural prosperity and grassroots level development by empowering farmers with information, knowledge and training.

ii. Developing a decentralized, localized and non-human based extension program with proper management and efficient delivery.

iii. Promoting, supporting and enhancing rural farm and non-farm enterprises.

iv. Mainstreaming ICTs in the areas of agricultural research, development, education, extension and training.

v. Establishing mechanisms to receive farmers' feedback for effective research, development, extension and training.

6. NOTABLE EXTENSION PROJECTS BY PUBLIC AGENCIES

6.1. Major Extension Reform Projects

In order to address the gaps, efforts were previously made through various extension projects. For example, the following projects approached holistic farm management in agricultural extension advice:

i. Agricultural Support Services Project (ASSP, 1992-1999),

ii. Agricultural Services Innovation and Reform Project (ASIRP, 1999-2002),
iii. Agricultural Diversification and Intensification Project (ADIP, 1997-2004),
iv. Thana Cereal Technology Transfer and Identification project (TCTTI, 1995-2000),
v. Integrated Soil Fertility and Fertilizer Management Project (SFFP, 1993-2000) *etc.*

6.1.1. Agricultural Support Services Project (ASSP, 1992-1999)

In Bangladesh, the Agricultural Support Services Project (ASSP) succeeded the T&V System in 1992. The project emphasised a number of issues hitherto not given adequate attention. (see Agricultural Extension Manual, 1995, Ministry of Agriculture, Government of Bangladesh). These included the following:

- The basic unit of agricultural production is the rural household and its farm where various activities are performed. The extension staff, therefore, needs to take the 'whole farm system' into account when transferring technology.
- Homestead, storage and processing are important parts of the production system. As women are responsible for much of these activities, both men and women should have access to extension services.
- Resources and circumstances of farmers vary. Larger farmers have different possibilities for developing their farms than the small and the resource-poor farmers. Therefore, the extension staff should be aware of the variation and provide appropriate information and services to all categories of farmers based on their needs and circumstances.
- Farmers adopt technology and farming practices that benefit them. Often the extension staff encourages farmers to adopt changes that may meet the strategic agricultural needs of the country but not essentially meet the immediate needs of the farmers.

Based on these considerations, the ASSP developed five principles of extension:

i. Responsiveness to farmers' needs – this requires a comprehensive Farmers' Information Needs Assessment (FINA) by Block Supervisors and other field staff to develop extension plans in Upazilas and districts.
ii. Working with groups – this would increase contacts with greater number of farmers (than was the case of working with individual farmers).

iii. Decentralisation of extension planning - Upazila and Districts develop their own extension plans while the DAE headquarters will concentrate on monitoring and supervisory functions.

iv. Targeting all categories of farmers - extension plans are carefully targeted to all categories of farmers (big, medium, small, marginal and landless, and also women farmers).

v. Using a range of extension methods - methods would include not only training and visits to farmers and demonstration, but also group meetings, motivational tours, folk media like songs and drama, *etc.*, (Agricultural Extension Manual, 1995, Ministry of Agriculture, Government of Bangladesh).

The ASSP sought to utilise a 'process' rather than a 'product' approach (Report of the ASSP Evaluation Mission, 1998). Its achievements are summarised as follows:

- Introduction of a New Agricultural Extension Policy (NAEP) in 1997 based on the principles outlined above.
- Establishment of linkages among related agencies involved in agricultural and rural development (Departments of Crops, Livestock, Fisheries and Forestry, related parastatals, NGOs and the private sector).
- Decentralised, bottom-up and demand-led extension planning.
- Partnership of Government Departments and NGOs.
- Establishment of a regular monitoring and evaluation procedure.

In the Bangladesh context, these were significant achievements that opened up avenues for participatory, situation-based technology development and transfer that can mobilise local creativity, energy and experience, and capitalise on diversity.

6.1.2. Agricultural Services Innovation and Reform Project (ASIRP, 1999 – 2002)

This is a follow up project based on the achievement made and the organisational changes initiated during the ASSP (1992 - 1999). The project aims to improve agricultural extension services which in turn will contribute to rising agricultural productivity and so declining rural poverty. The goal of the project was: To improve the capacity of all categories of farmers, especially landless, marginal and small farmers, to optimise their use of resources on a sustainable basis. The purpose level of Objectively Verifiable Indicators (OVIs) of the project include, among other things:

i. Accelerated implementation of the New Agricultural Extension Policy including a significant increase in the number and scope of partnership activities. This aims at cross-sectoral partnership (*e.g.* crops, livestock, fisheries and forestry, *etc.*) in order to address farmers' needs. In other words, the project aims to address issues related to holistic farm management.

ii. Innovative approaches to agricultural extension piloted and evaluated. This aims at addressing holistic needs of farmers through integration of various extension services (crops, livestock, fisheries, forestry, *etc.*,). This also addresses farm management issues.

6.1.3. ADIP : Agricultural Diversification and Intensification Project (ADIP, 1997-2004)

This project aims to:

(i) Enable small farmers to intensify and diversify crop production as well pursue other rural enterprises.

(ii) Enable landless and marginal families to increase their incomes and improve their nutrition through poultry and fish enterprises, agro-processing and other income generating activities and homestead gardening.

The project seeks to achieve above objectives through creating favourable environment of technology transfer, market and land development, using group formation and community development.

6.1.4. TCTTI : Thana Cereal Technology Transfer and Identification Project (TCTTI, 1995-2000)

This was aimed at sustainable intensification of production and productivity of rice under irrigated conditions. In order to achieve its objectives, the project identifies appropriate varieties and location specific package of technologies that permit crop diversification and development of sustainable intensive rice based farming system.

The project adopted "farmer's first approach" for technology transfer and established linkages between researchers, extension officers and farmers for technology identification to suit farmers' needs under specific ecosystem niches. By compact block demonstration using farming system approach and encouraging need-based crop diversification the project addressed issues of farm management. In addition "Community Nurseries", "Integrated Rice-cum- Fish Culture" and promotion of grain legumes and or green manuring in the cropping system helped ensure sustainable intensification of rice crop production.

6.1.5. Integrated Soil Fertility and Fertilizer Management Project (SFFP, 1993-2000)

The objectives of the SFFP project are to improve soil fertility by assuring that farmers adopt improved farm management practices in relation to soil fertility and fertiliser management. The extension messages focus not only on a balanced use of inorganic fertilisers, but also on integrated soil improvement approach including the use of organic fertilisers such as green manure, compost, crop residue, farmyard manure, biological nitrogen fixation, *etc.*, It also aims to develop new practices in soil fertility management that are subject to test by research and to establish better linkages between extension and on-farm research.

The project uses field demonstrations of cropping patterns in relation to balanced fertiliser application, green manure, farmyard manure, biological nitrogen fixation with grain legumes and use of inoculum, block demonstration, training, posters, pamphlets, manuals, *etc.*

6.2. e-Extension Initiatives

In Bangladesh, the foundation for leveraging ICTs for agricultural capacity building and marketing had existed for a long time. Long before computers arrived and the cell phones made their way to farmers, other types of easily accessible ICTs such as newspaper, community public address system, radio and television were leveraged to deliver vital agricultural information. One of the longest running, state-run television shows on Agriculture "Mati o Manush" (People and Soil) is still being telecast 5 days a week and has had around 760 episodes aired thus far over the course of 18 years. This TV show changes the traditional outlook on Agriculture by focusing on Crop Diversification non-farm sectors for income generation and is still the most watched show by the rural farmers. Subsequently "Shyamol Chaya (Green Shadow)", "Hridoye Mati O Manush (People and Soil at Heart)", "Nodi O Jibon (Riven and Life)" are a few of the privately run Agricultural TV Shows, which demonstrate crop, poultry and fishery management technologies on a hands-on basis, share farming success stories and offer a platform for farmers to speak to agricultural experts and policy makers. "Desh Amar Mati Amar", a state-run agriculture radio program is also broadcast everyday for 25 minutes. This program answers hundreds of letters from farmers on issues related to pest control livestock disease *etc.* However, the postal delay and inability to articulate the question in a letter cause inefficient and ineffective response.

With the coordination of Agricultural Information Service Regional Offices of the Ministry of Agriculture (MoA), other local farm programs are also being broadcast regularly from the respective regional radio station.

Having witnessed the popularity of these programmes, Government of Bangladesh and non-government players are now introducing interactivity in these programs such as phone-in option, talk shows for farmers and extension workers to interact with agriculture policy makers and experts. In the recent years, the Ministry of Agriculture of Government of Bangladesh, through its various agencies has started harnessing ICTs more effectively to strengthen extension system, perform need-based research and deliver information and services to the farmers. The Department of Agricultural Marketing (DAM) of Ministry of Agriculture website, currently in its pilot phase, provides wholesale and retail prices of around 200 commodities from 30 of the 64 districts on the web. Periodic reports are also available from the website but cannot be modified by user. Regional DAM offices use this data and hand-write it on a notice board of the local market. In the second phase, the proposed commercial system will collect up-to-date market information (*via* cell phone or computer) fed into a database managed on a SMS Server, which would then be accessible to clients requesting price information for agricultural products *via* text messaging. The text messages would both request and receive price information in Bangla language.

e-Krishi communication strategy must ensure that communication mechanisms are two-way. In the present scenario, the proposed ICT solutions are often driven not by empirical evidence from studies of how people use the technologies, but by ICT developers and providers who are usually a great distance from the local users and contexts in which their tools are to be used. Careful content development is critical to ensure the suitability of knowledge and information for the local client needs in rural communities and agriculture societies. With that in mind, the Department of Agricultural Extension, under its project "The ICT Development of DAE" maintains a website (www.dae.gov.bd) which stores and disseminates field level information from and through extension workers.

The Soil and Land Resource Information System (SOLARIS), another agency of Ministry of Agriculture has developed a database of soil data using primary information from 460 districts. SOLARIS is a customized GIS software SOLARIS-GIS maps soil data based on classification (soil texture, landtype, landform, drainage, slope, surface water recession) and condition (crop suitability, land zoning, nutrient status and fertilizer recommendation). This GIS system can analyze data at the upazilla level, district level and finally at the national level. Other public organizations like Local Government and Engineering Department (LGED), Bangladesh Agricultural Research Council (BARC), Bangladesh Meteorological Department (BMP) and Water Resources Planning Organization (WARPO) are also trying to incorporate Geometrics knowledge in their planning process (BARC, September, 2001).

These initiatives have not only created possibilities for public-private partnership in extension and marketing, but also have set the stage for exciting farmer-oriented policy reform. Mobile phone companies are now collaborating with public agencies and disseminating alerts and warning messages to the masses *via* SMS. During the 2008 epidemic of bird flu, cautionary or preventive messages were sent to millions of mobile phone subscribers. Grameen Phone CellBazaar has developed an online market place, where various items including agricultural retail and wholesale items are listed, bought and sold, by simply using a mobile phone thereby making it accessible to over 16 million users. Several NGOs are running telecenters all over the country to provide agricultural information and services to grassroots communities through computers, the Internet, and other digital technologies.

6.2.1. Agriculture Information and Communication Centre (AICC)

There are about 20,000 farmers' clubs around the country, working with the welfare of farmers in mind. The Integrated Pest , Crop Management Clubs (DANIDA), Special Programme Food Security Clubs (FAO),

Farmers Information and Advice Centers (World Bank) are to name a few. The latest trend towards union-parishad based information centers sheds a new light into the conventional information delivery mechanism. Providing appropriate information and communication tools to these information centers and or clubs could potentially enhance and improve the efficiency and effectiveness of service and information delivery to a great extent.

Farmers Field School in Group Activities

Strengthening Plant Protection Service (SPPS) Project funded by DANIDA introduced Farmers Field School (FFS) as a strategy for implementing Integrated Pest Management (IPM) in Bangladesh. Many of these FFSs have been promoted to IPM Clubs and some have moved to form Community Based Organizations (CBOs). An external review of the project remarked: "IPM Clubs give voice to their members in the context of their communities. Clubs gain access to financial support from a wide variety of sources, including politicians, loan societies, banks and individuals. They conduct activities that strengthen agriculture in their villages. Their status has been enhanced. Non-IPM trained farmers are anxious to become part of these IPM Clubs, not only to learn about but also because of status."

DANIDA extended the FFSs concept to another of its project, Local Initiatives for Farmers' Training (LIFT), implemented by CARE Bangladesh. LIFT uses FFS for building capacities of Primary Participants to apply, test and adapt innovations associated with bio-intensive (vegetable) gardening. Field Schools were generally composed of around 25 participants in a cluster. LIFT's target group is the poorest segment of the farming community. The reviews of the project remarked:

- *The project had a remarkable impact on increasing production, consumption and sales of vegetables among the participants.*
- *FFS encouraged the formation of association of members (samity) and there was enthusiasm among participants for continuing contacts with other FFS members.*
- *Gender development was integrated in the project. The majority of the members were women.*
- *The project contributed the strengthening local rural organizations, especially those that are controlled by the poor.*
- *Targeting capacity building among the poorer segment of the rural community strengthens the position and empowered the poor.*
- *Raised poor farmers' awareness on key entitlement issues (associated with homestead production) and ensuring that the gains (made through homestead production) are not lost. These included land access and tenure, nutrition and factors that increase women's vulnerability (such as dowry, physically threats and violence, divorce, etc.)*

ICTs can boom their capacity as a rural informal organization that will enhance production and livelihood improvement.

AICCs aim to leverage the existing infrastructure of IPM/ICM clubs and or other CBOs to set up integrated information centers for farmers. These community-based information centers would provide a platform for local public and private organizations that are working in the field of agriculture to come together and offer coordinated support and service to the agricultural community. This could also be a venue for male and female, young and old farmers, fishermen and livestock owners to interact and learn together. Although seemingly negligible, these would afford the villages a voice that can be heard at a distance, and if sustained, carry prospects for altering previous relationships with extension workers, traditional communication systems and modes of learning that are usually provided top-down. Within the AICC network, farmers from the different pilot sites would be able to share indigenous information, serve as each other's "experts" and learn from each other. Some of the perceived benefits farmers could receive through an AICC are listed below.

6.2.1.1. Access to Information

- Use of SMS, MMS or web cam to access Sub Assistant Agriculture Officers (SAAOs), field workers, veterinary officers and or other experts to receive real-time solution to crop, fisheries and livestock management issues.
- Better access to local and international market information through website and cell phones would enable farmers to become more adept at negotiating with middlemen and provide a mechanism for bringing farmers and their products to the attention of markets.
- Technology transfer through films, documentaries, radio and TV programs. These programs may be developed by Agricultural Information Service(AIS), private partners and or procured from private TV channels.
- Training in agro-processing will increase the market for rural products and keep farmers informed and in-sych with national and international market demand and guidelines.

6.2.1.2. Coordination

- Most farmers own livestock and or fisheries. Farmers' information and service needs are therefore not confined to crop management issues. Coordination among agriculture extension workers and livestock and fisheries field workers along with research organization is critical to providing meaningful service to the farmers.
- Coordination with local NGOs and private sector players working in the field of agriculture to avoid confusion and duplication of service and information is imperative. Public agencies can also play a role in verifying the authentication of private sector data and information.
- Support to the small enterprises by establishing linkages with NGOs and private sector. These would play a vital role in women empowerment and promotion of rural agri- businesses.
- Establish farmers' network to broaden their understandings and build bonding, bridge and link capital.

- Greater participation in communities is assumed to contribute to stronger social capital within the community at the local, state, national and global levels and hence contribute to improved economic and social outcomes. This "whole-of-community" perspective on the potential benefits of ICT provides a possible focus for future research into the dynamics of learning and creative communities.

6.2.1.3. Input Dealership

- *Agro-Production Input*: The information center could also have a small sales center for production input. This will not require a license, rather the coordinator will buy seeds, fertilizer and pesticides in wholesale and sell to the local farmers.
- *Agro-Processing Input*: Raw materials needed for agro processing or packaging such as plastic bags, glue and tape may also be purchased here.

Through AICC established networks, farmers may sell their products directly to the market and receive a higher price for their crops than they would if they sold them through the middleman system. In addition, AICC will provide farmers with lower prices for inputs and other goods they need as well as higher yields through the education and information services offered.

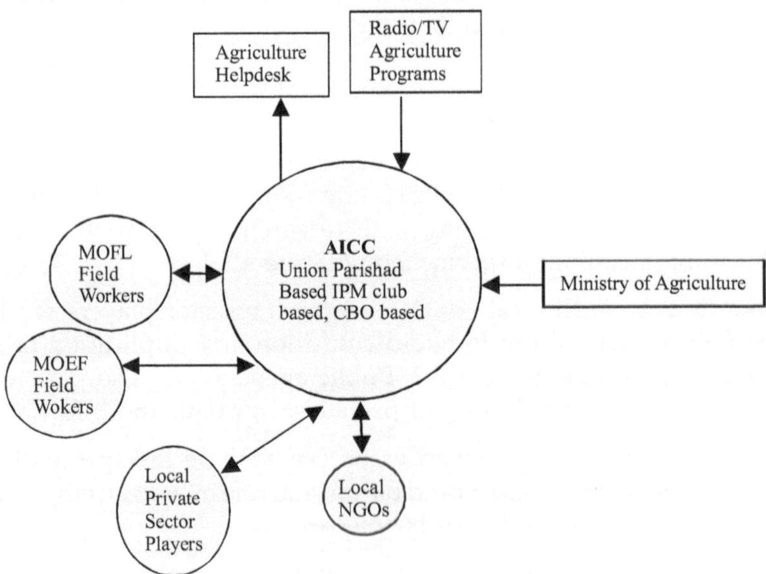

Figure 1: AICC Network

Below are some of the services that AICC will provide.

- Information Dissemination
  - All Levels (Pre, During and Post Production) through traditional methods along with film, documentaries, fillers *etc.,*
- Diagnostic Service
  - Soil Test
  - Pest and diseases (Preventive and Curative)
  - Poultry, Livestock and Fisheries (Preventive and Curative)
- Provide Market Information
  - Input Information (Availability, Alternatives, Pricing, Marketing)
  - Output (Pricing, Market *etc.,*)
- Weather and Precautionary tips
- Pos-Disaster Recovery Information
- Phone-In to Radio and TV Program and Helpdesk
- Networking with farmers, private sector players and NGOs

AICCs will be set up a common access point in the community for getting livelihood information and services using information and communication technology. The members of the club or groups will run the center. DAE extension officials and AIS officials will act as facilitators in conducting trainings and different kind of operations (computer, multimedia *etc.,*). These centres will provide livelihood-based information (ICT, print and audio-visual contents) on agriculture, fisheries, poultry and livestock, education, non-farm initiatives, appropriate technology, human rights, employment, disaster management *etc.* The centers are supposed to be registered by the social welfare department.

From the AICC, several Farmer's Communicator (FC) will be developed. The farmer's communicator will act as a local trainer for the community member, thus farmers' training facility will be improved at the doorstep of the community member. The executive member of the AICC will be responsible for the gradual development of the AICCs under the facilitation of Agricultural Information Service (AIS).

e-Governance "Quick-Win" Initiatives by "Access to Information"

More efficient, accountable and transparent administration and delivering critical services at the citizens' doorsteps are the cornerstones of e-Governance. To achieve the true benefits of e-Governance, a top-down push combined with institutional ownership is paramount. With this goal, the Prime Minister's Office of GoB had requested the "Access to Information" programme to organize a series of workshops for all Secretaries of GoB. Accordingly, 52 Secretaries of various Ministries and Divisions participated in three separate e-Governance Service Identification Workshops at the Prime Minister's Office on May 12, June 2, June 3, 2008. Parallel group discussions were held, where each Secretary identified at least one citizens' "quick-win" service, which would incorporate ICT in the delivery mechanism. The 'quick-win' phase of the project is meant to quickly showcase the impact of the newly engineered service delivery mechanism to the citizens, receive their feedback and allow government the necessary preparation time for nationwide implementation. The Union Information Centres (UICs), the Agriculture Information and Communication Centers (AICCs), and the Fisheries Information Centers (FICs) were three of these quick wins, which were respectively proposed by the Local Government Division, Ministry of Agriculture and Ministry of Fisheries and Livestock of Bangladesh.

The AICCs will take initiative for the launching of suitable income generating activities which will help to improve the income status of the members and the institution itself. With the establishment of AICC, a participatory problem solving forum will be developed at the community level leading to the empowerment of the rural community member and eventually, this will lead to the improvement of food security and livelihoods of the rural community member.

Agriculture Information Service (AIS), one of the agencies of Ministry of Agriculture (MoA), will be the implementing agency of this project. The officers concerned (Regional and Head quarter) under the leadership of the Director of AIS will oversee the activities of the AICCs.

6.2.2. Union Information Centres (UIC)

In 2007, UNDP, in collaboration with Local Government of Bangladesh, had developed two pilot Community e-Centers (CeCs) in two union parishad offices, in Madhainagar (Sirajganj) and Mushidhat (Dinajpur). The goal of this initiative was to create a comprehensive framework for establishing a model of sustainable CeCs which would provide e-Services to grassroot communities and in the longer term become knowledge hubs for communities to which they belong. The primary objectives of this initiative were to:

- Develop an e-Centre model consolidating all best practices of all existing models and centres that would be owned by community people and would ensure financial and social sustainability.
- Conduct primary research to determine a localized framework of both generic and specific information delivery required by the community.
- Generate and aggregate generic contents and develop a content development framework which can be disseminated through the already existing network of CeCs.
- Mobilize and sensitize local community about the purpose and benefits of CeCs and enhance awareness among the stakeholders.
- Develop a capacity development framework for conducting training for the infomediaries and or CeC operators.

The unique feature of the CeCs is a "4P" theory, ensuring Public, Private and People's Participation. As Union Parishad is still the most trusted entity for meeting people's information and service needs at the grassroot level, UNDP leveraged the existing goodwill and infrastructure of Union Parishad and housed the CeCs in their office buildings. Contents were procured and obtained through consultations and agreements with extension workers, NGOs and private companies. People's participation was ensured through mobilization workshops, involvement of young local volunteers and one-on-one consultation.

Both centers were highly appreciated by local beneficiaries and enjoyed a great publicity and recognition from the highest level policy makers. As continuation of the pilot phase, the Local Government of Bangladesh has now undertaken a project to develop 30 such centers at various union parishads all over the country with technical assistance from the "Access to Information Programme" of the Prime Minister's Office ("Access to Information (A2I)" is a UNDP funded project at the Prime Minister's Office, which supports development of new e-Governance projects and programmes and works to mainstream ICT into the national development policies and provides assistance to develop a national e-Governance Vision and strategy). These centers will be called "Union Information Centers (UICs)".

6.2.3. Fisheries Information Center (FIC)

In order to widen the extension coverage of the Department of Fisheries (DoF) following the LEAF concept (Local Extension Agent for Fisheries (LEAF) is a farmer extensionist and is developed to supplement the shortage of the field level extension manpower of the DoF. A pilot implementation of the LEAF concept was started in June 2004, in 90 villages of 30 upazillas in Fourth Fisheries Project (FFP), the Ministry of Fisheries and Livestock has decided to set up pilot Fisheries Information Centres (FIC) under the Department of Fisheries. This initiative is intended to extend support to meet the acute shortage of manpower and facilitate participatory and need-based fisheries extension service. Primarily these centres will be set up at 10 upazillas where some structural facilities are already in place, LEAFs are working well. A few non LEAF areas will also be the target of this initiative to make unbiased assessment of the impact upon completion of the pilot phase.

The FICs are intended in delivering information on aquaculture and fisheries management with which they are equipped, and which they have access to and are readily available with them through ICT, in particular saved in computer and on internet. To render their service V-FICs will take assistance from the ICT equipments, *viz.* computer, internet, camera, *etc.* To respond to some specialised problem of immediate need (*e.g.* disease diagnosis, de-shape fish) V-FICs will make Video conference with U-FICs. To address some periodic nature problem video or digital still camera could be used for taking picture and showing those to the U-FIC personnel for diagnosis and remedies (*e.g.* water quality).

The FICs will work in a three tire communication system. The core FIC (1st tire) will be established at village level (V-FIC). The V-FICs will be fed and updated with innovated and appropriate information on aquaculture and fisheries management by the Upazilla level FIC (U-FIC). The U-FIC (2nd tire) will be established in the Upazilla Fisheries Office and be co-ordinated by the Senior Upazilla Fisheries Officer/Upazilla Fisheries Officer. Activities of the U-FICs will be supervised by the District Fisheries Officer (DFO). One FIC will be established in each DFO's office (D-FIC). Countrywide activities of all the FICs and the whole course of information flow will be coordinated by the Extension Section of the Department (C-FIC).

6.2.4. One-Stop-Shop Telecenters

These telecenters are meant to provide e-services to grassroot communities and in the longer term become knowledge hubs for communities they belong to. Through these, people can access computers, the Internet, and other digital technologies that enable people to gather information, create, learn and communicate with others while they develop essential 21st-century digital skills. While each telecenter is different, their common focus is on the use of digital technologies to support community, economic, educational and social development, reducing isolation, bridging the digital divide and creating economic opportunities. They may be established by businesses, social entrepreneurs, nongovernmental organizations, governments, or development donor organizations. As of December 2007, there are

- 100 Gonokendra: Centre for Community Development (Dhaka Ahsania Mission);
- 05 Amader Gram Learning Centre (AGLC);
- 27 Internet Learning Centre: School based Telecentre (Relief International);
- 03 Youth Community Multimedia Centre (YPSA);
- Rural Technology Centres (Practical Action);
- 02 Ghat: Rural ICT Centres (DEN);
- 452 Community Information Centre (CIC) (Grameen Phone);
- 27 Pallitathyakendra (D.Net)

6.3. Other e-Initiatives

6.3.1. SRDI SOLARIS Data Interpreter

The Soil and Land Resource Information System (SOLARIS) is a data repository that was developed by CEGIS and later modified by Soil Resourses Development Institute (SRDI). This 2GB database stores soil data using primary information from Upazila Nirdeshika from 460 districts. A customized GIS software SOLARIS-GIS maps soil data based on classification (soil texture, landtype, landform, drainage, slope, surface water recession) and condition (crop suitability, land zoning, nutrient status and fertilizer recommendation). This GIS system can analyze data at the Upazilla level, District level and finally at the National level.

6.3.2. DAM Website

The DAM (Department of Agricultural Marketing) website provides commodity price information with the aim to involve all agricultural stakeholders, especially farmers in the decision making process. Currently in its pilot phase, the database provides wholesale and retail prices of around 200 commodities from 30 of the 64 districts. The price information is collected daily and sent to the DAM headquarter in Dhaka through e-mail. The headquarter populates the database with the received price information and prepares a report. The daily price reports are also posted on the website and are accessible in three different formats including an .xls file, which allows data to be downloaded and manipulated by user. Periodic reports are also available from the website but cannot be modified by user. Although this website is not currently being used for the original purpose intended, it is being used mostly for researching and studying price trends. A more farmer-Friendly format is needed.

6.3.4. BRRI Knowledge Bank

Bangladesh Rice Research Institute (BRRI), Bangladesh Rice Knowledge Bank (BRKB) is a treasure of rice knowledge. This is a dynamic source of knowledge that is updated regularly to keep consistency with the latest innovations and users' feedback. The BRKB contains rice knowledge to address the regional as well as national issues associated with rice production and training. It started with rice but extends promise to be expanded to non-rice technologies in future. The extension service providers are the immediate beneficiaries of the BRKB. However, ultimately farmers will be benefited from it (Agriculture R&D Institutes in Bangladesh).

6.3.5. BARI Computer Training Center

Bangladesh Agricultural Research Institute (BARI) had pulled resources from select National Agricultural Research System (NARS) institutes and put together a computer training center at BARI. This training center has 20 computers available for hands-on training for all BARI scientists, officers and staff. BARI has developed its own training modules, consisting of training on e-mail, internet usage and other BARI relevant activites. 180 scientists and officers starting from DG to CSO have participated in these trainings so far. These trainings have proven to be so effective that NARS has come to an agreement with BARI to train officials of all the other NARS institutes through BARI.

6.3.6. Bangladesh Country Almanac (BCA)

This project, jointly implemented by Bangladesh Agriculture Research Council (BARC), SRDI, BRRI and CIMMYT with financial support from USAID, was commenced in July 2002. The main objective of this project is to disseminate the natural resources, climatic and socio-economic information generated through earlier FAO/UNDP funded AEZ & GIS project at BARC and also the information from other organizations. In the BCA-CD, both spatial and non-spatial type of data are stored that also includes a GIS tools for use in agriculture and natural resource management activities for characterizing and analyzing geographic variation in relation to agricultural and environmental concerns.

6.3.7. DAE Website

In order to develop and strengthen the ICT department of Department of Agricultural Extension (DAE), Government of Bangladesh had initiated a project called "The ICT Development of DAE". The DAE website (www.dae.gov.bd) is one of the key components of the project. This website stores and disseminates field level information from and through extension workers.

7. ICT IN NAEP AND NAP

The DAE is the key implementing agency of the New Agricultural Extension Policy (NAEP). Although NAEP mentions use of mass media *i.e.* radio, television, newspaper *etc.*, for rapid extension of agricultural technologies, the current situation calls for a much more comprehensive information and communication strategy.

"The primary goal of the National Agricultural Policy (NAP) is to modernize and diversify the entire agricultural system, through initiation and implementation of a well-organized and well-coordinated development plan (National Agriculture Policy)". Agricultural policies worldwide, including those of our neighboring countries like Srilanka, India and Indonesia, include clearly pronounced ICT policies. Although the National Agricultural Policy of Bangladesh does not currently include a national integrated ICT strategy, most entities, however, within the Ministry of Agriculture and Ministry of Fisheries and Livestock (Training, Research, Development) have their own vision and strategy for ICT. At the top level, the importance of ICT development is recognized and supported. In order to put an e-Krishi enabling environment in place, the development and implementation of various e-Policies and e-Strategies should always remain supreme in the processes. This section discusses 5 key areas where inclusion of ICT policy is critical (National Agriculture Policy).

7.1. Human Resource Development

- Increase farmers' literacy and education through development and implementation of distance-learning curricula, adjusted to the needs of farmers. Public-private partnerships often lead to more timely and effective implementation of such programs and services. Promotion of vocational schools and community clubs is crucial for development of the farm and non-farm rural sectors.

- Train field extension personnel in a broader range of subjects not limited to technology and provide them with the resources they need to advise farmers on how they can obtain credit, add value to their agricultural products and find markets for their products and other marketing information.
- Empower women through access to information at home (TV, radio, Internet), which will enable self learning and increase self employment opportunities. Women must be allowed under the law to inherit land, join credit and savings clubs and water user groups, have access to extension advice, set up small enterprises, and be self-sufficient as household heads if the family unit breaks down (Hafkin and Odame, 2002).
- Promote women's networks, such as microfinance and livestock development programs, which can be effective advocacy groups for women's rights while strengthening the social capital of women and supporting them in new activities.
- Future growth will depend on increased researcher funding combined with significant capacity building. The share of scientists with advanced degrees needs to increase, as do the number, quality, and frequency of international staff exchanges (both into and out of Bangladesh) for training and capacity building. Through global scientific linkages, core research staff must be trained to take advantage of advances in bio-technology and ICT.

7.2. Research and Development

- Agricultural research investments should be increased, the continued application of conventional agricultural techniques and recent developments will have considerable improvement in crop and livestock yield growth in all agro-ecosystems.
- Linkages with international research institutes, universities should be established. Coordination efforts should lead to improvement of the highly fragmented research and development sector.
- Through collaborative research, joint ventures, alliances and networks with the private sector, public research institutions can access advanced scientific information and technologies held by the private sector; mechanisms for developing, processing, marketing, and distributing final products to farmers and consumers and financial resources, which are increasingly difficult to obtain.

- Private firms, on the other hand, can access untapped or emerging markets, opportunities to participate in formulating regulatory systems, and prospects to improve corporate profiles (Agriculture R&D Institutes in Bangladesh).

7.3. Need Based and Inclusive Extension System

- There should be a multi-stakeholder steering committee which would develop linkages with other relevant ministries, agencies, NGOs and private sector players. This will ensure better coordination amongst extension service providers.
- Community based information center could emerge as a system that would enable farmers to receive information and communicate with regional and national counterparts. This could also be the platform for all public and private sector information providers to share and exchange information with other stakeholders for better coordination.
- Privatization of extension through contracting, for example, with seed companies; success is more likely when extension is linked to the delivery of a specific technology and to larger, more homogeneous groups of farmers; if private extension services cannot be self-supporting, the Government must continue providing assistance and training (Agricultural Extension Manual, 1999).

7.4. Agribusiness

- Policies must promote agricultural diversification commercialization, and vertical integration and coordination. Priorities are (i) effective legal, regulatory, and institutional environments for trade, commerce, and manufacturing; (ii) rural financial reforms; and (iii) infrastructure investments to reduce marketing margins and transport costs (www.bdagribusiness.com).
- Redraft key regulations and documents with provisions to facilitate agri business entrepreneurs.
- Ensure participation of women, farmer groups, associations, chambers, and existing and potential entrepreneurs in relevant seminars and consultations.
- Strengthen the capacity of DAM, MoA and MoFL in agribusiness governance and adjust their roles as facilitators in linking primary producers with the market.
- Contractual and regulatory frameworks must give smallholders access to high-value commodity production and equitable contractual

arrangements with agro-processing companies. It is also critical to provide technical expertise to identify regulatory impediments to private sector operations in agribusiness.

- Train the staff of Ministry of Agriculture agencies to deal with NGOs and private sector, as well as addressing gender issues in agribusiness development (ADB-Report, 2004).

7.5. Market Access

- Assess current situation of public market management and impediments to farmers' access to market, and suggest mitigation measures.
- Web-based trading platforms offering one-stop shop facilities are emerging, especially for main commodities. The Department of Agriculture Marketing should include policy provisions in their mandate to accommodate these services.
- There should be provisions to broadcast market information *via* rural radio, TV or mobile phone, thereby creating a 'level playing field' between producers and traders in a region.
- In recent years, short message and text services have taken up and effectively deliver prices and trading information *via* mobile phone to farmers. The sustainability of these systems requires attention, with an important role for the private sector and organized producer groups.
- Agro-processing information should be obtained, offered and delivered to the farmers. Separate 'Center for Agricultural Market Research, Intelligence and Certification' should be established by restructuring and renaming DAM with additional facilities and trained manpower.

8. CONCLUSION

Agricultural extension is back on the global development agenda, and reform measures are being actively pursued in many parts of the developing world. With changing environment of agricultural extension, institutional pluralism and bottom up participatory approaches are necessary to address new challenges. The public sector may still need to play an important role in providing agricultural extension services due to its 'public good' nature, but its role should be changing in the face of increasing role of private and NGO sector and additional responsibilities of extension services. Entry of actors such as the private sector and NGOs in delivery of such services should be relaxed and creation of innovative public-private partnerships (PPP) in extension should be facilitated and promoted. But the process is thus far incomplete, not only in implementation but also in policy analysis. We have endeavoured to identify some important gaps in analysis and some key issues that still need to be addressed. It is important to remember that the impact of ICT on agriculture and rural economy depends not only on access to ICT and information, but also on the status of other infrastructure (such as roads and electricity), services (like education and health) and presence of a National Agricultural Policy with particular focus on e-Krishi. This is a time for agricultural policymakers to reflect afresh on the unmet demands, implicit and explicit, for provision of agricultural extension services to all the deserving farmers of Bangladesh.

References

ADB-Report, 2004. Agribusiness marketing and post-harvest systems in Bangladesh. *Interim Report*, AGRIC limited, June 2004, Ministry of Agriculture and Asian Development Bank, New Zealand.

Agricultural Extension Manual, 1999. Agricultural extension manual, Department of Agricultural Extension, Ministry of Agriculture, Government of the People's Republic of Bangladesh.

ASIRP, 2003a. Agricultural extension with the Department of Agricultural Extension (DAE) Bangladesh, A Ten Years Review, Volume-1, 1992-2003, Agricultural Support Services Project (ASSP) and Agricultural Services Innovation and Reform Project (ASIRP), November 2003, Agricultural Extension with the Department of Agricultural Extension (DAE) Bangladesh.

ASRIP, 2003b. Performance of extension service providers in Bangladesh. Quality or Quantity of Service; ASRIP, Bangladesh.

BARC, 2001. National coordinating institution for the National Agricultural Research Systems in Bangladesh, Fifth Edition, September 2001, Bangladesh.

Hafkin, Nancy J and Odame, Helen Hambly., 2002. Gender, ICTs and agriculture. August 2002.

Hoque, Enamul., 2007. Crop diversification in Bangladesh. Ministry of Agriculture, Government of the People's Republic of Bangladesh.

Hossain, M. Gul., 2004. Dhaka, Thematic Reports on Crops Sub-Sector (Introduction, Agriculture Sector Overview, Agricultural Extension, Policy and Legislations (Crops), Review of DANIDA Interventions (Crops), Water Management and Irrigation, Biotechnology, Options for Support, Capacity of Implementing Agencies for Recommendations Made-as inputs to Sector Analysis Report- December 11, 2004).

Hossain, M. Gul., 2005. Bangladesh Agriculture: A Critique on performances and the challenges of tomorrow.

MoA, 2004. Actionable policy brief and resource implications. Agriculture sector review (Crop Sub-Sector), Ministry of Agriculture, Government of the People's Republic of Bangladesh, Dhaka.

Yuan Zhou, (n.d.). Reinventing agricultural extension to smallholders. Syngenta foundation, Bangladesh, Dhaka.

Reference Websites

Bangladesh Agriculture University
http://agri-varsity.tripod.com

Bangladesh Agriculture Research Council
http://www.saic-dhaka.org/saic/BARC.htm
www.bdagrbusiness.com

Agriculture R&D Institutes in Bangladesh
http://members.tripod.com/~UTTAMDEB/res.html

Ministry of Agriculture
http://www.bangladeshgov.org/moa/moa.html

Bangladesh Agriculture Facts
http://www.nationmaster.com/country/bg/Agriculture

Overview of Agriculture Sector
http://countrystudies.us/bangladesh/67.htm

Bangladesh Livestock Research Institute
http://www.blri.gov.bd/

Bangladesh Fisheries Research Institute
http://www.fri.gov.bd/

Food and Agriculture Organization: Bangladesh Country Profile
http://www.fao.org/countryprofiles/index.asp?lang=en&iso3=BGD

CARE: Rural Livelihoods Program
http://www.careinternational.org.uk/cares_work/where/bangladesh/projects/project.php?id=97

Comparative Advantage and Crop Diversification in Bangladesh
http://wbln0018.worldbank.org/lo%20web%20sites/bangladesh%20Web.nsf/0/370f3af793c7c41546256711001bb7c1?OpenDocument

National Agriculture Policy
http://www.sdnbd.org/sdi/issues/agriculture/national-policy/review_agricultural_policy.htm

Department of Agricultural Marketing
http://bdagribusiness.com/

ICT for Disaster Management/ICT for Disaster Recovery
http://en.wikibooks.org/wiki/ICT_for_Disaster_Management/ICT_for_Disaster_Recovery

Author Information

Ms. Rubaiya Ahmad

Policy Associate

Access to Information (A2I) Programme (A UNDP Funded Programme)

Prime Minister's Office, Bangladesh

Old Sangshad Bhaban, Tejgaon

Dhaka-1215

BANGLADESH

Phone No (O): (880-2) 914-4848 (R): (880-2) 883-1038

Mobile Phone No.: 01718643497

Fax No: (880-2) 8159895

e-mail: rubaiya.ahmad@undp.org; rubaiyaa@gmail.com

URL: www.pmo.gov.bd

Ms. Rubaiya Ahmad, a graduate of the University of Texas, has over six years experience in the field of Information Technology and two years in ICT4D. She was one of the key members of the "National e-Governance Vision" team of the Prime Minster's Office, Bangladesh and the key investigator of "e-Krishi (Agriculture) Vision 2025". She has also co-drafted various "e-Extension" project proposals for the Ministry of Agriculture and Ministry of Fisheries and Livestock, which are currently being implemented.

❑❑❑

Chapter 2

CARIBBEAN NATIONS

Wayne G. Ganpat and Claudette De Freitas

1. INTRODUCTION

There are many countries surrounded by the waters of the Caribbean Sea. These countries vary widely in size, topography, ethnicity, language spoken, agricultural production systems, main sources of income, income levels and styles of governance, to name a few.

This analysis covers only the countries where English is spoken as the first language. These countries are linked through a grouping of Caribbean Communities, called CARICOM, an organization of 15 Caribbean nations and their dependencies. It starts with Jamaica in the North to Trinidad and Tobago in the South, but also includes Belize in Central America and Guyana and Suriname in South America. Although French speaking Haiti and Dutch speaking Suriname belong to CARICOM, they are not discussed because of the language barrier. The other French speaking islands (Guadeloupe and Martinique) and the Spanish speaking islands (mainly the Dominican Republic, Cuba and Puerto Rico) have different administrations for agricultural development. CARICOM countries stretching from the tip of Florida, USA, in the North to Venezuela in South America. CARICOM includes Belize in Central America and Guyana and Suriname in South America.

2. COUNTRY AND AGRICULTURE SECTOR OVERVIEW

The importance of agriculture as a major economic sector has declined within the Caribbean Community (CARICOM) countries over the past two decades. Agriculture, based on traditional commodities such as rice, sugar and bananas, which once dominated the region's economies, has been replaced by tourism and services in a number of countries. Significant challenges resulting from loss of preferential markets for traditional commodities, poor competitiveness of Caribbean farm enterprises in a global trading environment and high levels of risk associated with natural disasters (particularly hurricanes), have also contributed to great losses to farmers and producers. Changes in trading arrangements at the global level, climate change, new phyto-sanitary standards and issues relating to intellectual property rights, have created major challenges for Caribbean farmers. In addition, other factors such as an aging farmer population, use of outdated production technologies, high costs of inputs and unavailability of labour, among others, have contributed to the decline of regional agriculture.

The farming community comprises a multiplicity of farmers operating on small holdings, small defined as less than two hectares, is characteristic of most of the islands. Production and marketing systems exist for the traditional commodities, whereas those for root crops, herbs, vegetables, small ruminants and other non-traditional commodities are not well defined. Many of these challenges have been summarised into ten key binding constraints in *the Jagdeo Initiative* - the current regional agricultural policy framework adopted by the CARICOM Community in 2005. The policy identifies a series of recommendations for implementation to improve the competitiveness of the region's agricultural sector (Antoine, 2009; Bourne, 2008).

In the Caribbean region, the role of agricultural extension services has moved from that of passive transfer of information and has been evolving to that facilitator, service provider and information broker. The Information and Communication Technology (ICT) revolution of the past two decades has certainly had an impact on extension services in the Caribbean region.

This chapter will provide an overview of extension services and systems and highlight the challenges. It will also trace some key developments in ICT use in the agricultural sector, with special reference to extension services. Challenges and recommendations for ICTs in Caribbean agricultural extension have also been identified.

A brief overview of the general economy, with highlights of the agriculture sector is presented below for each country. The main source of information for the country description detailed is the website of the Caribbean Agricultural Research and Development Institute (CARDI, 2009), a regional Research and Development organization serving CARICOM countries. Population statistics and economic data are from websites of the United Nations (UN, 2008).

Antigua and Barbuda

The larger of the twin island state, Antigua, is 281 square kilometers in area. Barbuda covers 161 square kilometers and lies forty-eight kilometers due north of Antigua. Together, they have a population of 83,000. The economy is service-based, with tourism, financial, and government services representing the key sources of employment and income; tourism is the leading foreign exchange earner. Agriculture contributed 3 per cent of GDP in 2007. The main agricultural products are cotton, fruits, vegetables, bananas, coconuts, cucumbers, mangoes, sugarcane and livestock, mainly focused on the domestic market. There is an organized Extension system as part of the Ministry of Agriculture. This division faces a number of challenges such as staffing and equipment.

Belize

Belize is located in Central America, bordering the Caribbean Sea to the east, with Mexico and Guatemala bordering on the northwest and southwest, respectively. Belize covers a total area of 22,960 square kilometers and includes many small uninhabited islands called cayes. The tourism industry is the primary foreign exchange earner and is becoming the country's second most important industry, next to agriculture. The agricultural sector accounted for 28.4 per cent of GDP in 2007 with the chief crop, sugar, generating almost half of the exports. There are just about 20,000 farmers in Belize and while the government is the main extension service provider, a number of other organizations, including several NGOs also provide extension services.

Barbados

Barbados is the eastern most island of the Lesser Antilles, situated 160 kilometers east of St. Vincent, with a total land area of 430 square kilometers. It has a population of 295,000. Historically, the Barbadian economy was dependent on sugarcane cultivation and related activities, such as cotton and vegetables. However, the economy has shifted from an emphasis on agriculture towards the provision of mainly tourism, business, finance and general services, and manufacturing. As part of the county's "sugar adaptation strategy", a sugar-based ethanol industry is being developed together with non-sugar agriculture and manufacturing. The agricultural sector accounted for 4 per cent of the GDP in 2007. There are just about 3,500 farmers in Barbados and the government is the main provider of Extension services.

Commonwealth of Dominica

Dominica is the largest and most northerly of the Windward Islands. Geographically, Dominica is distinctive in many ways. The country has one of the most rugged landscapes in the Caribbean, and is also among the earth's most rain-drenched lands. The island, home to rare species of wildlife, is considered by many as a beautiful, unspoiled tropical preserve. The country is approximately 751 square kilometer in area and has a population of about 70,400. There are less than 10,000 farmers. The agricultural sector accounted for 18 per cent of the GDP in 2007, mainly from bananas, citrus, coconuts, cocoa, herbal oils and extracts. Historically, banana production has employed over one-third of the labour force, directly or indirectly. However, the sector has been facing challenges with the loss of the preferential European Union markets for bananas. There has been diversification into other crops such as citrus fruits, vegetables, coffee, patchouli, aloe vera, cut flowers, mangoes, guavas, and papayas. Increasingly the Government is promoting the island as an "ecotourism" destination. Agriculture Extension services are provided mainly by the government, but other special interest organizations and groups also provide extension services.

Grenada

Grenada and its largely uninhabited outlying territories are the most southerly of the Windward Islands. The Grenadine Islands chain consists of some 600 islets. Grenada and its territories occupy a total area of 433 square kilometers and are known as the Spice Isle because of its production of nutmeg and mace. Grenada is the second largest producer of nutmeg in the world after Indonesia, and even after devastation by Hurricane Ivan in 2004, remains a significant producer of mace, cinnamon, ginger and cloves. The economy of Grenada is dependent on agriculture and tourism, and while tourism is the main source of foreign exchange, historically, the island has been an agricultural society. In 2007 the agricultural sector accounted for 6 per cent of GDP. Agricultural export products are nutmeg, mace, cocoa, bananas, other fruits, vegetables and fish, while agro-processing is becoming more important. There is an organized Extension system coordinated by the Ministry of Agriculture, pursuing mainly commodity based extension programs.

Guyana

Guyana is located in northern South America, bordering the North Atlantic Ocean, between Suriname and Venezuela, covering an area of 214, 969 square kilometers; with a population of 736,100. It is a sparsely populated country. Over 90 per cent of the population reside in the narrow coastal plain that comprises only five per cent of the total landmass. Agriculture in Guyana contributed to 31 per cent of GDP in 2007. The major industries are sugar, bauxite, rice, timber, fishing (shrimp), gold mining, and diamonds. The government provides extension along with commodity associations, farmers' groups and other organizations.

Jamaica

Jamaica is the largest island of the Commonwealth Caribbean and the third largest of the Greater Antilles. The country lies 145 kilometers south of Cuba and 160 kilometers west of Haiti. Its capital city, Kingston, is about 920 kilometers southeast of Miami. Jamaica occupies an area of 10,911 square kilometers. Jamaica's economy is heavily dependent on services, which account for 70 per cent of GDP. The main income earners are tourism, the bauxite industry, funds from remittances and agriculture. At five per cent, agriculture was the lowest contributor to GDP in 2007, but it employs 20 per cent of the workforce. Major commodities produced include sugar, bananas, coffee, cocoa, citrus, pimento, yams, vegetables, poultry and goats. Jamaica has several commodity extension programs for traditional commodities, for example: banana, cocoa, coffee, and sugar. Extension services are provided by an autonomous government authority, the Rural Agricultural Development Authority (RADA), which has been fairly successful. RADA coordinates the delivery of services to rural communities along with providing advice and technical support to farmers.

St. Kitts and Nevis

The islands of St. Kitts and Nevis are part of the Leeward Islands group of the Lesser Antilles. A narrow strait of 3.2 kilometers wide separates the two islands. The total land area is 269 square kilometers, and together both islands have a population of 46,000. Sugar was the traditional mainstay of the St. Kitts economy until the 1970s when the commodity became increasingly unprofitable. The Government of St. Kitts and Nevis embarked on a programme to diversify the agricultural sector and stimulate the development of other sectors of the economy. By 1987, tourism had surpassed sugar as the major foreign exchange earner for St. Kitts and Nevis, with export-oriented manufacturing and offshore banking also playing key roles. The agricultural diversification programme has continued with emphasis on crops such as hot pepper, white and sweet potato and vegetables, in addition to livestock - dairy, beef, mutton, pork and poultry. Agriculture contributed 3 per cent of the GDP in 2007.

Saint Lucia

St. Lucia lies to the north of Martinique, south of St.Vincent and north-west of Barbados. It covers a total land area of 616 square kilometers. St. Lucia's economy depends primarily on revenue from banana production and tourism with some input from small-scale manufacturing. The contribution of agriculture to the GDP has declined over the years. However, although agriculture contributed only four per cent of the GDP in 2007, it still accounts for a significant number of jobs, with the majority related to the banana industry. Other agricultural products of economic importance are: cocoa, coconut, citrus fruits and livestock. The banana industry is faced with low prices and a new, less favourable, set of rules for the European Union market which has made economic diversification increasingly important. Efforts to this end include the establishment and promotion of fruit crops such as mangoes and avocados; a variety of vegetables are being produced also for local consumption. This diversification programme is supported by the extension services which is largely provided by the Ministry of Agriculture. There are also a number of important commodity associations which provide extension services to their members.

Saint Vincent and the Grenadines

St. Vincent is the largest of a chain of small islands, known as the Grenadines, which extends to Grenada. St. Vincent covers a land area of 344 square kilometers. The islands altogether, they have a population of 121,000 persons. The economy of St Vincent and the Grenadines is based largely on agriculture, which contributed seven per cent of the GDP in 2007. Banana is the main crop and accounts for over 50 per cent of the work force and exports. There is a growing service sector which includes tourism, telemarketing and offshore financial services. The Government has been implementing a series of structural reforms within the agriculture sector as part of a diversification programme, including promotion of private sector growth and restructuring of the banana industry. Among the agricultural products are: coconuts, sweet potatoes, eddoes, dasheen, arrowroot starch, spices, as well as small numbers of cattle, sheep, pigs, goats and fish; the output for plantains, pineapples, yams and peppers is increasing. There are about 7,000 farmers and while extension service is provided mainly by the government, a number of other agencies exist to provide service to the farmers.

Trinidad and Tobago

Trinidad and Tobago, a twin island state, is situated nearer to mainland South America than any other Caribbean island and comprises an area of 4,828 square kilometers. Tobago is located thirty kilometers northeast of Trinidad, and separated by a channel thirty-seven kilometers wide. Trinidad and Tobago has a population of 1.3 million with one of the highest per capita incomes in the Caribbean. The country is a producer of oil, natural gas, steel, and petrochemicals. Agriculture contributed 0.6 per cent of the total Gross Domestic Product in 2007, and the value of crops grown for the domestic market is believed to be considerably greater than that of the export crops. Since 2003, with the divestment of the industry, sugar has declined as the main export crop cultivated on large estates employing a large number of persons. Trinidad is also world famous for its "fine or flavor cocoa" which fetches a premium price on the world market. In recent times, rice, citrus, corn, cassava, peanuts, and pigeon peas are being cultivated on large estates both private and state managed, as part of a programme to diversify agricultural output and promote food security and sovereignty. The Ministry of Agriculture has historically been the main provider of Extension services to farmers. However, in recent times a number of state-assisted agencies and organizations have begun to provide specialized extension support to farmers.

3. MAIN EXTENSION PROVIDERS AND SYSTEMS

In most countries, the extension service is attached to the Ministry of Agriculture or a similar ministry. They are national in scope, cover most agricultural commodities and in some instances, Forestry and Fisheries. Jamaica, however, has an alternative national system, known as the Rural Agricultural Development Authority (RADA) which is a semi-autonomous statutory authority of the Government. Limited services are provided by commodity associations, farmers associations and other special service groups. Seepersad (2002) described in detail the variety of Extension systems in the English-speaking Caribbean and this publication is used extensively in the section below.

3.1. Ministry of Agriculture (MoA) Extension Services

3.1.1. Structure

Various models are observed in various countries. For example, in St. Kitts and Nevis extension officers are located within the crops and livestock units respectively. In Barbados, the extension service was once merged with the Research Division. They are now separate. Trinidad and Tobago (TT) has attempted decentralization of extension and other services of the MoA through the use of regional offices. However, this resulted in some problems with the overall management of the extension staff and coordination of extension programs. In Grenada and Barbados, the Extension Division is further subdivided into crops extension and livestock extension. In St.Vincent and the Grenadines, there has been a move in recent times to have extension services pursue a commodity focus. Most countries have information or agricultural communication units attached to the Extension division for the production of publications in all instances, radio programs in some instances, and audio-visual material in few cases.

3.1.2. Scope of Extension Programs

Most MoAs have a national mandate for almost all commodities and the main focus of the Government's programme over the years has been on increasing production and productivity of the agricultural sector. Extension programmes provide services for the most part to the vast majority of small farmers producing a variety of crops and some livestock as part of the farming system. Commodity extension services exist for main export crops like; cocoa banana, and sugar. Such services are usually provided by Commodity Associations and extension officers are not heavily involved in Technology Transfer and other support services related to that commodity. The poultry industry is dominated by contract farming, and the processors provide the contract farmers with advice, inputs and technical assistance. However, the MoA has continued to provide information for those farmers who are outside the contract farming system along with a host of other commodities. Farmer training programmes are also conducted as a means to provide technical information and farming techniques to farmers.

Extension staff often has an understanding of rural development issues although most organizations, with the exception of RADA in Jamaica, are not set up for delivering rural extension or rural development services. Even so, RADA's role is limited to coordinating services and or liaison to ensure better delivery of services offered by various agencies involved in rural development.

There are limited arrangements for Subject Matter Specialists (SMS) to provide backstopping and technical assistance for extension officers. Trinidad and, to some extent, St. Lucia have experimented with such arrangements. In other countries, staff members of Research Divisions serve in a limited way as SMS.

3.1.3. Role of the Extension Officer

Frontline Extension officers operate as "generalists" in most countries and address a range of problems of the farming community - from providing technical advice to serving as a channel for matters which require the wider attention of the Ministry. Thus from time to time they are required to carry out responsibilities which are important, but are not strictly informational or educational. Across countries duties may include assessment of flood damage, collection of data, administration of subsidies, and regulation of land issues, to name a few. The extension officer is usually seen as the representative of the Ministry and even of the Government since he or she is the first line of contact with the public. This means that they could be asked to carry out functions that are needed that are not carried out by any other agency. Often these activities take precedence over information and education delivery. In a number of cases, therefore, where there is stated dissatisfaction with Extension services, it may be a reflection of general dissatisfaction with the response of the Ministry of Agriculture as a whole, rather than with the Extension officers themselves.

3.1.4. Staff Qualifications

All frontline officers in the MoAs in Jamaica, Guyana, Barbados, and Trinidad and Tobago have at least a two year diploma in agriculture at the tertiary level. In Trinidad and Tobago, a number of the frontline officers have undergraduate (first degrees) in agriculture. This is so because the Faculty of Agriculture of the University of the West Indies is located in Trinidad making access for Trinidadians much easier. In most other countries, the majority of staff have been trained in agriculture science *via* two-year diplomas from various national agricultural colleges, with the ultimate aim of training everyone at that level. The remaining staff have Certificates in agricultural science, acquired after two years of training. All the staff at the senior levels have at least undergraduate (first degrees) in agriculture, while others also possess post graduate qualifications including Master's degrees in various agricultural disciplines. While a number of Extension Officers also have specialized qualifications in Extension at the diploma level, there are very few persons trained in Extension at the post graduate level.

3.2. Commodity Extension Services

Commodity associations bring together all the functions for one particular commodity *i.e.* research, extension, input supply, marketing and training under one umbrella. The Extension function is usually carried out by a separate unit which is normally referred to as a commodity extension service (CES). This is perhaps the second most important type of Extension service in the region. Commodity associations still exist for most of the traditional commodities such as banana, sugar, rice along with coffee, cocoa and coconut.

3.2.1. Banana

All the countries that export bananas – the Windward Islands (Dominica, Grenada, St. Lucia, St. Vincent), Jamaica and Belize have commodity associations for this commodity. Extension is mainly for the purpose of technology transfer in that specific practices are promoted for higher yields and improved quality. Overall farm management is also addressed. Extension staff is also involved in regulatory functions *e.g.* inspections related to fruit quality and certifying farmers. The Banana Extension Officer is also responsible for recommending additional inputs when warranted and ensuring proper marketing arrangements.

3.2.2. Sugar

The sugar industry is the largest agricultural industry in the Caribbean with large companies (State owned and otherwise) responsible for all the processing and most of the production. Guyana's sugar industry is the largest agro-industry in the Caribbean producing over 300,000 tons of sugar, most of it being produced by the state. In Jamaica, sugar cane commodity extension service is operated by the Sugar Industry Research Institute (SIRI) which has field staff throughout the producing areas. They carry out technology transfer and assist with providing technical services to farmers. In Belize, there is also some level of support for sugar by dedicated extension officers.

3.2.3. Rice

Guyana is by far the Caribbean's largest producer and exporter of rice, which is produced by both large and small farmers. The industry is managed by the Guyana Rice Development Board which runs a research station and a rice extension service. Extension Officers are posted throughout the rice growing areas and are engaged in technology transfer as well as working with seed growers. In Belize, there is also some level of support for Sugar by dedicated extension officers. As with sugar, there also some level of support for rice farmers by dedicated extension officers.

3.2.4. Poultry

Contract farming is practiced largely in the broiler industry. It is probably the second largest agricultural industry in the Caribbean and is certainly the fastest growing industry. The industry is dominated by vertically integrated businesses involved in feed manufacturing, processing and production. The contracted farmers are serviced by technical representatives who are responsible for technology transfer, monitoring supplies of feed and other inputs and providing other forms of support. It is a very intensive system and technical representatives have to visit the farms at least twice a week.

3.2.5. Other Commodities

Jamaica has commodity extension services for coffee, cocoa, and coconuts. Trinidad has, within the last five years, started to provide a specialized extension service for cocoa in an effort to revitalize the industry. Grenada also has a cocoa commodity extension service in addition to banana, and another Commodity association responsible for the nutmeg industry. With agricultural diversification away from traditional commodities, a number of small commodity associations or groups are being formed across the region.

3.3. Extension by Farmers' Associations

3.3.1. National Associations

The most prominent national associations over the years are the Agricultural Societies – the Jamaica Agricultural Society (JAS), the Barbados Agricultural Society (BAS), and the Agricultural Society of Trinidad & Tobago (ASTT). They are all over 100 years old, and at their peak wielded considerable influence on agriculture. They tend to command the most attention from governments as the voice of farmers. These groups receive funds or a subvention from the national governments but may sometimes also get funds from various other agencies. JAS receives considerable funding from government and is the only one that has several field staff in the various communities. ASTT has minimum field staff. The Agricultural Societies get information to members and the public in various ways such as agricultural exhibitions, field trips and various types of publications and strong advocacy.

3.3.2. Commodity Based Associations

A number of commodity- based organizations exist in the various countries that represent the interest of the farmers in their various enterprises or commodities, assisting them to better manage challenges with a common front. They have also been important for sharing information especially where the "official" knowledge systems had not been able to provide access to relevant information or mechanisms to advance the industry. For example, the Horticultural Society of Trinidad and Tobago, which is almost 100 years old, has been credited with providing information and services which support the development of the floriculture industry.

Some other farmers' associations that have field staff are the Guyana Rice Growers Association, the Toledo Cocoa growers association in Belize and to a lesser extent the Grenada Cooperative Nutmeg Association and the Citrus Growers Association in Trinidad. They liaise mainly with farmers and do provide some limited extension support.

3.4. Other Extension Providers

3.4.1. Research and Development (R&D) Institutions

In general, there are limited technology transfer opportunities associated with on-farm trials, workshops and publications, and research institutions have little or no extension staff. However, some projects have "Technology Transfer and Adaptation Specialists" or similar staff who may interact with farmers on an ongoing basis, through their research programs. Generally, they are few in number and can only reach a limited number of farmers. Some R&D institutions also have information and communication units which produce publications such as factsheets, technical bulletins, manuals and other types of materials in print or electronic format, to disseminate research findings. The main R&D institution in the region is the Caribbean Agricultural Research and Development Institute (CARDI), which provides on-farm demonstration and on-station demonstrations and trials, through principal scientists. In the few countries where national Research Stations exist, staff may occasionally provide some limited extension service to farmers.

3.4.2. Input Suppliers

Agricultural Input suppliers, also called "agro-shops" - those who sell various types of agricultural chemicals, animal feed, machinery and equipment - now promote themselves as going beyond selling products. They consider themselves as "friends" of the farmer who have genuine concerns for overall agricultural development. They participate in and help to sponsor agricultural events. Some do R & D as well but their activities are largely technology transfer on the products they sell. They are generally quite willing to collaborate with government, the University of the West Indies (UWI), and other organizations in the agricultural sector.

3.4.3. NGOs

The importance of NGOs in agriculture in the region has been growing, especially those with an emphasis on rural development. However, they are not yet widespread and the reach of those that exist is generally small. In mainland countries such as Belize and Guyana there are greater numbers of NGOs than other countries. While NGOs have special mandates, they often have extension staff who work with target clientele (farmers, women and children).

3.5. Fisheries and Forestry Extension

Fishery and Forestry extension are relatively new in the region. Some units have had to "borrow" staff trained in agricultural extension to develop their services. However, increasing emphasis is now being placed on fisheries and the extension units are developing regionally. Forest conservation is now on the agenda in the region and it is expected that extension capacity will improve.

4. CHALLENGES OF EXTENSION SERVICES IN THE CARIBBEAN

de Freitas (2006), noted that extension involved empowering farmers to identify and articulate their needs and to obtain the required knowledge and skills necessary to negotiate and manage their environment. Hence, Extension Officers are expected to become facilitators in a relationship between the client/producer. They are charged with the responsibility of addressing not only issues relating directly to farming, but also supporting the delivery of other services to facilitate the overall development of the farm family and rural community. Several challenges face the Caribbean Extension community, some are historic and some are new issues arising from more recent developments:

- High expectations due to changed view of agriculture. Extension officers are now expected to deal with the production cycle as well as food and non-food products along the entire value chain of a range of commodities.
- Weak linkages between agricultural research and extension.
- Large numbers of small farmers with a wide range of commodities and farming systems.
- Varying levels of functional literacy and low levels of computer literacy among farmers.
- Decentralisation of extension services suffering from poor coordination and management.
- Extension infrastructure, systems and approaches not yet reconfigured to embrace a new culture of participation and empowerment.
- Limited access to and application of new and appropriate farming techniques for the region.
- A general trend of poor human and financial resources for Extension Services throughout the region making for low contact levels with clients on a timely basis.

It is with respect to this last challenge that the advent of ICTs has most meaning, providing greater opportunities for interaction, more in keeping with current extension practice. Furthermore, with more actors involved in the development process, Extension Officers are no longer the sole voices reaching farmers. Thus modern tools and techniques must be adopted that will enable extension to be the source that clients turn to in quick time to provide timely information, clear up conflicting information, validate information sourced elsewhere *etc.*

5. NEED FOR ICTs IN CARIBBEAN AGRICULTURE

The role of Information and Communication Technologies (ICTs) is popularly held to be very critical to economic development, with potential to be used as both an engine of economic growth and for delivery of products and services. ICTs facilitate collection, organisation, storage and dissemination along with communication and exchange of data, information and knowledge between and among groups. While the "old" or traditional technologies such as the fixed line telephones, radio, television, print, video and film are still widely used, there has been increased, albeit slow, adoption of "new" technologies such as computers and telecommunication tools including the Internet and Mobile telephony.

ICT is a major cross-cutting issue in development. So important is the issue that Caribbean Community (CARICOM) Ministers responsible for Information and Communication Technology agreed "to an aggressive and focused strategy to accelerate the adoption of ICT for development" (CARICOM, 2004). A Regional ICT Steering Committee was formed in 2007 and has secured funding from the European Development Fund as part of a project to develop the Caribbean Information Society (CARIB-IS). The CARIB-IS project is programmed to involve key Information Society (IS) stakeholders including ICT/IS personnel, ICT and telecommunications regulators and operators, thematic stakeholders, the private sector and civil society (Eurostrategies, 2004). To date, both Trinidad & Tobago and Jamaica have begun to implement national IT policies. In the Organisation of Eastern Caribbean States (OECS), a sub-grouping within the CARICOM, countries operating within the framework of the Eastern Caribbean Telecommunications Authority (ECTEL) have initiated but not yet completed work on ICT policies.

The use of ICTs in agricultural extension in the Caribbean region, similar to many other developing regions, will be influenced by its use within the wider society and particularly in the agricultural sector. All the countries have recognized the value of an information based society and the opportunities it provides to citizens in the areas of business, education, personal development and entertainment. Definitely, the work of Agricultural Extension in the region has been enhanced over the years with the introduction of new ICTs.

There are however, some challenges which impact on the rate and extent of ICT adoption and use in the Caribbean region, and to the extent to which these challenges are not addressed, farmers, rural producers and indeed overall agricultural development will be affected. de Freitas (2007) outlined the following challenges faced by the ECTEL states, challenges which are also present in the wider Caribbean region:

- Lack of comprehensive quantitative data on ICT use at national and regional level
- Low levels of ICT penetration
- Absence of a well coordinated approach to ICTs both nationally and regionally
- No single entity responsible for ICT
- Significant shortfall in deployment of ICT

These challenges must be addressed so that the region and the Extension services in this case, can reap greater benefits from the application and use of ICTs. The following section provides an overview of the use of ICTs within agricultural extension in the Caribbean region presently.

6. USE OF ICTs WITHIN AGRICULTURE AND EXTENSION

Campbell (2003), in a paper prepared for the CTA ICT Observatory, noted that information about use of ICTs has not been widely researched to date. This situation still exists today; hence information for the purposes on this paper was obtained based on presentations made by Extension staff and other Ministry staff at workshops, reports, feedback from colleagues in Extension across the region and participation in joint projects executed with the involvement of Extension agents. Experiences and outcomes from several studies spearheaded by CTA and completed with support from CARDI were also used extensively in this chapter (CTA,1997; CTA,2005(b); CTA, 2008(b)).

Since the 1990s, Extension Divisions or units, usually located within the Ministries of Agriculture, have moved from a situation of very limited access to hardware and software to a position where there is increased emphasis on the acquisition and use of ICTs on a day-to-day basis. Extension agents within the respective Ministries of Agriculture in the region are often in charge of Information or Communication Units, and are responsible for dissemination of information. These officers now use ICTs to provide support for the preparation of factsheets, brochures, for mounting exhibitions, in providing training, delivering public lecture presentations, and for the development of radio and television programmes. These units are good examples of the integrated use of old and new ICTs where more traditional printing and reprographic technologies are still in operation.

Several countries have benefited from donors and technical support from equipment supplied *via* projects through agencies such as the Commonwealth of Learning (COL), the CTA, the Inter American Institute for Cooperation on Agriculture (IICA), and other technical support agencies (COL, 2002). Local governments also continue to provide ICT tools as part of basic printing and communication equipment such as desktop computers, laptop computers, printers and photocopy equipment. Digital scanners, cameras and video equipment are less popular.

6.1. Country Status

Extension officers in *Antigua & Barbuda* have been providing technical assistance and manpower to facilitate training through courses offered by a local NGO, the Gilbert Agricultural Research and Development Centre (GARDC). GARDC has a fully equipped networked computer lab, but very limited full time staff, while the Extension Officers do not have access to the training facilities and ICT equipment available at the GARDC. Courses are conducted in use of ICTs for record keeping, for the preparation of business plans and for accessing information relevant to the business *via* the Internet. The Centre also hosts a Website that can support presentation of courses, and materials that can support agri-businesses or community projects. This present situation provides an ideal opportunity for collaboration by Government Extension Officers and GARDC, all to the benefit of the local farmers. Extension division uses e-mail for correspondence and presentation software for training, and has the potential to place information on the website as part of a Government of Antigua & Barbuda gateway. A bibliographic database provides access to information held in the GARDC Library. In 2007, the Government of Antigua & Barbuda agreed "to invest in development of traditional media (radio, TV) and modern ICT infrastructure, which is widely accessible", as part of efforts to build their national capacity in technology and to promote "Antigua as the Regional Centre of Information Communication Technology Excellence". Plans are in place to reduce the digital divide and provide ICT access at the community level (CARICOM, 2009).

Barbados's Ministry of Agriculture used ICTs mainly for accessing information, for correspondence, compilation of trade and production data as well as for information dissemination. These activities are carried out in collaboration with other stakeholders in particular the Barbados Agricultural Management Co. Ltd. (BAMC), using customized software called *AGRICAST*. The Website is used as a conduit to the Ministry's Question and Answer Service (QAS) and for online searches of its Library (Ministry of Agriculture and Rural Development, 2009).

In *Belize* ICTs were used mainly to access technical information from the Internet, for word processing, for messaging and reporting between supervisors and staff and to access information from the agricultural library's database. Presentation software is becoming popular for use in training and seminars. A limited number of NGOs, farmers' and commodity associations maintain websites which provide access to information and news for their members. The Ministry of Agriculture Website, still in development, may be accessed through the gateway for the Government of Belize (Government of Belize, 2009).

The Ministry of Agriculture in *Dominica* has developed a plan called Agricultural Information and Management System (AIMS) designed to support the provision of a reliable agri-business intelligence service to key stakeholders in the agricultural sector, including improved management and production of local agricultural statistics. Implementation of the plan has been limited by funding and access to skilled personnel. At this time, ICTs are used for development of databases, for presentations, to provide information to farmers, and for publication of newsletters and fact sheets.

Grenada's Ministry of Agriculture has continued efforts to use ICTs for the establishment of databases, including a marketing database for analysis and dissemination of user friendly marketing information. An Information Resource Center was organized in 2008, while the Ministry initiated policies designed to capture important local data and reports. A Public Relations Officer was employed to disseminate information. The Communication Unit in the Ministry has acquired more advanced equipment, and a decision has been made to improve staffing levels and skills in ICTs. This will contribute to the Ministry's efforts to work in collaboration with the local marketing agency to develop a single platform or Agri-business Information System (AGBIS) to provide timely and accurate production, market and trade information. Information on weather, e-learning opportunities, events, technical information related to livestock, crop and fisheries sub sectors is planned, along with a discussion forum. This forum aims to bring the farming community, scientists, policy makers and extension workers together onto a single platform to exchange ideas.

In *Guyana*, more traditional ICTs are mainly used. However, there is increasing use of presentation software such as PowerPoint by extension for training, fairs and exhibitions. The St. Stanislaus College Farm has put a number of agriculture-related information on CDs and these are being used as teaching tools for both farmers and students. Students in the Faculty of Agriculture at the University of Guyana are being taught the use of presentation software for use in Extension and other outreach activities. ICTs were used to assist in the running of small scale-enterprises and to support the preparation of business plans, and the production of articles, brochures, reports and photographs. The Government of Guyana has developed Agri-Net - Guyana's Agriculture Information Network (GAIN), in collaboration with the IICA, to provide information to users including farmers and extension officers (Agri-Net Guyana, 2009).

Jamaica's Rural Agricultural Development Authority (RADA) established a website which is a "One-Stop-Shop" for the provision of business counseling, technical and marketing services, and the facilitation of investment. It features an Agricultural Business Information System (ABIS), described as "a web based information system to provide information on crops, livestock and agricultural production, markets and stakeholders to support the business of agriculture" (Ministry of Agriculture Jamaica, 2009). ICTs are indeed at the heart of its implementation approach and support activities such as providing continuous training for farmers, providing marketing intelligence services to farmers and intermediaries, and facilitating liaison with research and development stakeholders to deliver technology to farmers. The RADA is also making use of ICTs to update the farmers' database. Some frontline extension agents are provided with Blackberry mobile phones through which the officers can communicate with farmers and provide timely responses to requests for information and advice to farmers. Some Extension staff also have some experience with GPS for collecting and recording farmer data. Other government agencies such as the Land and Surveys Division use ICTs for land registry thus making it easier for farmers to better manage land. At the local level, Internet cafés are accessible *via* district offices, *e.g.* post offices, facilitating Internet access at a community level.

St Kitts/Nevis is one of the countries which benefited from the training provided to Extension Officers in the use of cameras and audio/visual reproduction equipment under a Commonwealth of Learning Media Empowerment (COLME) programme. Studies by the COLME project team, local extension staff and farmers, recognized that video was the best medium since many of the farmers already owned television sets and VCR/CD players. Use of multi-media technology greatly extends the reach of each extension officer and allows the capture of local content.

ICT use in *St Kitts and Nevis* is mainly centered on print and broadcast, with a relatively slow emergence of e-mail, websites and electronic networking by Extension agents. Standard radio and TV programmes are aired (Walker, 2003). The Public Library in Nevis makes local factsheets and other technical information available to Extension agents and farmers through searches on its bibliographic database.

Extension services in *St Lucia* are well serviced by the Information Unit, comprised of the Documentation Centre and Communication Unit, which has a reasonably well developed capacity of print and multi-media, with less capacity for managing text, maps and databases. The basic hardware and software for developing information products is available, however access to facilities for dissemination of information - reprographic and audio-visual equipment - is limited. Human resource constraints are also experienced due to under-staffing and a limited range of in-house skills in the use of ICTs. The Extension Division is developing a model to use the Ministry's website to feature local multi-media training material developed specially for farmers. ICTs are expected to enhance efficiency and effectiveness, facilitate monitoring and evaluation, assist with programme planning, and support training for users. Efforts continue to upgrade the Information Centre and use the Website as an interactive portal for information dissemination and training to farmers and other stakeholders (Ministry of Agriculture, Forestry and Fisheries, 2009).

In *St Vincent and the Grenadines*, the Ministry uses ICTs for traditional information sources such as technical advisory services, reports, statistics, publications and databases. Dissemination of information is *via* conventional media, print and extension services. Within the last five years there has been increasing recognition and integration of new and 'old' ICTs. The use of radio call-in programmes for farmers, coupled with providing answers to questions *via* mobile phones has been proposed. An innovative model of farmers' organizations partnering with Extension services to digitize and publish factsheets has been successful. ICTs have also been used to support the implementation of a Question and Answer Services (QAS) in collaboration with CARDI and CTA. The bibliographic database allowed easy access to information, which was used to answer questions put to the Agricultural Information Resource Centre.

Trinidad & Tobago is one of the few countries in the Caribbean that has a National ICT Policy. The Ministry of Agriculture has developed a national initiative to integrate its information into a one stop shop, as part of the national e-Government Programme. Within this Ministry and among key statutory agencies, ICTs are used for database development, including farmers' databases, training, and managing state lands. Software development has been customized for various institutional activities *e.g.* to manage contact lists, prepare and maintain farmer databases and collect, organize and disseminate production, marketing and other statistical data. The Website presents information on upcoming farmer training courses and provides access to technical information for farmers (Ministry of Agriculture, Land and Marine Resources, 2009). The marketing aspect of the NAMDEVCO programme is ICT driven, where price and volume information is collected on a daily basis with the use of Palm Pilots. The information is then computerized and disseminated close to real time on a daily basis, for the benefit of farmers and consumers. Since 2005 an information system called the National Agricultural Marketing Information System (NAMIS) has provided this information to all stakeholders in the agricultural sector *via* their website. The Extension Division has a Mobile Learning Unit which is essentially a modified fully air conditioned 25-seat bus, fully powered and equipped with computer and presentation equipment, which goes to rural areas. A facilitator uses prepared multimedia

CDs on various topics to provide training to communities on board the bus. Recently, in Trinidad, a curriculum has been developed for the training of farmers to enable them to access information *via* the internet and training is due to begin shortly, since many farmers have access within their homes mainly for their children's use.

The Extension divisions are all making use of both old and new ICTs in delivering current services to a greater or lesser extent based on resources invested in the technologies. Integrating old and new technologies is the current practice, but with low penetration of hardware and software, use of new ICTs is not ideal. Some 50 per cent of countries presented have websites dedicated to the agricultural sector, while the Ministry has a presence as part of the Government gateway. To date, the Extension Services in Jamaica appear to have the most targeted use of ICTs for the purpose of communicating with their clients. St Lucia and Trinidad & Tobago feature Extension Services as part of the Ministry of Agriculture Website. Barbados and Guyana have dedicated links to sources of agricultural information and data. There is also some use of ICTs in related areas which will ultimately contribute to more effective extension services such as production and marketing agencies, trade and bi-lateral entities, research and development organizations and other government agencies.

7. TRACKING INFORMATION, COMMUNICATION AND TECHNOLOGY NEEDS AND CAPACITIES

The information revolution has resulted in a proliferation of better information access, tools, and facilities. Hence, according to Gachie and Ruault (2006), a larger volume and range of information is available through portals, websites and gateways. Publications can be done by anyone with access to the hardware and software. Communication is available in real time allowing easy networking, sharing and exchange of information. Within the agricultural sector, ICTs have been used to support activities in the following broad areas (IICD, 2006):

- Facilitate a conducive policy environment
- Obtain increased profitability of small farms
- Provide increased market access
- Support improved production efficiency
- Enhance and empower farmers and the rural community in general

They have been used specifically to:

- Coordinate and develop systems for managing various projects and activities within the agricultural sector
- Disseminate technical information relevant to agriculture
- Post prices and other market information
- Identify markets and sell online
- Manage agricultural and information resources
- Better communicate and network between and among stakeholder groups

The use of ICTs should be determined by needs. These include direct and indirect information needs, communication needs as well as the need to solve problems and manage challenges. Information needs of key stakeholders including farmers and extension agents have been identified *via* evaluation at workshops and seminars, feedback at meetings, direct requests from respective stakeholders, formal surveys, consultations and trend analysis.

ICT was identified as a major cross-cutting issue affecting agricultural development and thus was seen as a means to an end and not an end in itself. A long standing partnership between the Caribbean Agricultural Research and Development Institute (CARDI and the Technical Centre for Agriculture and Rural Cooperation (CTA), made significant contributions to meeting information needs and providing policy support linked to Information and Communication Management (ICM). A two-pronged approach emphasized improving both access to relevant information products and services, and developing the capacities of regional stakeholder groups in information and communication management.

Historically (CTA, 1997), extension agents highlighted the need for further technical support for the delivery of information and advice to farmers. They pointed specifically to the need for information on pests, diseases and marketing, and for training in photography techniques, radio broadcasting, script writing and mounting exhibitions. Farmers indicated they required access to information on pests and diseases management, marketing and trade. The farmers also preferred information delivered *via* on-farm visits or personal contacts, although they also utilized radio, television, video and printed materials. This 1997 agricultural information needs survey confirmed that there was a general lack of formal mechanisms for organizational collaboration, accessing and managing information, and not much use of electronic information sources for disseminating and communicating information. In addition, there were no strategic marketing and communications plans for developing and targeting existing agricultural information products and services.

Campbell, (2003) recorded that 'agricultural extension relied on face to face, print and radio communication' (CTA, 2005b). A report on Agricultural Information Needs (CTA, 2005; (a)), identified "word of mouth", personal connections, extension visits and publications as the preferred channels of communication for users. Meetings, workshops, email and press releases were also used to exchange information. However, radio programmes, TV and multi-media were still not well utilized as key information sources. The report also showed that although the use of the Internet had greatly increased, there was still a dearth of information relevant to the Caribbean farmers available on the Internet. A few agricultural libraries continued to provide access to technical information mainly in print formats, although the organization of some in-house collections had improved to some extent. Even in 2008, stakeholders continued to indicate a preference for personal communication to obtain information. While there was more significant use of the Internet, email and online information sources, information gaps still existed in relation to current technical agricultural information in a variety of subject areas, particularly local information and data (CTA, 2008;(a)).

7.1. Farmers' ICT Needs, Capacities and Use

ICT capacity, information needs and other challenges of Caribbean farmers are also important since farmers are the main clients of the extension services. In 1999, CARDI worked with the German Agency for Technical Cooperation (GTZ) and the Jamaica Agricultural Society (JAS), to pilot the use of Information Cafés to provide information to farmers in rural communities. The results of that exercise revealed that more use was made of the Information Café's services by students and teachers than by farmers. Even when farmers used the service, it was to request non-farming information and or obtain services such as photocopying and printing for other purposes.

A number suggestions were offered based on these findings, including the need to provide assistance to access the information - either to guide in the use of the computers and Internet for searching, or for interpretation and analysis of the volume of materials accessed by farmers in the Internet cafés. Locally relevant information with advice on specific commodities and conditions most often requested, was not available on the Internet. There were also difficulties in accessing information which was not packaged in a manner that would facilitate easy understanding by farmers (CAIS, 2004).

Seepersad (2003) noted that while small farmers were making limited use of the internet, the highly educated floriculturists were using it regularly for their development. Users of a CTA/CARDI-supported Question and Answer Service (QAS) across six Caribbean countries comprised approximately 7-9 per cent of farmers and 5-13 per cent extension officers over the period 2005-2006 (CTA, 2005a; 2006b).

The report on the Survey of over 150 farmers' organizations and rural NGOs in 13 Caribbean countries highlighted the ICM/ICT status of groups ranging from less than 100 members to those with upwards of 10,000 - 50,000. Approximately 76 per cent of the groups owned telephones, while 68 per cent had computers, with less than 50 per cent with cameras, VCR/DVDs, televisions, radios, or photocopy machines. Over 50 per cent of employees in farmers associations had communication skills in computer science, audio-visual use and journalism. The main sources of information for those farmers' groups were meetings/conferences, followed by Extension Officers, newsletters, magazines and the Internet. Information exchange was identified as one of the main purposes for setting up a regional network of farmers' groups. Marketing information, information on training and education, as well as access to information on general agricultural production, was stated as the top three priorities. Problems linked to communication capacity of the respective organizations related to lack of proper infrastructure such as office space, basic communication equipment, insufficient human resources - numbers and range of skills, along with limited finances (Rajack-Talley and Lawrence, 2004).

7.2. Regional Responses to Support ICT Use – Training and Use of Networks

Based on the results of the various needs analyses, efforts were made to create an entity called the Caribbean Agricultural Information Service (CAIS), which was designed to improve capacity for information and communication management (ICM). ICTs were identified as only one aspect of ICM with efforts to support network development, design and development of information on products and services, as well as provide technical assistance in ICM. The CTA/CARDI partnership organized and hosted several courses in the last few years, which provided regional training courses and seminars to build capacity in ICM. Training included topics such as:

- Web design and development
- Strategic management of ICTs
- Communication management

- Public awareness
- Management of electronic networks
- Management of question and answer services (QAS)

One course was especially for Extension agents in the field of communication - electronic production and publishing of agricultural extension materials. Over twenty extension agents from across the Caribbean were exposed to using ICTs to design and develop customized solutions to meet information needs and empower farmers. The concept of using a participatory approach and involving farmers/clients in the design of the products was promoted (CTA, 2006a).

Other institutions such as the Inter American Agency for Cooperation on Agriculture (IICA) spearheaded the development and hosting of online and CD-ROM training courses for farmers and technicians in areas such as computer literacy, farm management, and production of selected commodities. National governments also launched training programmes in computer literacy and use of basic software and hardware. Staff members of several Ministries of Agriculture, NGOs, farmers associations, and research and policy institutions were able to improve skills in ICM and identify potential use of ICTs in their day to day operations.

Some 12 commodity and thematic networks were set up over the period 1998-2004 to encourage networking, including *via* face to face, print and electronic means, and for ongoing exchange of ideas and information. A 2003-04 evaluation of PROCICARIBE - the Agricultural Science and Technology Networking System (established in 1996) - pointed to weak capacity for research and development and a poor culture of networking among the reasons for the failure of all but a few of these networks (Asiedu, 2005). Many Extension Officers and farmers were included among the stakeholder groups in these networks.

8. CHALLENGES FOR ICT USE IN EXTENSION SERVICES

Studies on Agriculture, Science, Technology and Innovation (ASTI), spearheaded by CTA, determined that the Caribbean region, as with African Caribbean and Pacific (ACP) countries, "do not sufficiently and adequately apply science, technology and innovation to facilitate meeting the challenges associated with improvements in agriculture" (CTA, 2006c). This also applies to the use of ICTs which have already been identified as having great potential to support development and poverty alleviation.

On a more global level, while ICTs have been used to support the development of industry and delivery of services, there has been very limited application in development matters such as poverty alleviation and rural development. However, pre-requisites for ICT use are not only the access to the equipment and tools, but also the knowledge and skills for application and use of same. Relevant institutional frameworks, together with appropriate strategies and policy frameworks are also critical for success.

The ICT Development Index, a statistical tool designed to measure the relative application and use of ICTs on a comparative global scale developed by the International Telecommunications Union (ITU), revealed that while the Caribbean region is among the top 50 per cent of countries in the area of ICT access and infrastructure, they fall below the expected level of ICT use based on their current GDP levels (ITU, 2009). From a global perspective, the region is lagging behind its potential position in becoming an information society. This is reflected in the relatively slow adoption of ICTs at regional and national level in both business and development projects.

Insufficient attention has been paid to targeted training to increase the skills of the users overall. Issues relating to the sustainability of systems, work flow processes and how the activities within respective institutions would or could change with the introduction of new technologies, are not greatly considered. There is at present low capacity to promote greater application and use of ICT for productive development. This speaks to the need to increase human resources and equipment, strengthen institutional and policy framework, further develop local content, strengthen the integration and networking among key stakeholder groups and address socio-cultural issues such as 'information insensitivity', indigenous knowledge and the use of more participatory approaches (Henry, 2009). All of the above must be addressed at a national or even sectoral level.

9. THE WAY FORWARD FOR INCREASED ICT USE IN CARIBBEAN EXTENSION

There is a real risk that in times of global food and financial crises, poverty and hunger could be experienced in developing countries including the Caribbean region. At both the regional and international level, there has been recognition that in response to the global crisis in food security and energy, agriculture must change (World Bank, 2007). Chesney posits that the Caribbean region must respond appropriately and quickly to meet potential opportunities, along with the challenges that accompany them. He recommended a "new agriculture", which should examine activities along the value chain and view agriculture as both food and non-food items (Chesney, 2007).

Results from studies by the International Institute for Communication and Development indicated that ICTs can increase empowerment and income of participants in the agricultural sector, based on its experience over two years (IICD, 2006). ICT interventions provide support for the development of the agricultural sector in areas such as improving knowledge on policy environment of the agricultural sector, increasing market access and productivity efficiency of farms, and increasing empowerment and social inclusion of rural communities.

Barker, (1997) suggested that production on small farms increased if more attention was paid to improving the quality of the frontline extension worker. His research pointed to the need to intensify support from research, involve personnel from other relevant agencies that support the implementation of extension programming, and the need to establish co-ordinating mechanisms at the local, national and regional levels. Based on this preamble, several suggestions to enhance ICT use in improving the impact of extension on farming are detailed below.

9.1. Reconfigure Extension

Agricultural Extension is changing worldwide, with emphasis on innovation (Saravanan, 2008) and increased involvement of other rural groups besides farmers. Access to information is critical to rural communities in this new global environment, if they are to constructively participate in new opportunities and successfully manage the challenges. The "new agriculture" being promoted in the region requires a different set of extension, educational and research skills and programmes, and is organically linked to other economic sectors including tourism, education, health, transport, trade and finances (Chesney, 2007).

In the context of ICT use, we embrace and promote an extension service that is configured to facilitate the transfer of information from both the global knowledge base and local research, to farmers and rural communities, with the objective of development. Such development includes empowerment of the local communities and equipping them with tools and skills to manage their growth and maximize the use of their resources (Richardson, 2007).

The region therefore needs to reexamine and restructure Extension if it is to serve the needs of all of its clients in the context of the new global environment. A situational analysis might be useful to support this. Such an analysis should include an inventory of ICTs for extension, and must detail the skills, infrastructure, resources, capacity of farmers and officers that are needed to transform the sector. Extension workers must be empowered through technology to:

- Be able to reach out and work more closely with each other and also with other agricultural professionals, especially those in Research organizations at national, regional and global levels.
- Design and develop appropriate learning experiences to reach and educate farmers.
- Be in direct link with specialist officers in the country and elsewhere in the region to respond quickly to farmers' problems and queries.
- Continue to assist in development and repackaging of materials such as videos, factsheets, technical manuals, and in supporting services such as databases, skill banks and question and answer services, specifically tailored to the needs of clients.

Moreover, extension officers require an expanded skill-range for them to better understand ICTs and their potential for use to provide innovative solutions to the host of challenges being faced in the community. Extension agents must also develop the capacity to act as conduits to bring ICT tools to their communities. Training for Extension officers should include the use of multi-media, alternative media, and other ICTs relevant and appropriate to their needs such as preparation of CDs/DVDs, participatory video, web 2.0 tools, Internet Radio, streaming video, and other evolving technologies. Skills training must also cover methods to wisely integrate old and new technologies and use that effectively and appropriately given the content and the audience.

This is a medium term to long term development as most countries will require some time to move forward given the present ICT situation in each country. In the short to medium term however, in the extension institutions, experts are needed to repackage, analyse and make information more locally relevant to the various audiences, in particular, farmers and rural communities. Skills in advocacy and sensitization, networking and collaboration will be necessary since a multi-disciplinary approach will be necessary to address the range of needs and audiences. This will assist the institutions in extension to become better users of ICTs, while raising awareness of the potential the technology provides to meet the objectives of the work programmes and empower farmers and the rural community.

9.2. Develop Policy and Institutional Frameworks

Much of the effort required to address the broader issues of communication and knowledge exchange falls beyond the remit of researchers, extension and the Ministries of Agriculture in the region. An holistic approach is needed. Given the cross-cutting nature of the ICTs, Government policy on ICT in each country must be designed to ensure that all benefit; individuals and farmer groups, NGO's, and state assisted organisations must all have the opportunity to move ahead. Extensive country-consultations followed by regional consultations with all potential users, to identify needs and current uses of ICT in agriculture, is a good starting point, even for countries which already have national ICT policies. Ministries of Agriculture and Extension Services in particular should 'piggy-back' on existing projects and activities such as the CARICOM Regional ICT Steering Committee. This committee will support the development of appropriate policy frameworks for the region through use of networks, development of appropriate ICT and information society indicators, and also promotion of information-sharing, discussion, sharing of experiences and fostering innovation (CARICOM, 2007).

9.3. Improve Access and Build Capacity

Ministries of Agriculture and other statutory agencies are still major repositories of information useful for agricultural production and farmers. There is also only one major research institute (CARDI), governed by CARICOM which has a mandate to work in collaboration with respective national governments and other regional and national research and development stakeholders. While some information resides with Commodity Associations and other special interest organizations, the present situation is clearly insufficient to meet ICT needs of the region.

According to the ICT Development Index, the Caribbean region has relatively high scores in ICT access and skills, but somewhat lower than expected application and use in business, education and development. It will also be necessary to ensure that users of modern ICTs use these tools in pursuit of their agricultural development goals. Further, ensuring that users are able to use and or reach relevant information through intermediaries or combination of old and new ICT technologies is vitally important. This will truly facilitate better access to information by those most in need.

There is a need to fuel innovation by improving the capacities of persons to retrieve and interpret information. The region should strive to go beyond computer literacy to reach information literacy, where users are better able to access, evaluate and use information in their everyday life. In addition, ICTs should be examined for options to improve systems, processes and business models currently in existence.

9.4. Promote a New Culture

Cultural changes, including the development in the region of "information sensitivity" - an appreciation of the usefulness of information - and willingness to think outside of the box, will support innovation (Henry, 2009). Critical thinking, participatory approaches and feedback mechanisms are also to be encouraged. Such innovative approaches are necessary, along with the design and development of modern information and communication management strategies that would identify the best policies, frameworks and resources to support implementation of Extension programmes. And, while, as this paper shows, general capacity issues are known, to address the problem requires more explicit information. Information and Communication management (ICM) strategies should be prepared to support the implementation of Extension plans in the context of Ministries' strategic programmes.

9.5. Build on Present Strengths

The Internet is a key source of information for extension agents and farmer associations. As such, the development of local and Caribbean content suitable for the Internet should be encouraged. Much of the present documentation, technical information and data relevant to the Caribbean should also be collated, displayed and digitized for access. In addition, recognizing that the use of the mobile telephone has become so wide-spread, it can become a cost-effective tool for accessing information by the population including rural persons. Increased efforts must be made to develop ways to provide relevant information and structured networking *via* this method.

There are a few organized interventions using ICTs for extension support. For example, the use of Blackberries to collaborate with farmers in Jamaica, the use of PDA's to capture and transmit data by the NAMDEVCO, and the use of a Mobile Learning unit for farmer training in Trinidad and Tobago, have been identified as more formal and institutionalized examples. There also seems to be a number of new approaches and pockets of innovation being implemented or championed either individually, at unit, or department level that may not yet be mainstreamed at the institutional level. Some examples include the exploration of mobile phones for supporting question and answer services and the linking of QAS to radio talk shows focusing on agriculture in St Vincent and the Grenadines, use of interactive multi media training materials on the Web in St Lucia, use of interactive CDs for training farmers in Trinidad and Tobago, and the partnership between the GARDC and Extension Services in Antigua & Barbuda. All these initiatives require appropriate institutional frameworks, policy support, adequate funding and human resources. These are certainly examples that can be shared and perhaps replicated elsewhere in the region where appropriate, through networking of Extension Officers.

9.6. Encourage Youth

Encouraging youth may assist the use of ICTs in the agricultural sector, and at the same time encourage or attract the addition of a younger pool of farmers to a group where the average age is over 60 years. As elsewhere in the world, youth and ICT use seem to be almost natural. ICTs can therefore be used to encourage wider participation of youth in agriculture, not only to attain technical, professional and business skills in agriculture, but also to promote opportunities for innovation by youths to assist the development of the agricultural sector. Such opportunities are available at the UWI, and various colleges of Agricultural education in Trinidad, Jamaica, Guyana and Belize, as well a range of Youth in Agriculture Programmes in the region *e.g.* the Youth Apprenticeship Programme in Agriculture (YAPA) in Trinidad and Tobago, which targets young people 17-25 years of age and the host of 4-H programmes which targets a wider age category and is particularly strong in Jamaica.

9.7. Develop and Use Networks and e-Forums

Communication is necessary to share and exchange information and ideas, and to support innovation. Therefore, the encouragement of linkages and exchange of information/discussions both within extension groups and also across stakeholder, cultural, country and institutional barriers must be encouraged (Gandelsonas, 2002). A number of national and regional networks already exist; however, it is recommended that a network or Community of Practice for Caribbean Extension Officers be established. This network should position itself to link with other existing networks.

Local, national and regional producer networks such as the Caribbean Farmers Network (CaFAN) and the Caribbean Association of National Fisherfolk Organisations (CANUFO) can provide good fora for community learning, social interaction, collaboration and resource sharing (de Freitas *et al.*, 2006). Other groups such as the Caribbean Agricultural Forum for Youth (CAFY), the Caribbean Network of Rural Women (CANROP) and the Caribbean Agri-business Association (CABA), all have the potential to become increasingly important as vehicles for sharing information among and between stakeholders groups such as producers, youth, women and business groups. Networks and groups could also serve as important channels for training using the application of ICTs in the rural areas. e-Forums can become important means by which expertise between and among extension in the region and already existing producer/stakeholder networks, can be shared.

The Food and Agriculture Organisation (FAO) has developed a model known as the Virtual Extension and Research Communication Network (VERCON) for promoting networking and communication between agricultural research, extension, farmers and their associations, and policy makers (O'Farrell, 2003). The Caribbean may examine this existing model for experiences and relevance to the region. While policy frameworks are being developed, the Extension agents can begin to network, support each other and also learn from one another. The consequent benefit expected is improved human resource capacities and a stronger interface with farmers and rural communities.

10. SUMMARY AND CONCLUSION

The use of ICTs in Extension services in the English speaking Caribbean in the future will be based on the vision for agriculture in the region. IICA (2007) noted that agriculture must be seen as more than helping poor farmers and must be viewed as a strategic sector in Caribbean development; that there is sufficient market demand for the output of the sector to make it potentially profitable. Bourne (2008) stated that public policies which address structural constraints on agricultural performance and on the marketing and distribution of agricultural output are required. Extension Services should make every effort to build on the current strengths and experiences from application and use of ICTs both within the agricultural sector and other sectors in society. The existing capacity and infrastructure with respect to equipment, education and literacy levels puts the region in a position of readiness to make good use of ICTs in its development programme. However, greater emphasis must be placed on raising awareness, improving the relevant skill sets and improving 'information sensitivity', so that both extension officers and their potential clients can benefit from the use of ICTs in helping meet key objectives within the agricultural sector of the region.

References

Agri-Net Guyana. 2009. *http://www.sdnp.org.gy/minagri/* (retrieved 27.03.09)

Antoine, P., 2009. *Implementing a CARICOM Community Agricultural Policy (CCAP): Issues, Options and Process.* IICA, Port of Spain.

Asiedu, F., 2005. *PROCICARIBE Mechanism: Evolution, Achievements and Future Directions.* 4th FORAGRO Meeting, Panama, 13-15 April 2005.

Barker, S.C.P., 1997. The impact of farming systems extension on Caribbean small-farm agriculture: The case of St Kitts and St Vincent. *Tropical Agriculture,* 74 (1), Pp: 58-63.

Bourne, C., 2008. Perspectives on enhancing sustainable growth and development of Caribbean agriculture. Keynote Address by the President, Caribbean Development Bank, *Forty-Fourth Annual Meeting of the Caribbean Food Crops Society (CFCS),* 14 July 2008, CFCS. Miami, Florida.

Campbell, D., 2003. The relevance for and experience made in the Caribbean region in using ICTs for agricultural extension for the benefit of small scale farmers. *CTA/ICT Observatory for ACP agricultural development – "ICTs – transforming agricultural extension,* Wageningen, Sep 22-25, 2003. *http://www.cta.int/observatory2003/keynote_papers/Keynote_Caribbean.pdf* (Retrieved on 28.03.09).

CARDI, 2009. Caribbean Agricultural Research and Development Institute (CARDI) website, *http://www.cardi.org/* (Retrieved on 30.01.09)

CARICOM, 2004. CARICOM Ministers adopt aggressive ICT strategy: Communiqué issued at the conclusion of the third meeting of Ministers responsible for Information and Communication Technology. *Press release: 16,* October 2004, Barbados. http://www.caricom.org/jsp/pressreleases/pres164_04.htm

CARICOM, 2009. Information and Communication Technology for Development website. http://www.caricomict4d.org/index.php?option=com_content&task=view&id=14&Itemid=95 (Retrieved on 05. 04. 09)

CARICOM, 2007. Press Release 99/07, *http://www.caricom.org/jsp/pressreleases/pres99_07.jsp* (Retrieved on 25.03.09)

CAIS, 2004. *Status Report [Internal Report],* Caribbean Agricultural Information Service (CAIS). CARDI. St Augustine.

Chesney, A., 2007. An agricultural renaissance. *Spore* No. 131, October 2007. Pp:16.

CTA. 2008a. Annual *Report 2008,* Technical Centre for Agriculture and Rural Cooperation (CTA), Wageningen.

CTA. 2008b. Regional Workshop: Information and Communication Management Policy and Strategy in the Caribbean. Technical Centre for Agriculture and Rural Cooperation and Caribbean Agricultural Research and Development Institute (CARDI), St. Vincent and the Grenadines. October 9 - 10, 2008.

CTA. 2006a. Regional Workshop: Training course on Electronic Production and Publishing of Agricultural Extension Materials/17-28 January *2005, Dominica, W.I.*.Technical Centre for Agriculture and Rural Cooperation, and Caribbean Agricultural Research and Development Institute.

CTA. 2006b. *Annual Report 2006,* Technical Centre for Agriculture and Rural Cooperation (CTA). Wageningen.

CTA. 2006c. Technical Centre for Agriculture and Rural Cooperation and Caribbean Agricultural Research and Development Institute (CARDI). *Regional Training of Trainers (TOT) Workshop for ACP Experts on Agriculture Science, Technology and Innovation Systems-* Course Notes. 28th August - 1st September, 2006, St Vincent and the Grenadines.

CTA. 2005a. *Annual Report 2005,* Technical Centre for Agriculture and Rural Cooperation (CTA), Wageningen.

CTA. 2005b. *Assessment of Agricultural Information Needs in Africa, Caribbean and Pacific States for CTA's products and services. Final Report: Phase 1 Caribbean.* Technical Centre for Agriculture and Rural Cooperation.

CTA. 1997. *Report on Regional Overview on Agricultural Information Needs.* Technical Centre for Agriculture and Rural Cooperation (CTA)/ Caribbean Agricultural Research and Development Institute (CARDI).

COL, 2002. Using Information and Communications Technology for Agricultural Extension. Commonwealth of Learning (COL), *http://www.col.org/resoruces/publications/trainingresources/knowledge* (Retrieved on 13.02.09).

de Freitas, C., 2006. Extension methodologies for effective technology transfer and sustainable adoption. *FAO Food Security Project National Coordinators/Focal Points Meeting,* Trinidad & Tobago, 3-4 July, 2006.

de Freitas, C, Phillips, T. and Greene, J., 2006. Exploring the potential of networks for empowerment of Caribbean agricultural producer groups through learning and collaboration. *Fourth Pan Commonwealth Forum on Open Learning Commonwealth of Learning (COL) and the Caribbean Consortium.* Jamaica. 30th October -3rd November, 2006.

de Freitas, C., 2007. Development of a National ICT Policy: ICT Policy in the ECTEL Member States. Caribbean Internet Forum, Eastern Caribbean Telecommunications Authority, St Lucia. *http://www.ectel.int/aboutECTEL.html* (Retrieved 08.03.09).

Eurostrategies, 2004. *Feasibility Study on the Future of the ICT regulatory and institutional framework and establishment and development of the Information Society in the Caribbean.* August 30 2004.

Gachie, I. and Ruault, L., 2006. *Facilitating and Managing Information for Rural Development: information service tools, methods and experiences.* Editor du GRET. CTA.

Gandelsonas, C., 2002. (Ed.) *Communicating for Development: experience from the Urban Environment.* ITDG Publishing, London.

Government of Belize , 2009. *http://www.governmentofbelize.gov.bz/* (Retrieved 5 April 2009)

Henry, R. 2009. The Food Crisis: a blessing in disguise? *The Financial meltdown, the food crisis and the way forward for agriculture and agricultural research,* CARDI Public Forum. Port of Spain.

IICD, 2006. *ICTs for agricultural livelihoods impact and lessons learned from IICD supported activities*. International Institute for Communication for Development (IICD), The Netherlands.

ITU, 2009. *Measuring the Information Society; the ICT Development Index (IDI) 2009*. International Telecommunication Union (ITU), Switzerland. ITU. ICT Statistics Newslog - Caribbean. At *http://www.itu.int/ITU-D/ict/newslog/CategoryView,category, Caribbean* (Retrieved on 13.02.09).

IICA, 2007. Inter American Institute for Cooperation on Agriculture. A New Vision for Agriculture in the 21st Century, Presentation at the *Caribbean Donor Conference*, June 2007, *http://www.caricom.org/jsp/community/donor_conference_agriculture/iica.pdf*, (Retrieved on 01.04.09).

Ministry of Agriculture, Forestry and Fisheries (St Lucia), 2009. http://www.slumaffe.org/Agriculture/Extension_Services/extension_services.html (Retrieved on 27.03.09)

Ministry of Agriculture and Rural Development, 2009. Barbados Website http://agriculture.gov.bb/default.asp?V_DOC_ID=814 (Retrieved on 30.03.09)

Ministry of Agriculture Jamaica, 2009. (Agricultural Business Information System- ABIS) Website http://www.abisjamaica.com.jm/ (Retrieved on 30.03.09)

Ministry of Agriculture, Land and Marine Resources, (Trinidad and Tobago), 2009. http://www.agriculture.gov.tt/applicationloader.asp?app=articles&id=1051 (Retrieved on 02.04.09)

O'Farrell, C., 2003. VERCON, *FAO's Virtual Extension Research Communication Network*. Presentation, CTA Observatory, Wageningen.

Rajack-Talley, T.A., and Lawrence, W., 2004. *Final Report on Survey of Caribbean Farmers' and Non-Governmental Organisations working with Farming Communities in the Caribbean*. CTA/CARDI.

Richardson, D., 2007. *How can agricultural extension best harness ICTs to improve rural livelihoods in developing countries. http://www.comminit.com/en/node/271587/306* (Retrieved on 03.03.09).

Saravanan. R. 2008. (Ed.). *Agricultural Extension-Worldwide Innovations*. New India Publishing Agency (NIPA), New Delhi.

Seepersad, J., 2002. *Research on Extension and Support Services: Towards sustainable agriculture in Latin America and the Caribbean*. FONTAGRO Project report.

Seepersad, J., 2003. Case study on ICT's for Agricultural Extension in Trinidad and Tobago. Paper prepared for CTA's ICT Observatory, ICTs- Transforming agricultural extension, Wageningen, September 23-25, 2003.

United Nations. 2008. National Accounts Main Aggregate Database. United Nations Statistics Division.*http://unstats.un.org/unsd/snaama/resultsCountry.asp? Country=28&SLevel=99&x=44&y=8&Selection=country* (Retrieved on 10.03.09).

Walker, D., 2003. Agricultural extension: COLME: Caribbean digital video instruction. *ICT Update*. [14], October 2003.

World Bank, 2007. *World Development Report 2008*. World Bank, Washington D.C.

Authors Information

Dr. Wayne G. Ganpat

Deputy Director

Extension Division

Ministry of Agriculture Land and Marine Resources

PO Box 389 Port of Spain, Trinidad, West Indies

Phone No.(O): 868-646 3778

Mobile Phone No.: 868- 741 3957

Fax No.: 868 642 6747

e-mail: waygan@flowrtinidad.net, etis@malmr.gov.tt

URL: www.agriculture.gov.tt

Dr. Wayne G. Ganpat has extensive experience as an extension officer working mainly with small farmers in Trinidad. His focus is primarily on Participatory Methods of Research and Extension. He also works in the wider Caribbean region as a consultant in the field of Agricultural Communications, mostly providing assistance with developing public Education Campaigns. He has presented several papers at International conferences, published several articles in International journals and co-authored two chapters in books on Extension.

Mrs. Claudette de Freitas

Head, Information Management

Technical Services Division

Caribbean Agricultural Research and Development Institute (CARDI)

P.O. Bag 212, The University of the West Indies, St Augustine Campus, St Augustine, Trinidad & Tobago.

Phone No.(O): 868 645 1206

Mobile Phone No.: 868 689 9883

Fax No.: 868 645 1208.

e-mail: cdefreitas@cardi.org; defreitas.c@gmail.com

URL: www.cardi.org

Mrs. Claudette de Freitas holds a B.Sc. (Agriculture) and post graduate diploma (International Relations) from the University of the West Indies; Masters Degree in Library and Information Science (MLIS) -University of Western Ontario and MA (Mass Communication) - University of Leicester. She has over 15 years of experience in information and communication management (ICM) for development - organising and providing support to more strategic approaches through technical assistance, training and designing information and communication products and services. Mrs. de Freitas has published several articles on various aspects of ICM in her capacity as the Programme Coordinator for key projects between CARDI and the Technical Centre for Agricultural and Rural Cooperation (CTA).

❑❑❑

Chapter 3

GREECE

Anastasios Michailidis

1. INTRODUCTION

Nowadays, any issue related to "information and communication technologies in agriculture (ICTs)" is extremely interesting and it belongs to the modern subject-matters of the agricultural economics science. Unambiguously the above phrase is the buzz remark for the agricultural extension scientists considering that the process of communication is fundamental to extension, training and passing on information. Thus, learning processes, dissemination of innovations and social change cannot be explained without reference to communication (Albrecht *et al.*, 1989).

An efficient communication strategy, especially when combined with a wide range of available ICTs, utilises the advantages of both personal and impersonal media communication. This can be achieved by building up, promoting and using communication networks and modern technologies. Under the term "network" we understand the complex of communication channels which ensure that all members of the target group receive the message more than once, both from various personal contacts and media. This calls for the use of existing communication relationships, the dissemination of information by multipliers and influential individuals and a range of media which are appropriate to the type of information and for reaching the target groups. As well thought out, communicative approach can partly compensate for the often lamented shortage of advisers provided that one is in a position to communicate a message in various ways, to evaluate feedback and then to take action accordingly.

However, knowledge is an increasingly significant factor of production in modern agriculture and ICTs can accelerate agricultural development by facilitating knowledge management (Michailidis, 2007a). Landholders in rural areas are continuing to expand their use of ICTs and specifically internet. This is mainly due to increased availability of hardware, software and communications infrastructure at reasonable cost. Nevertheless, it is unclear if all the benefits of adopting a new communication technology are simply cost reduction or might include potential gains to production and improved marketing opportunities. This chapter is concerned with the experiences description of ICTs adoption-diffusion in Greek agriculture. In addition, national agricultural scenario is presented to show the need of ICTs in agricultural extension. Taking into account the great importance of ICTs as the principal driver of change in agricultural areas, as well as the drastic increase in ICTs adoption over the last decade, a study determining the adoption parameters of ICTs can prove singularly valuable.

The rest of the chapter is organized as follows: Section 2 provides the Greek agricultural scenario including agricultural statistics, land use, production, employment, farming systems and agricultural policy implications. In section 3 the national extension systems are presented with emphasis on historical changes, scopes and importance. In section 4 the need of ICTs in agricultural extension is drawn and the relationship between agriculture and society is commented. In section 5 the national policy on ICTs, with emphasis on agricultural extension, is introduced. In section 6 the adoption-diffusion theory including the literature review is sketched out. Section 7 presents a case study as example, related to ICTs application for agricultural extension, in North-West Greece. Finally, the chapter ends with the concluding section in which recommendations and policy implications are drawn.

2. GREEK AGRICULTURE

The main aim of this section is to describe the Greece agricultural scenario including agricultural statistics, land use, production, employment, farming systems and agricultural policy implications. Traditionally, Greece belongs to the main agricultural countries of the European Union (EU). However, against all efforts that have been taken in the last 30 years, Greek agriculture is based on small-sized family-owned dispersed units. On the other hand, the extent of cooperative organization stays at rather low comparative levels. Nowadays, according to the NSSG (2008), Greek agriculture employs only 530,000 farmers namely 12 per cent of the total labor force. The Greek agricultural product represents only 3.6 per cent of the national Gross Domestic Product (GDP) that is about €13.5 billion annually. A large number of the country's immigrants (mainly from Albania and Bulgaria) are employed in the agricultural sector of the economy, as well as construction and public works. Employment in the agricultural sector slipped throughout the latter part of the 20th century and into the 21st Century. In 1981 the agricultural workforce was estimated at 972,000 and had fallen to 873,000 by 1991 and recently to 530,000.

One of the major problems of the Greek agricultural sector is that it suffers from a lack of many natural resources (Michailidis, 2007b). In particular, more than 70 per cent of the total land cannot be cultivated as it is covered by forests or because of poor soil. On the other hand, there are many factors that hold back the growth of the Greek agricultural sector like low rainfall, rural land ownership system, and the emigration of the rural community into urban areas or abroad. Thereby, during last two decades, agriculture is oriented mainly to the northern regions of the country and especially to the plains of Thessaly, Macedonia, and Thrace where corn, wheat, barley, sugar beets, cotton and tobacco are traditionally harvested.

More specifically, the main agricultural Greek products are (NSSG, 2008):

- *Cotton and tobacco:* While agriculture is not a thriving economic sector, Greece is still a major EU producer of cotton and tobacco.
- *Cereals:* Cereals constitute very important crops mainly for local consumption. Older policies of the Greek government aimed at cereals production being in equilibrium with the country's annual needs. Frequent expansive and defensive ople's nutrition were two factors that galvanized the former policy.
- *Wheat:* Annual production of wheat reached 1.8 million tones in 2004 down from 2.1 million tones in 1964. Average yield increased from 165.3 tones per square kilometer in 1964 to 211.3 t/km^2 in 2004, while the total area harvested decreased from 12,631 km^2 in 1964 to 8,519 km^2 in 2004.
- *Barley:* Annual production of Barley reached 220,000 million tones in 2004 down from 242,000 million tones in 1964. Average yield increased from 144.91 t/km^2 in 1964 to 224.97 t/km^2 in 2004. The total area harvested decreased from 1670 km^2 in 1964 to 978 km^2 in 2008.
- *Olives:* Many of which are turned into olive oil which is the country's most renowned export product.
- *Grapes, melons, tomatoes, peaches, and oranges* are also popular EU exports.
- *Wine* is an export with promise, and the government has urged vineyard owners to produce higher quality wines to increase its popularity as an international export.

As well as, animal production constitutes a significant part of the total agricultural output of Greece. In particular, goat and sheep products (meat and milk) are popular and provide more that 6 per cent of the agricultural production. But, the most popular animal product is the renowned *feta* cheese which is made from Greek sheep milk and recently characterized as of *"Denominated Origin Product"*. Other important livestock are hogs, cattle, chickens, rabbits, beehives, and pigeons.

Given Greek's vast coastline and its numerous islands, it is natural that a fishing industry exists. However, it is not as vital to the economy as would be expected from a country with a rich maritime history. The main reasons for the above note are the over-fishing and the pollution in the Mediterranean. Specifically, over-fishing has lessened the impact of fishing revenues on the economy and pollution in the Mediterranean has also damaged the industry.

Regarding cooperative systems, Greece adopted a framework of farming cooperatives as early as 1915 to streamline farming efforts. These cooperatives are step-by-step unionized and have been supported by every government that comes to power. In particular, under the governments of the 1980s, the cooperatives were greatly enhanced, and they received a large percentage of agricultural loans.

Currently, Greek agriculture is heavily subsidized by the Common Agricultural Policy (CAP), with controversial results. Certain deductions of subsidies are planned within the next decade. Moreover, Greece produces a wide variety of crops and livestock products. Fisheries are also playing an important role while forestry plays a secondary role. Greece's climatic and topographic conditions allow for an extremely large number of crops ranging from tropical (*e.g.* bananas) to those more typical of northern climates (*e.g.* mushrooms) to be grown.

The EU has granted Greece a number of subsidies to bolster its agricultural sector, but it continues to perform poorly in the 21st century. To expand the market for Greek food exports, the Ministry of Agriculture established a private company, Hellagro SA, to assist Greek companies in selling their products over the Internet. Private stockholders will hold the majority share in Hellagro SA and financing will come from e-commerce, commission (money paid for performing a given act or transaction), investment opportunities, and joint ventures. The government is hoping this effort will help revitalize the struggling industry.

Although industry and services have recently expanded, agriculture remains one of the important sectors of the economic activity in Greece. The growth of the industry and especially the growth of services during the last decade were not followed by the agricultural sector. In particular GDP declined by 2 per cent in 2008, following a drop by 1.9 per cent in the previous year. This brought down agriculture's movements and lower farm output. Agriculture, however, is of much greater importance if you take into consideration that it occupies a large percentage of the active population and the country is self-sufficient of food products (primary and processed) with the exception of some animal products (beef and milk).

Greek agriculture today is mainly focused on the production of high quality, competitive, certified, labelled products produced by innovative procedures compatible with sustainable development principles. Improved quality and the resulting increase in value of Greek agricultural products will be achieved through technological innovations resulting from focused and well-coordinated agricultural research.

3. AGRICULTURAL EXTENSION IN GREECE

This section mainly aimed to present the Greek agricultural extension systems with emphasis on historical changes, scope and importance. The Greek Extension service was first established as a directorate (Agricultural Extension and Education) of the Ministry of Agriculture in 1950 under the framework of the Marshall Plan (Koutsouris *et al.*, 1996). As a consequence of the Second World War and the Civil War, the Greek agriculture was essentially left ruined. Therefore, its immediate reconstruction became a top priority. During the 1950's the Extension Service undertook this developmental task with great success, mainly due to the massive and well co-ordinated mobilisation of its staff on the local level and on the basis of well planned extension programs. As an outcome the Greek agriculture was given a developmental spirit and the first sign of modernization were shown. In particular, productivity was remarkably increased and innovative practices (fertilizers, tractors and pesticides) were used by the majority of the farmers.

By the end of the next decade (1960's) the priorities of the Greek agriculture presented a significant change as a result of an outstanding shift in the agricultural policies. In particular, the development priorities changed from the increase of productivity to the restructuring of the agricultural production according to marketing considerations (Koutsouris *et al.*, 1996).

A very important time point for the Greek agriculture was the entrance of the country in European Union, in 1981, which signified the total bureaucratisation of the extension service since the administrative burden of the Common Agricultural Policy implementation was designated to its staff. As a result the Greek agronomists were transformed into the controllers of various policy measures regarding the provision of subsidies to farmers (Koutsouris *et al.*, 1996). However, since recently, the two main development parts of the Greek extension services cannot exist harmonically. More specifically, the Greek public agronomists are neither effective controllers nor efficient educators or advisers of the farmers. The main reason for the above finding is probably the insufficient institutional and organizational division between the two dimensions of the extension service.

Nowadays, the Greek agricultural extension services try to find a new and efficient role in a consecutive changing environment. Under these upheaval circumstances and taking into account the great importance of ICTs, as the principal driver of change in agricultural areas, as well as the drastic increase in ICTs adoption over the last decade, the determination of the adoption parameters of ICTs can prove significant interesting issue for the modern agricultural extension services. In the next section the need of ICTs in agricultural extension services is drawn and the relationship between agriculture and society is commented.

4. NEED OF ICTs IN AGRICULTURAL EXTENSION

By all odds, agriculture is a part of society. This simple fact can be interpreted in a simple way. Agriculture gets from society certain inputs and on the other hand it offers food or a nice landscape (van Woerkum, 1996). So, agriculture and society are intertwined in a complicated manner which creates a lot of tension on both sides and a lot of questions about animal welfare, healthy food, environment, nature, political behaviour and recently biotechnology. However, all these issues must be investigated in connection with a growing feeling of distance. More and more Greek consumers, for example, do not have any direct experience with agriculture and ultimately they do not get the feeling that their food is depending on the activities of some other fellows from other cities, countries or continents. In a big Europe and generally in a global perspective, the psychological feeling of dependency on own farmers is rapidly decreasing (van Woerkum, 1996).

The growth in the economics of many developed countries, over the past decade, can be partly explained through the contribution of ICTs (Jorgenson, 2001). Advances have occurred through improved productivity (particularly in the services sector), and changes in the labour and capital markets. These advances occur in part because of the integration of computing hardware and software into the production processes (allowing for greater productivity and better use of labour and capital), the development of new services and products (including the internet), and the improved linkages between businesses and consumers (including e-commerce).

According to Oliner and Sichel (2000), the majority of related studies to productivity changes in an economy focus on the non-farm sector. Specifically, Brynjolfsson and Hitt (2000) support that the greatest benefits from adoption of ICTs have tended to be at the manufacturing and service industry levels mainly because applications of ICTs allow not just direct productivity changes, but also organisational changes, marketing and consumer relationship changes, and better control over supply chain management.

On the other hand, agriculture is generally assumed to have fewer gains to make from ICTs. However, throughout the 20th century a striking array of new agricultural technologies has profoundly affected the nature of farming and the role of farmers (Cochrane, 1993). The adoption of a wide spectrum of technological advances in soil chemistry, fertilizers, hybridization, and animal health practices, has altered the farming practices and the structure of agriculture (Lasley *et al.*, 2001). Today's breakthroughs in cell biology and ICTs will offer new challenges in farming and extension.

Besides, the availability to farmers' options of new ICTs opens new channels for delivering information and thus extension services face new opportunities and challenges in meeting the information needs of farmers (Lasley *et al.*, 2001). New ICTs can replace traditional information systems of information delivery and training as currently practiced by extension services. Thus, as farmers become increasingly able to "info-graze" around the world, the thus far known role of the extension agent may be undermined.

Ramkumar *et al.*, (2007) describe the modern ICTs like the tools which provide powerful 'force multipliers' to conventional extension. As well as Akpabio *et al.*, (2007) suggested that ICTs will increase the effectiveness and efficiency of extension work and ultimately transform extension officers into catalysts who will access and transfer relevant information and assist farmers to utilize such information to solve pressing problems.

Furthermore, research in Texas revealed that only 10 per cent of the agents were using or had used extension by 2007 (Harder and Lindner, 2008a & 2008b). Such results indicate that extension agents are not prepared to promote extension to their peers or clientele and confirm previous findings that while ICT related skills are becoming an essential job-related skill and there is a lack of training 'beyond the basics level' which would support agents to learn more in-depth skills (Gregg and Irani, 2004). Besides, in 2006, Cooperative Extension publicly launched an online information resource known as extension with the goal for 2007 being to have at least 75 per cent of all Extension employees as registered users (Harder and Lindner, 2008a). Finally, research on extension also suggests that there is a role for the traditional one-to-one relationship with Extension clients, even with online clients (Herring, 2008).

However, a review of the available literature reveals that the opportunity for ICTs in extension is clouded with many challenges in both developed and developing countries (Akpabio *et al.*, 2007; Akpabio, 2007; Annor-Frempong *et al.*, 2006). Constraints affecting ICTs utilization for extension mainly refer to ICT training, infrastructure, financial resources and technical backstopping to enhance the use of ICTs in extension. While such constraints may be expected in developing countries, findings from the US Cooperative Extension Service, the world's largest informal adult education and development organization (Seevers, 1995), are illuminating. Despite the advances in ICTs over the past decades and the proliferation of distance-based learning and on-line resources, printed information sources have been found to remain the most preferred delivery systems for Extension and state agency natural resource professionals in Ohio (Rodewald, 2001).

Research findings in both developed and developing countries confirm the need to approach ICTs with caution. In many developing countries constraints relating to the issue of physical access, such as poor infrastructure and high costs, is a quite common one also complemented by the topic of the dissemination of unrelated information (Akpabio *et al.*, 2007; Annor-Frempong *et al.*, 2006). Furthermore, a research study in Turkey proves that there are positive relations between adoption and socio-economic characteristics and communication behaviours supporting the generalizations of diffusion theory (Demiryurek, 2006). Finally, Park *et al.* (2007) verify the findings of previous studies indicating that modern communication strategies (ICTs) are less preferred by the farmers and ascertain that farmers are constrained by time and finances, commitment to family and jobs, and responsibilities in the operation of the family farm or business.

In the case of the United States of America (USA), Abbott and Yarbrough (1992) in their studies on ICTs adoption among Iowa and New York farmers note that new communication technologies will probably never be adopted by many farmers as well as that adoption and use have created substantial inequalities. Furthermore, as early as 1986, Lasley and Bultena, (1986) found that 20 per cent of the surveyed 2,400 Iowa farmers opposed the use of personal computers. They also found that, on the one hand, scale of farming operations and farmers' skills determine the perceived potential benefits and thus adoption of ICTs, and, on the other hand, there is no overall pattern of all ICTs. As well as, research among farm leaders in southeast Alabama, northwest Florida, and southwest Georgia led to the conclusion that the dominant factors in PC and Internet adoption are type of farm production and age, with education being a supporting factor. The researchers also argue that a large segment of the farm population has not adopted the PC and an even larger segment is not connected to the Internet, a fact attributed to the lack of infrastructure availability in rural areas (Hall *et al.*, 2003). In Michigan, farmers' preferences for computers and Internet were related to gross annual income level, age and education (Howell and Harbon, 2004). A study among North Carolina forest landowners identified associations between delivery method preferences and various characteristics of landowners, including socio-demographics, land ownership, and management experience (Bardon *et al.*, 2007). Recent research examining internet technology use among flower growers in Hawaii affirms that the factor firm size determines whether to have a Web site as well as that among those who do not have the main reason cited is a deficiency in training (Burke and Sewake, 2008). Finally, there is evidence that GIS-GPS-RS technologies, in spite of having created many new tools for Extension, have widened the digital divide thus leaving many with little understanding of the technology and potential applications (Milla *et al.*, 2005).

Recent studies show that the internet and electronic mail are of less use to producers than other sources. In this respect, swine producers in Illinois have been found, on the one hand, not to use computers as a source of information despite the fact that they have access to on-farm computers and, on the other hand, that they rely on popular publications, corporate representatives, extension newsletters, and current users of new technologies to advance within the swine industry (Brashear *et al.*, 2000). Lasley *et al.*, (2001) underline that Iowa farmers regardless of technological sophistication (including communication technologies) desire a wide range of information delivery channels and show a strong preference for personalized communication. In line with results from many studies over the past 50 years (Vergot *et al.*, 2005) research in Michigan showed that landowners prefer traditional written communication strategies as compared to ICTs (Howell and Harbon, 2004) while printed information was also the channel most often used for information among beef cattle producers in north-west Florida (Vergot *et al.*, 2005).

In Ireland, surveys show farmers as one of the most marginalized sectors in the 'information society'. Issues such as low ICT skills and lack of appreciation of ICT benefits continue to be barriers to farmers taking up the new technologies (Morrow *et al.*, 2004). Furthermore, research shows that less than two-thirds of the farmers with access to ICT used this for farming purposes, that farming households with off-farm employment are more likely to have adopted ICT than full-time farmers. On parallel, farmers engaged in specialist enterprises as well as those with medium-size farms were more likely to have adopted ICT (Wims, 2007). Research has also found that dairy farmers are more involved in ICT use than less profitable farmers, mainly cattle farmers and has pointed to a growing digital divide in Ireland (Leveque *et al.*, 2007).

Moreover, 'access' is nowadays understood as a multi-layered and complex issue, not to be reduced simply to the physical availability of the technology; effective access should ensure that information technology-based services are usable, useful and economically viable thus pointing to issues of (information) literacy, content and economics (Koutsouris, 2006).

5. GREEK ICTs ENVIRONMENT

In the past decade Greece has witnessed significant growth in the ICTs sector, backed by an increase in higher education orientation toward the applied sciences and electrical and computer engineering disciplines. The decade has also witnessed a blooming of high-tech start-ups and spin-offs.

A remarkable index is the research and development (R&D) performed in Europe through collaborative projects co-funded by the European Union's dedicated framework programs. The project success ratio of Greek companies' driven efforts far surpasses that of EU countries with similar population demographics or GDP and compares only to highly-developed Western European countries that are bigger and industrially stronger.

Furthermore, the Greek ICTs environment presents some remarkable opportunities for rapid development:

- A significant number of highly-educated and proficient engineers graduate every year from Greek ICTs engineering schools. Many of these engineers go on to pursue Doctor of Philosophy (Ph.D.) and Master of Science (M.Sc.) studies in renowned USA and EU universities. A good portion of the students return to Greece to apply their expertise in their disciplines.
- The academic and research network infrastructure is already in the Gigabit broadband era, which constitutes an important element of ICT growth and enables the use of local human capital for remote partnerships and collaborations.
- The Ministry of Economy and Finance, the Ministry of Development, and the Ministry of Education oversee a regulatory framework that is highly supportive of emerging R&D efforts and acts as a key paradigm of the emphasis that the state places on advancing information and communication technologies.
- Greece lies at a strategic intersection and constitutes an excellent business vehicle for expansion into the Southeast European market.

Greece ranks 15th worldwide in terms of availability of skillful IT scientists and engineers, according to the most recent World Economic Forum's Global Information Technology Report 2004-2005. This human-capital capacity, coupled with a strong fluency in foreign languages, an excellent knowledge of the European and Southeast Mediterranean business environment, a modern European regulatory framework, top-notch telecoms infrastructure and a strong political will today to promote broadband, could prove vital in any business strategy which seeks to utilize resources efficiently on the basis of outsourcing.

One of the key reasons is that for the past 15 years universities and local small-medium enterprises (SMEs) participated in EU-funded industrial R&D programs, creating a pool of trained engineers versed in the latest technologies and full of creative ideas for competitive product solutions. The existence of technology companies, albeit small and medium size, along with the fact that Greece has enjoyed a decade of economic growth, attracted a strong wave of repatriated talent - successful engineers and operation managers that helped plant the 'seed' for the creation of the silicon valley for Southeast Europe and the Balkans.

Increased specialization and rapid technological change make focus on core competencies increasingly important. This is more than evident in the case of the semiconductor industry, where competition across all market segments forces specialty chipmakers to offer complete reference designs to their customers in order to facilitate shorter time-to-market schedules. In cases like this, the outsourcing of activities like software development (driver, tools) or board design and assembly can free valuable resources and redirect them to activities directly related to the core business. We strongly believe that information technology (IT) companies residing in Greece can successfully provide such services since, apart from the country-related advantages (geographical position, EU membership), the capabilities of Greek engineers in the main outsourcing areas of interest such as software development, are well appreciated internationally. Integrated circuits are gradually becoming more complex, creating increased demand for specialized Electronic Design Automation (EDA) tools and microelectronics companies are now turning to Greece in their quest for best-in-class solutions. The proliferation of wireless Systems-on-Chip (SoC) integrating sophisticated Radio-Frequency (RF) 'front-ends' along with digital gates, urges the EDA industry to come forward with new tools and higher degrees of design automation. Nanometer-scale semiconductor technologies tracking 'Moore's law' into the new millennium, are posing new problems for IC design that necessitate focused methodologies such as Design for Manufacturability (DfM) and Design for Yield (DfY). Greek scientific and engineering community is poised to address these challenges. Several local universities now feature IC design-related courses in their curricula, while EU-funded programs empower aspiring engineers and researchers with the latest in commercial EDA software. Coupled with several decades of tradition in RF and microwave education, the academic environment provides the ideal skill set for microelectronic engineers to pursue breakthroughs in analogue, RF, and high-frequency semiconductor design. And this is becoming more obvious as leading semiconductor companies from Europe, North America and Japan are sourcing solutions from Greek companies and university labs. Technologies on offer include electromagnetic simulation, analogue transistor modeling and automated semi-custom layout tools.

However, the above technology is being far away from the agricultural practices. In particular, landholders in rural Greek areas are just trying to expand their use of ICTs and specifically personal computers, internet, e-mail and GPS. Thus, taking into account the great opportunities of ICTs as well as the drastic increase in ICTs adoption over the last decade, any issue related to the adoption parameters of ICTs can prove singularly interesting. In the next section the adoption and the diffusion theory of innovations are sketched out including related literature review and issues about ICTs in agriculture.

6. ADOPTION OF ICTs IN AGRICULTURE

The basic idea underlying the adoption theory is that (potential) adopters of an innovation do not adopt the innovation independently, but instead influence each others' adoption decisions. In the context of farmers, for example, the more farmers have adopted an innovation, the higher the likelihood that the remaining farmers will also do so. The influence of early adopters on later adopters is often called "word of mouth communication" (Rogers, 1995), a term referring to a much broader set of phenomena than farmers simply talking to each other. For instance, a farmer might be influenced by another farmer simply by observing his/her behaviour (Kibwana *et al.*, 2001). Similarly, when the adopting units are firms, the uncertainty surrounding a firm's adoption of a technological innovation may be reduced when observing other firms' experiences (Ganesh and Kumar, 1996).

One of the first studies that sought farmers' opinions about new ICT's was carried out by Lasley and Bultena (1986). In this study, nearly 2,400 Iowa farmers voiced opposition to these technologies. Also, 40 per cent of the farmers expressed opposition to robotics (defined as computer-assisted machinery for on-farm use); 30 per cent opposed confinement livestock facilities and 20 per cent the use of personal computers.

In a following research, Bultena and Lasley (1990) introduced the principle of "the dark side of agricultural biotechnology". This phrase was first originated through the analysis of the responses from a large sample of farmers in Iowa. The results of this study show that, while the farmers are excited about the prospects of biotechnology, they are worried about the structural impacts caused by these new technologies.

Several other research outcomes have probed the acceptance of information technology, biotechnology and tele-communication (Ramirez,

2005; Rivera, 2000; Rogers, 1995). For example, Audirac and Beaulieu (1986) explored the adoption and diffusion of microcomputers in farming and identified a large number of personal and institutional constraints. Hoban *et al.*,(1992) examined the adoption of electronic communication technologies by farmers and the impacts of these technologies on the traditional extension information delivery approaches.

Access to information and improved communication is a crucial requirement for sustainable agricultural development. Modern communication technologies when applied to conditions in rural areas can help improve communication, increase participation, disseminate information and share knowledge and skills. It is being said that "Cyber Extension" would be the major form of technology dissemination in the near future.

However, it is observed that the rural population still has difficulty in accessing crucial information in order to make timely decisions. It is essential that information availability is demand driven rather than supply driven. The challenge is not only to improve the accessibility of communication technology to the rural population but also to improve its relevance to local development.

Improved communication and information access is directly related to social and economic development. However, it is observed that the rural populations still have difficulty in accessing crucial information in forms they can understand in order to make timely decisions. There is a concern that the gap between the information rich and information poor is getting wider.

New information and communication technologies are generating possibilities to solve problems of rural poverty, inequality, and giving an opportunity to bridge the gap between information-rich and information-poor and support sustainable development in rural and agricultural communities. However, remote rural communities still lack basic communication infrastructure. The challenge is not only to improve the accessibility of communication technology to the rural population but also to improve the relevance of information to local development.

One of the first large-scale studies exploring structural implications of the telecommunication age on agriculture was the Office of Technology Assessment report (O.T.A., 1987), predicting major shifts in information sources and channels. According to the study, biotechnology coupled with information-intensive technology would vastly restructure farming, leading to fewer and larger farms that would be vertically linked throughout the food chain. Farmers would be vertically linked within a commodity chain and receive all inputs, including information, within the chain system, reducing or eliminating outside suppliers, such as extension services. Farming in today's information-intensive world can be supported by obtaining information *via* an array of electronic communication technologies, personal computers, GPS, internet, fiber optics *etc.* Hence, "new age" technology may improve farm families' access to diverse information and result in better coordination between food processors and marketers.

Lasley *et al.,* (2001) have sought to discern farmers' opinions about new technologies. In their study, nearly 2,400 Iowa farmers voiced opposition to such technologies. Also, 40 per cent of the farmers expressed opposition to robotics (defined as computer-assisted machinery for on-farm use); 30 per cent and 20 per cent of them oppose to confinement livestock facilities and to the use of personal computers respectively.

6.1. ICT Application- An example

The ICTs adoption project, called "*wemaproject*", is projected to implement in the western Macedonia (north-west Greece) and destined for farm efficiency improvement purposes. The "*wemaproject*" includes several ICTs and destined in a representative farm framework of 600 farmers. Unambiguously, the implementation expenses of any project play an important role in the investor's decision, as they constitute a significant part of the total cost. The economic profitability of this project is also very important, notably in a world where funds available for agricultural investments are limited. In this work, a typical investment option was evaluated by applying DCF and a real options approach.

Cost projection estimates indicate that the "*wemaproject*" is expected to require an outlay of 750,000€ during the implementation phase. The project is required to provide 10 per cent of annual pre-tax revenue for payback during the operating stage. The annual operation cost (45,000€) includes salaries, materials, any conservation expenses and payments for several other services. On the other side, the estimates of total direct annual revenues are equal to 30,000€ including better quality, new markets, distribution canals, better marketing and generally farm efficiency improvement.

7. CASE STUDY FROM NORTH-WEST GREECE

This section presents a case study example, related to ICTs application for agricultural extension, in North-West Greece. In particular, this section, utilizing survey data, examines to what extent farmers have adopted electronic communication technologies. In addition, the section explores farmers' interest in precision farming in association with their willingness to adopt telecommunication technologies. Using indicated preferences, the analysis finds that the ability to receive information, *via* various means of telecommunication, reinforces farmer support for more one-on-one and personalized communication. The results also show that adoption of telecommunication technologies may supplement rather than replace traditional delivery methods of information; sequentially an increase of demand for education will be expected.

A range of communication technologies have been used to meet the information needs of the rural population. These include television, tone telephones, cellular telephones, personal computers, printers, answering machines, internet or e-mail capacity, fax machines, cable television, DSL internet, satellite television dish and global positioning systems as communication tools for mediating between rural people's needs and possible sources of information and expertise to respond to these needs.

The internet is emerging as a tool with potential to contribute to rural development. Internet enables rural communities to receive information and assistance from other development organizations; offers opportunities for two-way communication. It can also support bottom-up articulation of development needs and perceptions, and thus help in reducing the isolation of rural communities. It can facilitate dialogue among communities and with government planners, development agencies, researchers and technical experts; encourage community participation in decision making; and help agricultural researchers, technicians, farmers and others in sharing information.

A survey of ICTs use in Northern Greece was conducted by Dr. A. Michailidis in 2007. Rural ICTs users indicate that the ICTs and especially the internet provide them with a very convenient method for quickly accessing a large volume of information without being impeded by geographic barriers. They also report finding information of value from the ICTs in the form of new ideas, discussion groups, access to expert advice, continuing education resources, increased global understanding and cultural awareness. Social benefits include new opportunities to overcome geographic isolation, increased social interaction, opportunities to organize and advocate for social change, equalization of urban/rural disparities and new links between urban and rural communities.

The aim of this study is to examine to what extent farmers have been adopting ICTs in the Region of Western Macedonia. In particular, technologies which have already been adopted by farmers are explored. In addition, particular attention is given to the farmers' interest in precision agriculture such as Global Positioning Systems (GPS) and their preferences for extension services for the years to come.

The West Macedonian Region (Figure. 1) is located in the northwest part of the Greek state and represents the natural gate of Greece to the northwest borders and especially to Albania and to the Former Yugoslavian Republic of Macedonia (FYROM). The landscape of the region, accounting for 7.2 per cent of the country's surface, mainly consists of highlands (69.2 per cent), forest areas (26 per cent), rangelands (43 per cent) and cultivations or fallow lands (24 per cent).

The purpose of this research is to identify the extent to which farmers have been adopting ICTs and to determine the importance of agricultural extension as an information source in the study area. In particular, ICTs that farmers have already adopted in their farming practices are explored. In addition, particular attention is given on farmers' interest in precision agricultural methods. This focus is particularly important in Greece since, on the one hand, the country is geographically dispersed and, on the other hand, public extension has been trapped in a bureaucratic-administrative role leading to the provision of inadequate services to farmers (Koutsouris, 1999).

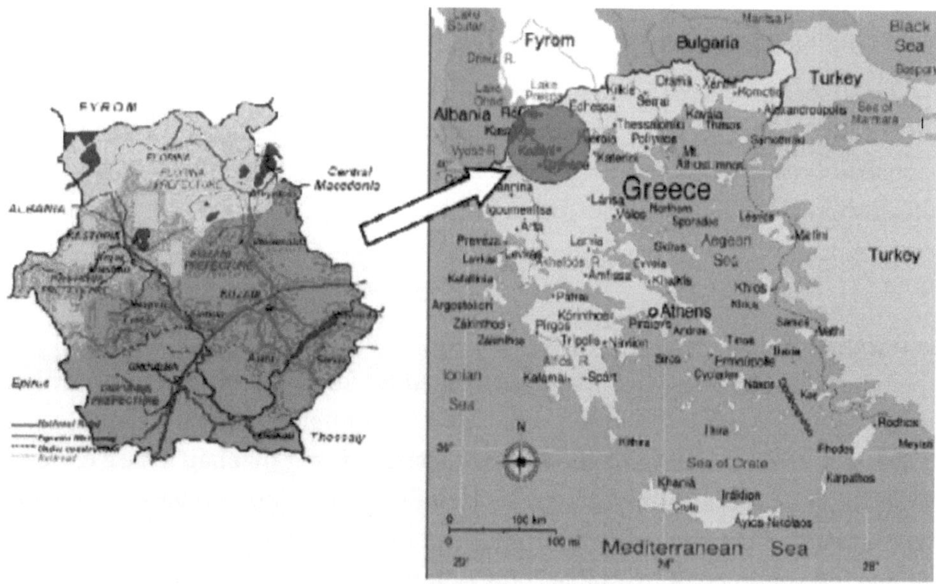

Figure 1 : A map of West Macedonian Region

The collected data relate to the farmers' use of ICTs and the nature of their enterprise. Demographic and attitudinal information was also collected. The key questions concern the value of ICTs use in their business according to a number of factors including the use of accountancy records as well as the use of online banking. To encourage participation and minimise the cognitive burden on respondents, most questions were framed with *Likert* scale intervals. The list of potential respondents was compiled from several sources. These comprised lists of members of cooperatives and lists of beef producers obtained by the regional authorities including producers with more than 20 heads. Participants were selected at random from the compiled lists

Data were collected through a mail-out/telephone response format. All surveys were mailed out in batches of 30 per week from January to March 2007. Respondents were contacted by telephone in the following week and asked if they would like to participate. Respondents could either complete the forms in their own time and return them by post, or respond over the telephone. By June 2007, 160 responses had been received from 500 questionnaires issued. There were another 34 per cent of respondents who indicated that they did not use ICTs and that the survey was not relevant to them, giving an overall response rate of 66 per cent.

Statistical analyses were performed using SPSS for Windows version 16.0 (SPSS, 2007). The focus of the analysis was on producing overall descriptive statistics as well as correlating the usage of ICTs with the characteristics of the respondents. The Categorical regression model was also used to handle optimally the transformed categorical variables and to predict future use of ICTs.

Reliability analysis (Bohrnsedt, 1977; Bohrnsedt, 1970; SPSS, 2007) for the twenty ICTs items was firstly performed to determine the extent to which these items are related to each other to get an overall index of the internal consistency of the scale as a whole and to identify items that had to be excluded from the scale. In fact, only two items (wireless connection system and pager) were excluded consequently from the total number of fourteen items. Table 1 shows the presence of available ICTs in farms.

Table 1 : ICTs on farms

	Responding "Yes"
Television	*96.87 per cent (155 respondents)*
Tone telephone	*88.75 per cent (142 respondents)*
Cellular telephone	*82.50 per cent (132 respondents)*
Personal Computer	*45.62 per cent (73 respondents)*
Printer	*30.62 per cent (49 respondents)*
Answering machine	*22.50 per cent (36 respondents)*
Internet or e-mail capacity	*13.75 per cent (22 respondents)*
FAX machine	*10.00 per cent (16 respondents)*
Cable television	*7.50 per cent (12 respondents)*
DSL Internet	*5.00 per cent (8 respondents)*
Satellite TV dish	*2.50 per cent (4 respondents)*
Global positioning system (GPS)	*1.25 per cent (2 respondents)*

The value of Cronbach's alpha (α) reliability coefficient was found equal to 0.88, indicating that the ICTs scale is reliable to accept. Friedman two-way analysis of variance, with x^2=2,631.69 (α=0.00) and Hotelling's T^2=1,069.29 (F=36.48 and α=0.00), indicated the significance in differences of item means. Having accepted the consistence of the twenty items, the average rankings for each respondent were used as numerical values of the dependent variable "ICTs" along with categories of nine independent variables shown in Table 2.

Table 2 : Selected independent variables

Independent variables	**Type**	**Categories**
Classification of farmers	Ordinal	1=no tech, 2=low tech, 3=medium tech, 4=high tech, 5=very high tech
Familiarity with precision farming	Ordinal	1=not familiar, 2=somewhat familiar, 3=very familiar
Interest of precision farming	Ordinal	1=not interested, 2=somewhat interested, 3=very interested
Use of precision farming	Ordinal	1=no plans to adopt, 2=plans to adopt within 5 years, 3=have already adopt
Gender	Nominal	1=male, 2=female

Contd...

Age	Ordinal	1=under 35, 2=over 35
Years of Education	Ordinal	1=none, 2=two or less, 3= from three to four, 4=from five to six, 5=seven or more
Annual income	Ordinal	1=less than 10,000€, 2=10,001€-15,000€, 3=15,001€-20,000€, 4=20,001€-25,000€, 5=more than 25,001€
Farming time	Ordinal	1=full time, 2=part time

Categorical regression (Kooij and Meulman, 1997) was used to handle optimally the transformed categorical variables. It yielded an R of 0.92 indicating moderate relation between the "ICTs" and the group of selective predictors. However, since R^2=0.8464, it is indicated that 84.64 per cent of the variance in the transformed "ICTs" rankings is explained by the regression of the optimally transformed variables used. The F statistic value of 7.42 with corresponding α=0.00 indicates that the model is performing well.

Standardized coefficients (Table 3) indicate that the transformed variables "classification of farmers", "annual income" and "gender" are significant in pointing out possible effects on "ICTs". In fact, from the zero order correlation coefficients between transformed predictors and the transformed response we get a better understanding of how these predictors are doing.

Table 3 : Categorical regression coefficients and other statistics

Independent variables	Standardized Coefficients		F	Correlations			Impo-rtance	Tolerance	
	Beta	St. Error		Zero-order	Partial	Part		After	Before
Classification of farmers	**-0.29**	0.05	22.80	-0.39	**-0.27**	**-0.25**	**0.12**	0.87	0.85
Familiarity with precision farming	0.10	0.07	2.20	0.36	0.08	0.07	**0.10**	0.66	0.64
Interest of precision farming	0.01	0.05	0.09	0.21	0.01	0.01	0.01	0.94	1.00
Use of precision farming	0.02	0.06	0.31	0.07	0.02	0.02	0.01	1.04	1.11
Gender	**0.14**	0.06	2.40	0.28	0.08	0.08	**0.09**	0.89	0.96
Age	0.05	0.06	1.47	0.10	0.07	0.05	0.02	1.13	1.22
Years of Education	0.04	0.07	0.05	0.24	0.00	0.13	**0.23**	0.68	0.72
Annual income	**0.21**	0.08	8.86	0.38	0.19	0.16	**0.27**	0.82	0.88
Farming time	-0.03	0.06	0.62	-0.13	-0.03	-0.03	0.01	1.12	1.19

Partial correlation coefficients indicate that, removing the effects of other variables, "classification of farmers" explains about 7.29 percent [or $(-0.27)^2$] of the variation in the "ICTs" factor. The rest of the variables explain smaller portion of variance if the effects of the other variables are removed. Moreover, by removing the effects of other variables from "classification of farmers" the remaining part of this variable becomes smaller, about 6.25 percent [or $(-0.25)^2$].

However, the relative important measures (Pratt, 1987) of the independent variables show that the largest importance to predict "ICTs" corresponds to "annual income" accounting for 27 per cent, followed by "years of education" (23 per cent), "classification of farmers" (12 per cent), "familiarity with precision farming" (10 per cent) and "gender" (9 per cent). The five variables' additive importance accounts for about 81 per cent. Finally, the tolerances of all variables are high enough to assure exclusion of the multicollinearity problem (Table 3).

According to the descriptive statistics analysis, the use of e-mail and social and recreational uses are rated highly by farmers, followed by weather, education and online banking (Table 4).

Table 4 : Different categories of ICTs use

	None (1)	Low (2)	Medium (3)	High (4)	Very high (5)	Median score
e-mail	2	1	3	6	10	**3.95**
Weather	8	2	2	4	6	**2.91**
Technical	12	3	4	1	2	2.00
Market information	10	4	3	2	3	2.27
Education	5	8	2	3	4	**2.68**
Online banking	9	5	3	2	3	**2.32**
Social and recreation	2	2	3	8	7	**3.73**
Buying	12	2	5	3	0	1.95
Selling	18	0	0	2	2	1.64
Own web site	20	0	0	1	1	1.32

Respondents were also asked to rate the advantages of ICTs use. The results (Table 5) show that the majority of respondents viewed the ICTs as being of low value to them. Among the reasons that ranked higher in the perceived advantage in ICTs use were "better information decision-making" and "greater clerical efficiency" (less paperwork)

Table 5 : Perceived advantage in ICTs use

	None (1)	Low (2)	Medium (3)	High (4)	Very high (5)	Median score
Better information	30	14	52	45	19	**3.06**
Less paperwork	67	24	38	17	14	**2.29**
Improved customer service	80	48	21	5	6	1.81
Faster supply of goods in	107	29	17	5	2	1.54
Better inventory control	128	14	18	0	0	1.31
Reduced costs	103	27	19	8	3	1.63
Service differentiation	71	45	21	19	4	2.00
Competitive advantage	112	24	10	12	2	1.55

The survey also included the following description of precision farming. *"Precision farming, or site-specific farming, is a way of looking at farms, fields, or specific areas within fields through use of information management systems."* Farmers were asked about their familiarity with precision farming, their level of interest in it, their intention to adopt it within the next years and potential benefits of precision farming. It was thus found (Table 6) that 30 per cent of the farmers are very familiar and 56.25 per cent of them are somewhat familiar with precision farming. Likewise, their level of interest in precision farming is reflected in their level of familiarity with technology. More than one-third (33.75 per cent) of the farmers indicated an interest in the technology and about one-fourth of the farmers (25.62 per cent) have already adopted precision farming.

Table 6 : Farmer's familiarity with precision farming

	Percent	Frequencies
Level of familiarity		
Not familiar	13.75	22
Somewhat familiar	56.25	90
Very familiar	30.00	48
Level of interest		
Not interested	22.50	36
Somewhat interested	43.75	70
Very interested	33.75	54
Level of use		
No plans to adopt	54.37	87
Plan to adopt within next years	20.00	32
Have already adopted	25.62	41

Furthermore, farmers were asked to indicate whether agricultural extension services should place "less emphasis", "the same amount of emphasis" or "more emphasis" on fourteen delivery approaches (Table 7). Farmers are slightly more likely to prefer local educational meetings or seminars (56.88 per cent). In addition, farmers are also likely to place emphasis on: involvement in applied research 55.63 per cent, printed bulletins 44.38 per cent and phone help lines 41.88 per cent.

Table 7 : Preferred educational delivery methods for farmers (indicating "more")

	Percent	Frequencies
One-to-one consultations	48	30,00 per cent
On farms demonstrations	62	38,75 per cent
Local educational meetings or seminars	91	**56,88per cent**
Prefectural educational meetings or seminars	28	17,50 per cent
Regional educational meetings or seminars	44	27,50 per cent
Farmer involvement on applied research	89	**55,63per cent**
News and reports *via* the farm media	43	26,88 per cent
Phone help lines	67	**41,88per cent**
Printed bulletins	71	**44,38per cent**
Training with farm supply dealers	58	36,25 per cent
Video tapes	11	6,88 per cent
Computer assisted instructions	36	22,50 per cent
Interactive video conferences	16	10,00 per cent

ICTs have helped to transform the non agricultural sectors of developed economies and drive real productivity gains. In recent years, agricultural enterprises and farms have adopted ICTs, suggesting that farmers are gaining real benefits from employing those technologies in their farms. However, these benefits have been difficult to identify and quantify.

In this section, survey information has been analyzed using categorical regression models and descriptive statistics. Neither cost and time savings nor production gains, resulting from access, are able to be identified by the farmers. However, the statistical analysis identified that high value rating for ICTs appeared to be associated with the use of e-mail, electronic banking, education, weather and social and recreational uses. There are also examples of individual producers who use ICTs in innovative ways to reduce paperwork and to obtain access to better information.

This section aims at contributing to a better understanding of the impact of adopting ICTs in farming. Findings support that farmers, though very apt at adopting new technologies, desire to maintain at their disposal a wide range of information delivery channels. However, farmers are reluctant to abandon any of the delivery methods provided by the extension services. Instead, it seems that their ability to receive information *via* ICTs may further enhance desire to enrich current delivery approaches.

Furthermore, findings suggest that regardless of the number of sophisticated ICTs available, there is a strong preference for personal communication. The preferred emphasis on local educational meetings and farmer involvement in applied research underscores the importance of personal communication. The proliferation of information channels could add to or complement personalized delivery approaches.

The analysis presented focuses on the examination of the extent to which farmers have been adopting ICTs in a specific Greek mountain region but has not included cross-region differences. Future cross-region or cross-country research on this issue would be a useful complement to the results presented here.

8. CONCLUSIONS

This chapter is concerned with the experiences description of ICTs adoption-diffusion in Greek agriculture. In addition, the national agricultural scenario is presented to show the need of ICTs in agricultural extension. Taking into account the great importance of ICTs, as the principal driver of change in agricultural areas, as well as the drastic increase in ICTs adoption over the last decade, a study determining the adoption parameters of ICTs can prove significant valuable.

In the 2nd section the Greek agricultural scenario was provided including agricultural statistics, land use, production, employment, farming systems and agricultural policy implications. The majority of the section is limited to statistical data including recent and historical numbers. However, the section in question also provides an overview of the national agricultural scenario ending with the finding that *"Greek agriculture today is mainly focused on the production of high quality, competitive, certified, labelled products produced by innovative procedures compatible with sustainable development principles. Improved quality and the resulting increase in value of Greek agricultural products will be achieved through technological innovations resulting from focused and well-coordinated agricultural research"*.

In section 3 the national extension systems are presented with emphasis on historical changes, scopes and importance. An interesting conclusion of this section is that, nowadays, the Greek agricultural extension services try to find a new and efficient role in a consecutive changing environment. Under these upheaval circumstances and taking into account the great importance of ICTs as the principal driver of change in agricultural areas, as well as the drastic increase in ICTs adoption over the last decade, the determination of the adoption parameters of ICTs can prove significant interesting issue for the modern agricultural extension services. In the next section the need of ICTs in agricultural extension services is drawn and the relationship between agriculture and society is commented.

In section 4 the need of ICTs in agricultural extension is drawn and the relationship between agriculture and society is commented. In particular a simple note was drawn that agriculture gets from society certain inputs and on the other hand it offers food or a nice landscape. So, agriculture and society are intertwined in a complicated manner which creates a lot of tension on both sides and a lot of questions about animal welfare, healthy food, environment, nature, political behaviour and recently biotechnology. On the other hand, agriculture is generally assumed to have fewer gains to make from ICTs. However, throughout the 20th century a striking array of new agricultural technologies has profoundly affected the nature of farming and the role of farmers. The adoption of a wide spectrum of technological advances in soil chemistry, fertilizers, hybridization and animal health practices, has altered the farming practices and the structure of agriculture. Today's breakthroughs in cell biology and ICTs will offer new challenges in farming and extension.

In section 5 the national policy on ICTs, with emphasis on agricultural extension, is introduced. However, the described technology is being far away from the agricultural practices. In particular, landholders in rural Greek areas are just trying to expand their use of ICTs and specifically personal computers, internet, e-mail and GPS. Thus, taking into account the great opportunities of ICTs as well as the drastic increase in ICTs adoption over the last decade, any issue related to the adoption parameters of ICTs can prove singularly interesting.

In section 6 the adoption-diffusion theory including the literature review is sketched out. The basic idea underlying the adoption theory is that (potential) adopters of an innovation do not adopt the innovation independently, but instead influence each others' adoption decisions. In the context of farmers, for example, the more farmers have adopted an innovation, the higher the likelihood that the remaining farmers will also do so. The influence of early adopters on later adopters is often called "word of mouth communication", a term referring to a much broader set of phenomena than farmers simply talking to each other. For instance, a farmer might be influenced by another farmer simply by observing his/her behaviour.

Section 7 presents a case study example, related to ICTs application for agricultural extension, in North-West Greece. In particular, this section, utilizing survey data, examines to what extent farmers have adopted electronic communication technologies. In addition, the section explores farmers' interest in precision farming in association with their willingness to adopt telecommunication technologies. Using indicated preferences, the analysis finds that the ability to receive information, *via* various means of telecommunication, reinforces farmer support for more one-on-one and personalized communication. The results also show that adoption of telecommunication technologies may supplement rather than replace traditional delivery methods of information; sequentially an increase of demand for education will be expected.

A further general finding is that ICTs adoption is significantly related to factors such as annual income, farmers' classification, familiarity with ICTs and education with gender being a supporting factor. Such empirical findings support Rogers (1995) socio-economic generalizations about early adopters. Additionally, they suggest the existence of a relationship of adoption to media use intensity and of a 'technology cluster'. Nevertheless, further research is needed along the lines of diffusion theory (addressing, for example, the role of personality traits and product attributes in ICT adoption) and its extensions (Mahler and Rogers, 1999; Lin, 2001; Li, 2004) as well as the technology acceptance model (Davis 1989) and its extensions (Venkatesh and Davis, 2000; Legris *et al.*, 2003; Yi and Hwang, 2003). Moreover, the successful ICT introduction is multidimensional and thus depends on complementary factors; in this sense, the exploration of issues such as social and economic structuration and personal circumstances (Selwyn *et al.*, 2005; Rhodes, 2004; Lamb and Kling, 2003; Uotinen, 2003) are also extremely important and has been proposed to be addressed within a Sustainable Livelihoods approach (Chapman and Slaymaker, 2002; Parkinson and Ramirez, 2006).

Finally, in line with previous research, it is shown that in Greece also regardless of the level of communication sophistication, farmers desire a wide range of information delivery channels, particularly on farm demonstrations and farmer involvement in applied research. A further finding was that 'high tech' farmers expressed a greater need for the intensification of most of the listed communication techniques and significantly higher preference for computer assisted instructions, video conferences and farmer involvement in applied research. In sum, these findings support, among others, Rivera and Qamar (2003) who state that the introduction of ICTs should adjust to the context of established communication systems as well as that ICTs should facilitate the work of extension - not replace it. In this respect, future research should also address the option for extension agents to undertake new roles, such as acting as information intermediaries (Souter, 2005), facilitators in the creation of 'local learning spaces' (Ramirez, 2001) and 'mediating commons' (Lagos, 2008) and so on.

References

Abbott, E. and Yarbrough. P., 1992. Inequalities in the information age: Farmers' differential adoption and use of four information technologies. *Agriculture and Human Values,* 9(2): 67-79.

Akpabio, I.A., 2007. Agricultural extension officials and utilization of information and communication technologies in the Niger Delta, Nigeria. *Journal of Extension Systems,* 23(1): 29-41.

Akpabio, I.A., Okon, D.P. and E.B. Inyang. 2007. Constraints affecting ICT utilization by agricultural extension officers in the Niger Delta, Nigeria. *Journal of Agricultural Education and Extension,* 13(4): 263-272.

Albrecht, H., Bergmann, H., Diederich, G., Grober, E., Hoffmann, V., Keller, P., Payr, G. and Sulzer, R., 1989. *Agricultural Extension: Vol. 1. Basic Concepts and Methods.* Wiley Eastern Ltd., Germany.

Annor-Frempong, F., Kwarteng, J., Agunga, R. and Zinnah. M., 2006. Challenges of infusing information and communication technologies in extension for agricultural and rural development in Ghana. *Journal of Extension Systems,* 22(2): 69-82.

Audirac, I. and Beaulieu, L., 1986. Microcomputers in agriculture: A proposed model to study their diffusion/adoption. *Rural Sociology,* 51(1): 60-77.

Bardon, R., Hazel, D. and Miller. K., 2007. Preferred information delivery methods of North Carolina forest landowners. *Journal of Extension,* 45(5). *http://www.joe.org/joe/2007october/a3.shtml* (Accessed on 20th January, 2009).

Bohrnsedt, G.W., 1977. Reliability and validity assessment in attitude measurement. In: Summers, G.F. (Ed.), *Attitude Measurement.* Kershaw Publishing Company, Ltd, London.

Bohrnsedt, G.W., 1970. Reliability and validity assessment in attitude measurement. In: Summers, G.F. (Ed.), *Attitude Measurement,* 80-99. Chicago: Rand-McNally & Co.

Brashear, G., Hollis, G. and Wheeler, M., 2000. Information transfer in the Illinois swine industry: How producers are informed of new technologies. *Journal of Extension,* 38(1). *http://joe.org/joe/ 2000february /rb4.html* (Accessed on 20th January, 2009).

Brynjolfsson, E. and Hitt., 2000, L.M., Beyond computation: Information technology, organizational transformation and business performance. *Journal of Economic Perspectives*, 14(4): 23-48.

Bultena G. and Lasley, P., 1990. The dark side of agriculture of agricultural biotechnology: Farmer's appraisals of the benefits and costs of technological innovation. In: Gendel S.M., A.D. Kane, D.M. Warren and F. Yates (Eds.), *Agricultural Bioethics: Implications of Agricultural Biotechnology*. Iowa State University Press, Ames, I.A

Burke, K. and Sewake. K.,, 2008. Adoption of computer and internet technologies in small firm agriculture: A study of flower growers in Hawaii. *Journal of Extension*, 46(3) *http://www.joe.org/joe/ 2008june/rb5.shtml* (Accessed on 20th January, 2009).

Chapman, R. and Slaymaker. T., 2002. ICTs and rural development: Review of the literature, current interventions and opportunities for action. *ODI Working Paper 192*, London: ODI.

Cochrane, W., 1993. *The Development of American Agriculture- A Historical Analysis* (2nd ed.), University of Minnesota Press, Minneapolis.

Davis, F., 1989. Perceived usefulness, perceived ease of use, and user acceptance of information technologies. *MIS Quarterly*, 13(3): 319-340.

Demiryurek, K., 2006. Distance education for rural people in developing countries: Turkish experience. *Journal of Extension Systems*, 22(2): 83-94

Ganesh, J. and Kumar, V., 1996. Capturing the Cross-National Learning Effect: An Analysis of an Industrial Technology Diffusion. *Journal of the Academy of Marketing Science*, 24: 328-337.

Gregg, J. and Irani, T., 2004. Use of Information Technology by county extension agents of the Florida Cooperative Extension Service. *Journal of Extension*, 42(3) *http://www.joe.org/joe/2004june /rb2.shtml* (Accessed on 20th January, 2009).

Hall, L., Dunkelberger, J., Ferreira, W., Prevatt, J.W. and Martin, N., 2003. Diffusion-adoption of personal computers and the internet in farm business decisions: Southeastern beef and peanut farmers. *Journal of Extension*, 41(3). *http://www.joe.org/joe/2003june/a6.shtml* Accessed 20th January, 2009.

Harder, A. and Lindner, J., 2008a. An Assessment of county extension agents' adoption of extension. *Journal of Extension*, 46(3). *http://www.joe.org/joe/2008june/rb1.shtml* Accessed 20th January, 2009.

Harder, A. and Lindner, J., 2008b. County extension agents' perceptions of extension. *Journal of Extension*, 46(3). *http://www.joe.org/joe/2008june/a2.shtml* (Accessed on 20th January, 2009).

Herring, P., 2008. Who's that knocking at our door? Characterizing extension's online clientele. *Journal of Extension*, 46(3) *http://www.joe.org/joe/2008june/a4.shtml* (Accessed 20th January, 2009).

Hoban, T., E. Woodrum and Czaja, R., 1992. Public opposition to genetic engineering. *Rural Sociology*, 57(4): 476-493.

Howell, J. and Harbon, J., 2004. Agricultural landowners' lack of preference for internet extension. *Journal of Extension*, 42(6) *http://www.joe.org/joe/2004december/a7.shtml* (Accessed on 20th January 2009).

Jorgenson, D.W., 2001. Information technology and the U.S. economy. *American Economic Review*, 91(1): 1-32.

Kibwana, O.T., M. Haile, L. van Veldhuizen and Waters-Bayer, A., 2001. Clapping with two hands: Bringing together local and outside knowledge for innovation in land husbandry in Tanzania and Ethiopia - A comparative case study. *Journal of Agricultural Education and Extension*, 7(3): 133-142.

Kooij, Van der, A.J. and Meulman, J.J., 1997. MURALS: Multiple regression and optimal scaling using alternating least squares. In: *Advances in Statistical Software 6*, (ed.) W. Bandilla and F. Faulbaum, F., 99-106. Stuttgart: Lucius & Lucius.

Koutsouris, A., 2006. ICTs and rural development: Beyond the hype. *Journal of Extension Systems*, 22(1): 46-62.

Koutsouris, A., 1999. Organisation of extension services in Greece. *Options Mediterranneenes*, Serie A: Seminaires Mediterraneenes, 38: 47-50.

Koutsouris, A., Panagiotou, A and Papadopoulos, D., 1996. Agricultural extension in Greece: New perspectives and development priorities. *Proceedings of the 12th European Seminar of Extension Education*. August 28-September 2, 1995. Edited by Siardos, G. and S. Androulidakis. Thessaloniki, 31-39.

Lagos, T., 2008. Mediating commons: Rural Greece. *New Media and Society*. 10(4): 565–583

Lamb, R. and Kling. R., 2003. Reconceptualising users as social actors in information systems research. *MIS Quarterly*. 27(2): 197-235/

Lasley, P., Padgitt, S. and Hanson. M., 2001. Telecommunication technology and its implications for farmers and Extension Services. *Technology in Society*. 23: 109-120

Lasley, P., and Bultena, G.,1986. Farmers' opinions about third-wave technologies. *American Journal of Alternative Technology*. 1: 99–110.

Legris, P., Ingham, J. and Collerette. P., 2003. Why do people use information technology? A critical review of the technology acceptance model. *Information and Management*, 40: 191–204.

Leveque, A.-E., Kelly, T. and Michelin. Y., 2007. Development of information and communication technologies: The impact at farmer level. Paper presented at the 6th Biennial Conference of the European Federation of IT in Agriculture (EFITA/WCCA 2007). *Environmental and Rural Sustainability Through ICT*, 2-5 July, 2007, Glascow: Glasgow Caledonian University.

Li, S.C., 2004. Examining the factors that influence the intentions to adopt internet shopping and cable television shopping in Taiwan. *New Media and Society*, 6(2): 173–193.

Lin, C., 2001. Audience attributes, media supplementation, and likely online service adoption. *Mass Communication and Society,* 4(1): 19-38

Mahler, A. and Rogers, E., 1999. The diffusion of interactive communication innovations and the critical mass: the adoption of telecommunications services by German banks. *Telecommunications Policy,* 23: 719-740.

Michailidis, A., 2007a. Adoption of Internet in Agriculture. *Journal of Extension Systems,* 23(1): 1-13.

Michailidis, A., 2007b. Agricultural Extension Services in the Mountain Areas of Greece. *Journal of International Agricultural and Extension Education,* 14(1): 71-81.

Milla, K., Lorenzo, A. and Brown. S., 2005. GIS, GPS, and remote sensing technologies in extension services: Where to start, what to know? *Journal of Extension,* 43(3) *http://www.joe.org/joe/2005june/a6.shtml* (Accessed on 20th January, 2009).

Morrow, L., Kelly, T. and Kirley, T., 2004. ICT – It's Potential as a channel for enhanced extension services. 20th Annual Conference AIAEE 2004. http://www.aiaee.org/2004/Accepted/064.pdf (Accessed on 24th January, 2009).

NSSG., 2008. Analytical Statistical Tables (1964-2008). National Statistical Service of Greece, Athens.

Oliner, S.D. and Sichel, D.E., 2000. The resurgence of growth in the late 1990s: Is information technology the story? *Journal of Economic Perspectives,* 14(4): 3-22.

O.T.A., 1987. Technology, public policy and the changing structure of American agriculture, U.S. Congress: U.S. Government Printing Office.

Palisade Corporation. 2000. @RISK Risk Analysis and Simulation Add-In for Microsoft Excel: A Software Package. Version 4, Newfield, NY.

Palisade Corporation. 1998. @BEST FIT: Distribution Fitting Software Package. Version 2, Newfield, NY.

Panigrahi, B., Panda, S.N. and Agrawal, A., 2005. Water balance simulation and economic analysis for optimal size of on-farm reservoir. *Water Resources Management,* 19, 233–250.

Park, D-B., Cho, Y.-B. and Lee, M., 2007. The use of an e-learning system for agricultural extension: A case study of the rural development administration, Korea. *Journal of Agricultural Education and Extension,* 13(4): 273-285.

Parkinson, S. and Ramirez, R., 2006. Using a sustainable livelihoods approach to assessing the impact of ICTs in development. *The Journal of Community Informatics,* 2(3) Special Issue: Telecentres. http://ci-journal.net/index.php/ciej/article/view/310/269 (Accessed on 22nd January, 2009).

Pindyck, R. S., 1991. Irreversibility, uncertainty and investment. *Journal of Economic Literature,* 29, 1110-1148.

Pratt, J.W., 1987. Dividing the indivisible: Using simple symmetry to partition variance explained. In: T. Pukkika and S. Puntanen, (Eds.) *Proceedings of the second International Conference in Statistics,* 245-260. Tampere, Finland: University of Tampere.

Purvis, A., W. G. Boggess, C. B. Moss, and Holt, J., 1995. Technology Adoption Decisions Under Irreveribility and Uncertainty: An Ex Ante Approach. *American Journal of Agricultural Economics,* August, 541-551.

Ramirez, R., 2005. Learning our Way into Communication: The Making of the Communication and Information Strategy with the National Agricultural Advisory Services Programme in Uganda. *Journal of Agricultural Education and Extension,* 11(4): 36-44.

Ramirez, R., 2001. A model for rural and remote information and communication technologies: A Canadian exploration. *Telecommunications Policy,* 25: 315-330.

Ramkumar, S., Garforth, C., Rao, S.V.N. and Heffernan, C., 2007. Design and formative evaluation of an information kiosk on cattle health for landless cattle owners. *Journal of Agricultural Education and Extension,* 13(1): 7-22.

Rhodes, J., 2004. South African, rural ICT implementation: A critical retrospective application of Latour's due process model. *Australasian Journal of Information Systems* 11(2): 46-56.

Rivera, W. M. and Qamar. K., 2003. *A new extension vision for food security: Challenge to change.* Rome: FAO.

Rivera, W. M., 2000. The Changing Nature of Agricultural Information and the Conflictive Global Developments Shaping Extension. *Journal of Agricultural Education and Extension,* 7(1): 31-41.

Rodewald, A., 2001. Delivery systems – is the "latest" technology the greatest? *Journal of Extension* 39(4) *http://joe.org/joe/2001august/tt2.html. http://joe.org/joe/2001august/tt2.html* (Accessed 20th January, 2009).

Rogers, E., 1995. *Diffusion of innovations* (4th ed.). New York: The Free Press.

Seevers, B., 1995. Extensionists as adult educators: A look at teaching style preference. *Journal of Extension,* 33(3) *http://www.joe.org/joe/1995june/rb2.html* (Accessed on 20th January, 2009).

Selwyn, N., Gorard, S. and Furlong, J., 2005. Whose internet is it anyway? Exploring adults' (non)use of the internet in everyday life. *European Journal of Communication,* 20(1): 5-26.

Souter, D., 2005. *The Economic Impact of Telecommunications on Rural Livelihoods and Poverty Reduction: A Study of Rural Communities in India (Gujarat), Mozambique and Tanzania.* Glascow and London: CTO and DFID.

SPSS. 2007. *SPSS Categories 16.0 and User Manual.* Chicago: SPSS Inc.

Uotinen, J., 2003. Involvement in (the Information) Society. *New Media and Society,* 5(3): 335-356.

Van Woerkum, Cees M.J., 1996. Communication and policy: From persuasion towards interaction. In: Siardos, G. and S. Androulidakis, Thessaloniki, (Eds). *Proceedings of the 12th European Seminar of Extension Education.* August 28-September 2, 1995. 113-119.

Venkatesh, V. and Davis, F., 2000. A theoretical extension of the technology acceptance model: Four longitudinal field studies. *Management Science,* 46 (2): 186-204.

Vergot, P., Israel, G. and Mayo, D., 2005. Sources and channels of information used by beef cattle producers in 12 Counties of the Northwest Florida extension district. *Journal of Extension,* 43(2). *http://www.joe.org/joe/2005april/rb6.shtml* (Accessed on 20th January, 2009).

Wims, P., 2007. Analysis of adoption and the use of ICTs among Irish farm families. *Journal of Extension Systems,* 23(1): 14-28.

Yi, M. and Hwang, Y., 2003. Predicting the use of web-based information systems: self-efficacy, enjoyment, learning goal orientation, and the technology acceptance model. *International Journal of Human-Computer Studies,* 59: 431-449.

Author Information

Dr. Anastasios Michailidis

Lecturer of Agricultural Extension
Department of Agricultural Economics, School of Agriculture
Aristotle University of Thessaloniki
541 24 (P.O. Box: 232), Thessaloniki, Greece.
Postal Code: 541 24
Phone No: (0030) 2310998783; (0030) 2310998827
Mobile Phone No: (0030) 6977353858
Fax No: (0030) 2310998828
e-mail: tassosm@auth.gr
URL: www.auth.gr

Dr. Anastasios Michailidis is a Lecturer of Agricultural Extension at Aristotle University of Thessaloniki. His research interests include agricultural extension, agricultural education, adoption-diffusion of innovations, water resources management and information-communication technologies. His education includes a B.Sc. in Agriculture, a M.Sc. in Agricultural Economics and a Ph.D. in Agricultural Economics. He has published more than 50 papers in international refereed journals, in collective volumes and proceedings.

□□□

Chapter 4

INDIA

R. Saravanan

1. INTRODUCTION

1.1. National Agriculture Scenario

Agriculture is the most important sector of Indian Economy, contributes 23 per cent of national GDP, accounts for 11 per cent of exports, and 50 per cent of billion depends on it. Interestingly India feeds 16 per cent of population with 2.4 per cent of global land.

In the last five decades, there has been a steady and spectacular transformation of Indian agriculture from the food deficit to food sufficient status. Diffusion and adoption of modern technologies, high yielding varieties, dedicated efforts of farmers, extension personnel and scientists and also programmatic support of central and state governments have all contributed significantly from 50 million tonnes in 1950-51 to land mark achievement of 230.67 million tonnes of food production in 2008-09.

1.2. Agricultural Sector Challenges

The challenges before Indian agriculture are immense. The agricultural growth rate is sluggish and stagnating. The sector needs to grow at a faster rate than the past to allow for higher per capita income and consumption. It is an accepted fact that the sound agricultural development is essential for the overall economic progress of India. Given its range of agro-ecological setting and produces, Indian agriculture is faced with a great diversity of needs, opportunities and prospects. The water scarce-rain fed areas, which account for 63 per cent of the cultivated land, exhibit low and also unstable yield and technology transfer gaps are much wider as compared to those of irrigated areas (Chatterjee and Prabhakar, 2009). National seminar on agricultural extension 2009 background note states that sustaining growth rate and achieving the required food grain production of 320 million tonnes by 2025 would be a Herculean task considering some of the challenges like non-expanding land, depleting soil and water resources, adverse impact of climate change, rising cost of production, diminishing agriculture labour availability and farmers reduced interest in agriculture (NSAE, 2009). If India is to respond successfully to these challenges and also to achieve accelerated growth needs, vibrant and innovative technology generation and delivery system are required. Greater attention will have to be paid to technology dissemination. To make farm information and technology transfer more effective, greater use will need to be made of modern information and communication technology among researchers, extension personnel, farmers and other stakeholders. Further, the agricultural extension requires paradigm shift from top-down, blanket recommendation of technological packages, towards providing producers with the knowledge and understanding with which they solve their own location-specific problems. Continuous two-way interaction among the farmers agricultural scientists and extension personnel is the most critical missing component of agricultural extension (Chatterjee and Prabhakar, 2009). To assist the farmer in these changing contexts, new strategies and innovative solutions are urgently required which in turn will require technological support (Website of National Agricultural Innovation Project - www.naip.icar.org.in).

2. NEED FOR ICT IN AGRICULTURAL EXTENSION

1. To accelerate agricultural growth

Recommendation of the Planning commission of India's working group on agricultural extension for XI five yea plan (2007-2011) states that the agricultural growth is stagnating and sluggish. Hence, there is an immediate need of vibrant, dynamic and innovative approach to be adopted for agricultural extension in order to achieve targeted growth rate and serve the farmers better. Integration of ICTs in agricultural extension will provide needed impetus to agricultural sector.

2. To expand knowledge resource

India feeds 16 per cent of world population with 2.4 per cent of global land. Land and water resources are almost reaching their limits, hence achieving food security heavily relies on "Knowledge Resource". In this scenario, ICTs can complement the traditional extension system for "Knowledge Resource" delivery to the millions of the farmers.

3. To facilitate better information access

Estimates indicated that 60 per cent of farmers do not access any source of information for advanced agricultural technologies resulting in huge adoption gap (NSSO, 2005). In this context, it is expected that convergence of ICTs with traditional extension system will improve the farmers' information access.

4. To supplement inadequate technical manpower

In India, there are about 120 million farm holdings and the number is growing year by year. It proposes to provide one village extension personnel for 800-1000 farm families than the requirement of field level extension personnel which is estimated to be about 1300000-1500000, against which the present availability is only about 100000 personnel (PC, GoI, 2007). In this scenario, inadequate technical manpower to be for some extent compensate by the extensive use of ICTs.

5. For stronger research-extension –Client system linkage

ICTs are required to facilitate stronger linkages with research- extension-client system. The feedback received through ICTs to be more accurate and faster.

6. To develop efficient feedback mechanism

Lack of efficient feedback mechanism in the research- extension linkage was identified as one of the weaknesses in the existing extension systems. Hence, it is believed that the media and ICTs will offer strong potential to improve linkage mechanism.

7. For cost-effective extension delivery

The ICTs tools such as: Internet and Mobile networks have the potential to provide agro-information services that are; affordable, relevant (timely & customized), up-to-date, high accessibility and farmer friendly.

8. To develop knowledge managers

The experience of rural centres shows that ICTs can help in enabling rural development workers to gather, store, retrieve, adapt, localised and disseminate a broad range of information needed by rural families. This, in turn, leads to the emergence of knowledge workers that will result in the realisation of a bottom- up, demand- driven paradigm for technologies generation, assessment, refinement and adoption.

9. To ensure gender equity in technology transfer process

Traditional extension is widely criticised for not concentrating women cultivators. Research evidences show that ICT enabled extension system offers equal opportunity to the farm women.

10. To empower small and marginal farmers

In India 77 per cent of cultivators are marginal farmers. Land holding declined from 2.28 hectare per family to 1.41 hectare. Empowering small and marginal farmers with the right information at the right time and place is essential for improving efficiency and vitality of small and marginal holdings (NPFF, 2007).

11. To serve the farm stakeholders beyond technology transfer role

There is a growing recognition that extension must go beyond transferring new food crop technology to farmers and focus on helping the rural poor by promoting agriculture diversification, increasing rural employment and helping farmer gain access to biotechnology and access to export markets and also environment awareness and rural health awareness. To perform this expanded role, extension systems should be equipped with ICTs.

3. ICT INFRASTRUCTURE SCENARIO

Strategic reforms in telecommunications sector since 1990's, facilitates strong ICT infrastructure in India. As on May 2009, 452.91 million fixed land line telephones, 415.25 million wireless and 6.4 million broadband subscribers were estimated by the Telecom Regulatory Authority of India (TARI, 2009). The tele density has reached 38.88 (number of telephone subscribers per 100 individuals). However, there is huge gap between urban and rural tele density, 64.48 and 9.03, respectively. Despite several policy initiatives to promote rural ICT penetration, growth in tele density continues to be skewed in favour of urban India. Total internet users are 49.40 million. In rural India, only 1.2 per cent people have internet access, whereas, it is 12 per cent in urban India. The overall urban and rural mobile penetration remains 43.88 and 4.92 per cent, respectively. It is estimated that by the end of the year 2010, India will have 500 million mobile phone subscribers.

4. NATIONAL POLICY ON ICT IN AGRICULTURAL EXTENSION

National policy framework for agricultural extension (2000) stated that information technology revolution is unfolding and has very high visibility. Harnessing information technology for agricultural extension will receive high point in the policy agenda. Extensive use of modern information technology will be promoted for communication between researchers, extension workers and their farmer clients to transfer technologies and information more cost effectively. Further, it emphasised IT application in marketing, wider use of electronic mass media for agricultural extension, farmer participation in IT programmes and support to the state government for using IT in agricultural extension, promoting IT based information kiosks and capacity building for use of IT (DoA&C, 2000).

National policy for farmers (2007) indicated that the potential of ICT would be harnessed by establishing gyan chaupels (Knowledge centres) in villages. Further, the Common Service Centres (CSCs) of the Department of Information Technology, Ministry of Communications and Information Technology, Government of India and those set up by the state governments and private initiative programmes will be evolved for inclusive broad-based development. Last mile and last person connectivity would be facilitated with the help of technologies such as broadband internet, community radio or internet-mobile phone synergies (NPFF, 2007).

Document of ICAR Framework for Technology Development and Delivery System in Agriculture (2008) outlined the need for the construction of Agri - India knowledge portal - A single electronic gateway to be developed through a peer review process with the help of 15 content accreditation centres from 15 agro - climatic regions of the country. Each accreditation centres will be coordinate with other Agricultural Universities and agricultural institutions in their region for development of content in regional language as well as in English and also do its validation, which will be collected in the central data warehouse integrated in the knowledge portal. The portal will also serve as a platform for facilitation of interaction among researchers and extension personnel in the KVKs through high speed server intranet (ICAR-FFTDDSA, 2008).

National e-Governance Plan indicated that the typical services envisaged in Agriculture as a Mission Mode Projects (MMP) to provide information to the farmers on seeds, fertilizers, pesticides, Govt. Schemes, Soil recommendations, Crop management, Weather and marketing of agriculture produce. Several projects such as ASHA in Assam, KISSAN and e-Krishi in Kerala and Krishi Maratha Vahini in Karnataka have been initiated by the Department of Agriculture and Co-operation (DoA&C), Government of India. To spearhead implementation of MMP in Agriculture, DoA&C has adopted twin strategy through AGRISNET & two portals AGMARKNET & DACNET (Mathur *et al.*, 2009).

5. REVIEW ON BEST PRACTICES OF ICT FOR AGRICULTURAL EXTENSION

5.1. aAQUA - almost All QUestions Answered

The *aAQUA* is effectively an online, yet archived, discussion forum accessible using a web browser, allowing members to create, view and manage content in their mother-tongue (Hindi, Marathi *etc.,*). It aims to incorporate innovations from the perspective of cross-lingual multimedia information storage and retrieval and intelligent databases. Thus, aAQUA empowers the members in a community to create their own content and provides an easy to use interface. The aAQUA eases the creation of content *i.e.,* text, images, short audio, video and animations, thus helping the users move from being passive consumers to active content creators. More than 90 per cent of the current English, Hindi and Marathi content of aAQUA was created locally by the community. Assistance in creating content for aAQUA is provided by the telecenter/ kiosk operator. This takes the technology to users who are unfamiliar to computers and often not literate. They can now experience the internet as well as participate through it. This is a great motivating factor for people as it showcases the benefits of using a computer in their everyday lives.

aAQUA provides language independent discussion services which are not limited to text and can have audio-visual elements to provide a simple, yet rich interface accommodating novice users. The user can formulate questions in a natural manner and is not forced to limit the query to a few words. This feature has been found to be popular among our rural community. It thus provides a communication framework for easy and fast access to reliable information (both through artificial agents and Human Experts from all over the world). The system performs "Meaning Based Search" through the available repository of information. Textual content is stored in a language independent fashion. This is intended to allow users to ask questions in their own language, access content in another language and view it in their own language (Ramamritham *et al.*, 2009).

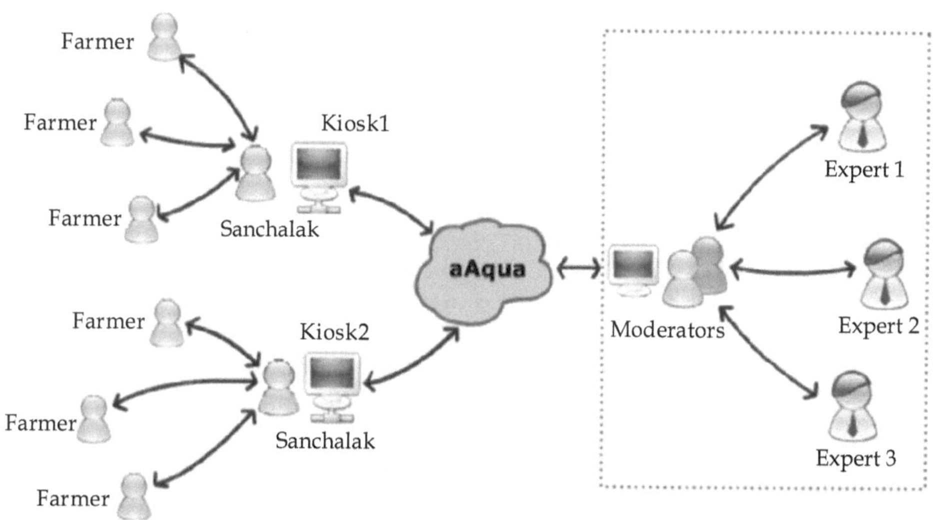

Figure 1 : Flow of information in aAQUA system (Source: Ramamritham *et al.*, 2005)

5.2. Digital Green – Participatory Video for Agricultural Extension

Digital Green (DG) is an ICT based research project for the production and dissemination of locally relevant agricultural through participatory video and mediated instruction. The digital Green farm based video has been interactively designed, deployed and evaluated among the small and marginal farmers of the Karnataka State of India. The project was initiated by the Microsoft Research, India in collaboration with Green Foundation, an NGO.

Farmer participatory video recordings facilitate the aggregation of scattered information into systematic and comprehensive format with a localized context. The video recordings are made by teachers of agricultural at grassroots level and expert reviewers ensure the accuracy, clarity and completeness of the context and guide the construction of a time and location sensitive video based curriculum. First, awareness meetings on DG services were conducted; interested farmers were identified and new content was recorded. Content was provided by local farmers with the guidance of experts. Selected farmers were deployed for field adoption, and content screenings were used to learn, adopt and innovate better agricultural practices. Community members act as mediators to conduct a minimum of three screening per week during suitable evening hours and mediators act as a facilitators, mediators are paid honorarium up to INR 1, 500 (USD 38) per month and they additionally supported by a full time extension staff either from the government or NGO. The Digital Green one year trial involving 20 villages (1, 470 house holds) in Karnataka, India increased the adoption of certain agricultural practices seven fold over a traditional extension system and ten times more effective per dollar spent (www.digitalgreen.org). DG system was able to multiply the value of a NGO's extension agents by a factor of 10 times per dollar spent (Gandhi *et al.*, 2008).

5.3. e-Arik (e-Agriculture)

The e-Arik (Arik means agriculture in the Adi tribal dialect of Arunachal Pradesh State of North-East India) project was implemented in Arunachal Pradesh, one of the most socio-economically backward states of North-East India. After assessing farmers' information needs, the project experimented single window system for the improved agricultural information and technology delivery by using computer, internet, phone, radio and television. Project provides all time expert consultation on agriculture production, protection and marketing aspects through ICTs. The e-Arik research project staff regularly undertakes field visits to observe crop condition, diagnosis the pest, diseases, and nutrient deficiency, physiological problems, and then field crop condition is digitally documented. To solve complex crop pest, diseases, nutrient deficiency and physiological problems, digital photographs are transmitted through e-mail from e-Arik-village knowledge centre to the farm scientists of Central Agricultural University and recommendations passed on to the farmers. Further, farm scientists undertake need-based field visits and provide expert advice to the farmers. Further, farmers training and demonstrations were conducted by the project staff and or Subject Matter

Specialists of Farm Science Centres, extension personnel from the developmental departments. Project portal (www.earik.in) provides information on crop cultivation, agriculture and rural developmental departments and their schemes, day to day market information and weather conditions, which is also displayed in the village knowledge centre notice board. Further, information on health, education, governance and other information for tribal farmers are available in the project portal. The village agricultural library at the e-Arik-village knowledge centre is having the collection of farm publications, multimedia CDs and daily news papers for the ready reference of the farmers and others. Farm input display unit at the e-Arik-village knowledge centre exhibits bio-fertilizers, organic pesticides and fungicide samples for the familiarization among the farmers. The ICT awareness lectures, regular trainings were conducted for the benefit of village children, students, village school teachers and villagers. The village advisory committee regularly reviews the progress of the project. Innovative approaches such as: farmer to farmer communication, local leadership and self-help approaches are employed for the agricultural technology transfer. The e-Arik pilot project covered 12 tribal villages, 500 registered farm families of East Siang district of Arunachal Pradesh State of North-East India (www.earik.in; Saravanan and Indradevi, 2008).

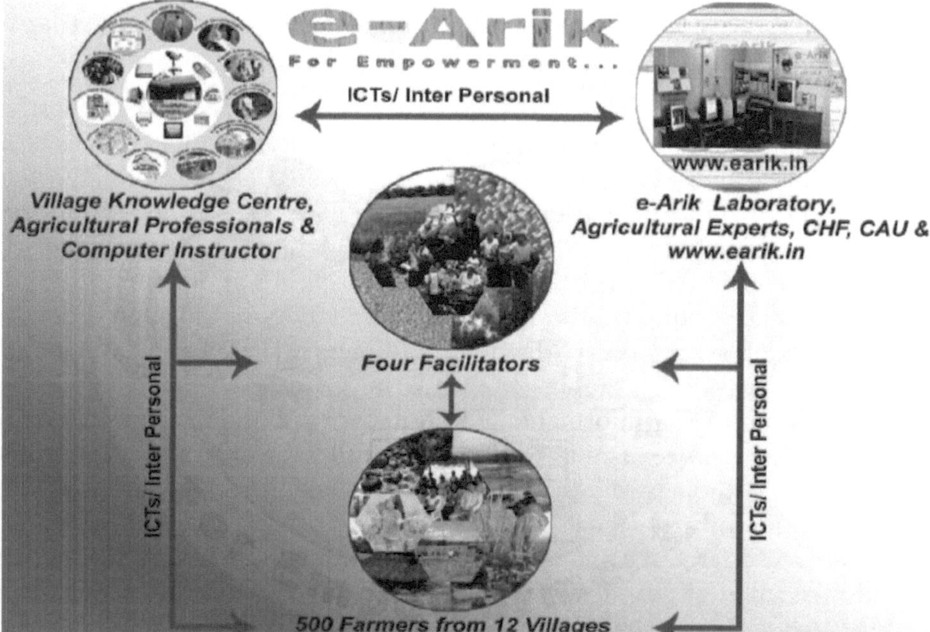

Figure 2 : Farm Information Dissemination in e-Arik: ICTs for Agricultural Extension Project

5.4. e-Sagu (e-Cultivation)

e-Sagu was a ICT based personalized agro- advisory system in which rather than on site visit of crops the agricultural scientists deliver the advice by getting the crop status in the form of digital photographs and other information. A co-ordinator who was associated with a group of farmers and who possesses agricultural experience and basic data entry skills, visits each designated farm on a weekly basis. The co-ordinator collects the registration information including soil data, water resources and capital availability from the farm and sends the crop details in the form of text and digital photographs through the communication system to agriculture experts. The agricultural experts gave appropriate recommendation with the help of computerized agricultural information system. Agricultural information system contained all the related information such as details of the farmer with corresponding soil and crop information and status of the crop. Also the details of agricultural technology related to various crop such as the level of pest resistance, requirement of water so on. They used research data, soil data, historical data, weather data and other information to generate appropriate recommendation and stored this advice in the system to improve crop productivity. The co-ordinator then explained the advice to the farmer. The feed back regarding the advice was sent along with photographs during the following weeks. The cycle was repeated every week for each farm (www.esagu.in).

A team of agricultural experts worked at the e-Sagu main lab, supported by an agricultural information system. One e-Sagu local centre consists of a few computers and a computer operator and covered a group of about ten villages. Educated and experienced farmers worked as coordinators. Depending on the crop, each coordinator was assigned a fixed number of farmers. He collected the registration details of the farms under him including soil data, water resources and capital availability and sent the information to the main e-Sagu system. Every day the coordinator visited a fixed number of farms and took 4-5 photographs for each farm. A CD was prepared with the photographs and other information and transferred to the main system by a regular courier service. The agricultural experts with diverse backgrounds (entomology, pathology, agronomy *etc.*) at the e-Sagu main lab analyzed the crop situation with respect to soil, weather and other agronomic practices and prepared a farm specific advice. At the local e-Sagu centre, the advice was downloaded electronically through a dial-up Internet connection. The coordinator collected the advice print out and delivered it to the farmer concerned. In this way each farm got the proactive advice at regular intervals starting from pre-sowing operations to post-harvest precautions (Ramaraju and Krishna Reddy, 2009).

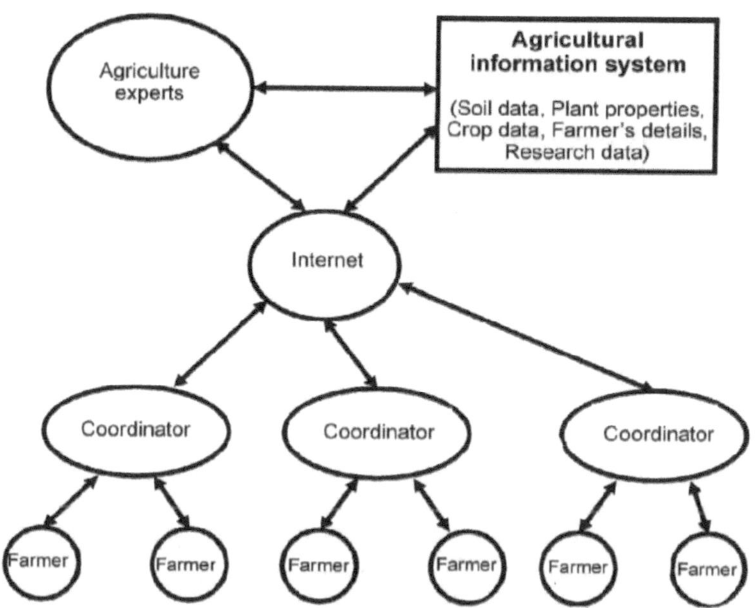

Figure 3 : Agricultural information dissemination system. (Source: Krishna Reddy and Ankaiah, 2005)

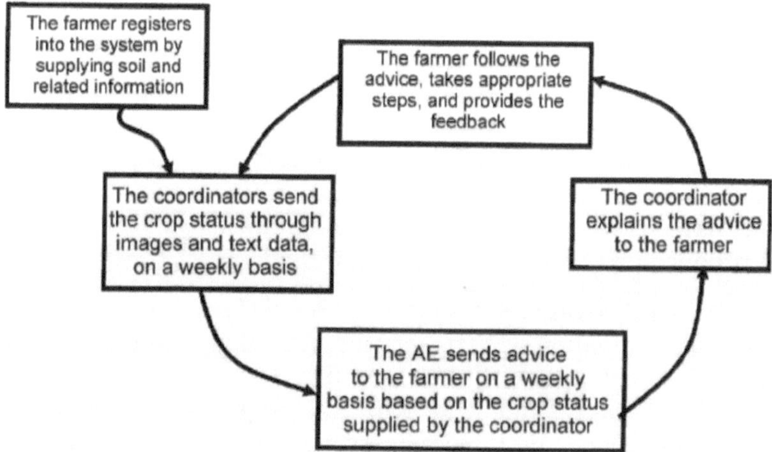

Figure 4 : Depiction of Agricultural Information Dissemination System (AgrIDS) Operation (Source: Krishna Reddy and Ankaiah, 2005)

The development of e-Sagu was started during Kharif season of 2004. The e-Sagu system was implemented by delivering advisory to 1051 cotton farms in three villages of Warangal district in Andhra Pradesh State of South-India. The experiment was successful. During 2005-06, as a scaled-up experiment, e-Sagu system was implemented for cotton, chilli, rice, groundnut, castor, red-gram and other crops for 4894 farms in 35 villages spread over six districts in Andhra Pradesh State. In addition, the e-Sagu system has evolved to deliver advisory service to fish-farmers. About 160 fish ponds were covered. By the end of 2005-06, there were eight e-Sagu local centres which had delivered 35925 advices covering 30 crops and fish rearing. About 260000 crop photographs and 8918 fish photographs were taken in the process. The average number of photographs taken per observation was five. The farmers are happy with the expert advice as it was helping them to improve input efficiency through the use of integrated pest management (IPM), judicious use of pesticides and fertilizers, *etc.* At the same time benefits accrue through higher yield. As per an impact assessment study carried out by social scientists, benefits accrued to farmers were (in 2004) Rs. 9435/- (USD 240) per hectare with cost benefit ratio of 1:3 and for the year 2005: Rs. 9491/- (USD 240) per hectare with cost-benefit ration of 1:4.1. Fish farmers have also realized significant benefits. In some centres the farmers have paid the subscription fee for the services and they are satisfied with the same. The e-Sagu project has successfully delivered advices for 5, 054 farms (4130 ha) of 3,035 farmers in 38 villages in Andhra Pradesh State of India (Ramaraju and Krishna Reddy, 2009).

5.5. KISSAN- Karshaka Information Systems Services and Networking

KISSAN is an innovative project by the Department of Agriculture, Government of Kerala. The mission of this project is to develop and deploy Information Systems, Network, Processes and Services for Agriculture in ways that lead to (i) amelioration of farmers' distress and enhance their welfare, (ii) increased farm productivity and (iii) better returns for farm produce and derived products.

The objective is to provide an effective knowledge management and smart information dissemination system that provides linkage among farmers, public research institutions, administrative and private entrepreneurs to share the information and knowledge. The KISSAN project is having a dedicated content processing and dissemination system. This involves the collection, analysis, classification, process various information from different sources. The project will establish an agriculture data center, which will cater for all the needs. High-end servers and other data processing equipments will back up the data center. All the information will be processed and disseminated through this data center. The project portal also provides Post Query for Expert Answers, Crop Information, GIS Based Agri Advisory, Farming Practices, Fertilizers & Pesticides, Kerala Agri Directory and Discussion Forum (http://www.kissankerala.net/kissan/kissancontents/about.jsp).

5.6. Lifelines India- Soochna Se Samadhan (Solutions through Information)

SOOCHNA SE SAMADHAN is an initiative to use the power of voice as the primary means of information dissemination. It facilitates the exchange of critical and timely information among marginalised communities so that it helps in improving their quality of life. It aims to provide connectivity, content and capability *via* a phone-based service. Specifically it will provide grassroots communities with access to a wide information and knowledge pool.

OneWorld South Asia (OWSA) with support from British Telecom and CISCO systems are piloting the service in North India in partnership with other civil society organizations. British Telecom and CISCO Systems have supported the initiative as a part of their Corporate Social Responsibility programme which assumes that access to information and communication technologies (ICTs) can improve people's lives and open doors to education, jobs, entertainment and interactions. The technology solution development has been supported by Tech Mahindra and WIPRO.

The benefactor calls a designated number using a land line/mobile telephone. The call first reaches the Interactive Voice Response System (IVRS) where she/he registers the query with the help of a voice menu. The query is stored as a voice clip in a database server. The knowledge worker (KW) logs in to the application through a web interface, views all the calls that are waiting for attention, and searches the FAQ database for the answers. If the KW finds the answer, it is retrieved and stored in the IVRS. If the answer is not found, the KW forwards the question to the subject matter experts. Once a response is received from any of the subject matter experts, the application alerts the KW, who examines the response and if appropriate, stores the answer in the database and makes it available for future queries. Voice Clip of the answer is played back to the benefactor when they call back.

The information can also be retrieved in text format from the Information Centre near the village. The farmers can send pictures along with their questions, and also voice clips to clarify on the issue (SOOCHNA SE SAMADHAN, 2009).

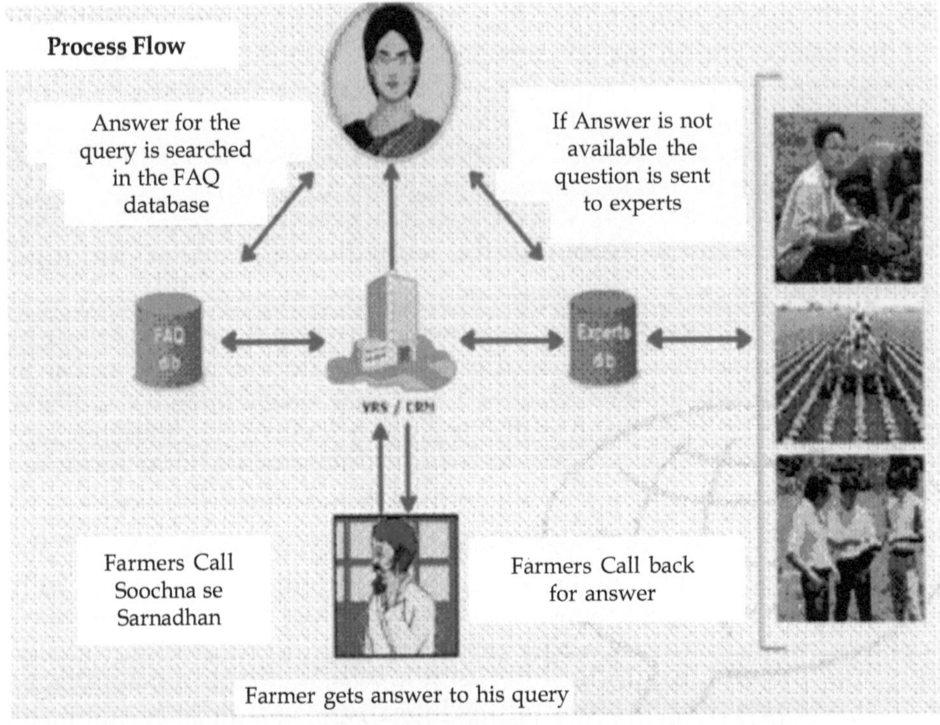

Figure 5 : Information Dissémination Pattern in Lifelines India (Source:www.solutionexchange-un.net.in/decn/cr/res05110601.doc)

Lifelines India which was launched in 2006 as an information delivery service at the grassroots was started in 700 villages in north and central India (Lall and Sahi, 2009), in partnership with Indian Society of Agribusiness Professionals and TARAhaat. Local volunteers will facilitate the use of the service by the farmers. The farmers can also visit the nearest information centre to access the offline database in local language, listen to the audio clips and also to send pictures of affected crops.

5.7. VASAT - VIRTUAL ACADEMY FOR THE SEMI-ARID TROPICS

The International Crops Research Institute for the Semi-Arid Tropics (ICRISAT) at Hyderabad, India has initiated the Virtual Academy for the Semi-Arid Tropics (VASAT), as a technology mediated extension and knowledge sharing programme (Navarro and Balaji, 2003; http://vasat.icrisat.org). As part of the VASAT activities, two field projects (one in India and another in Niger) involving the participation of rural community based organisations were launched in 2004. In Addakal region (South Central India), the VASAT project involved a local community based organisation, Adarasha Women's Welfare Organisation. In the experiment, a module prepared by ICRISAT experts was rendered into local language and the info-mediaries (volunteers from community based organisation) were administered the granules, which helped the volunteers to 'refine" farmers' queries before passing them onto experts. The video-conferencing and online forums in support of agricultural related questions and answers processes and also GIS derived tools were successfully experimented for micro-level drought preparedness among small and marginal farmers in drought-prone rainfed area (Balaji *et al.*, 2007).

5.8. Touch Screen Kiosk

The touch screen kiosk with information on cattle health was tried and tested under a United Kingdom Department of International Development (DFID) - Animal Health Programme (AHP) supported project on dissemination of animal health knowledge for development of landless dairy cattle owners in the peri-urban regions of Pondicherry, India (2002- 2004) by the Rajiv Gandhi College of Veterinary and Animal Sciences, Pondicherry, India in collaboration with the University of Reading, UK. The kiosks were installed in the veterinary centre and veterinary dispensary. The cattle owners bring their animals to the veterinary dispensary for treatment or Artificial Insemination (AI) and register the case. After that they will wait for their turn to come for service delivery. That waiting time is utilized for getting information from the kiosk (Ramkumar and Rao, 2004).

5.9. e-Extension (e-Soil Health Card Programme)

The Department of Agriculture, Gujarat State, India initiated office automation, networking and e-communication among department staff, digital data bank generation, management, storing and retrieving computerisation of farmer centric applications, web site, e-soil health card scheme. The crop estimation survey, online management and monitoring of Government schemes, online farmer accident insurance scheme applications and status, crop wise sowing area, crop and weather condition data will be entered online and state level weekly reports are generated automatically. It also provides online monitoring of individual beneficiary schemes and extension programmes (Krishi mahotsav) and online applications for natural calamity, *etc.* The e-Soil Health Card Programme, an e-Agricultural extension service programme, is one of the ambitious programmes which aims to analyse the soil of all the villages of the state and proposes to provide online guidance to farmers on their soil health condition, fertilizer usage and alternate cropping pattern (www.agri.gujarat.gov.in; www.shc.gujarat.gov.in).

6. VILLAGE KNOWLEDGE CENTRES

6.1. MSSRF - Village Knowledge Centres (VKCs)

The Village Knowledge Centre programme was started in 1998 in the Union Territory of Pondicherry in South India by the M S Swaminathan Research Foundation (MSSRF) in the east coast of South India with the generous support of IDRC and CIDA. In order to ensure access to all, the VKCs are located in public buildings (Panchayats, SHG building, Community Hall, School, Farmers and Fishermen Associations, Women Self-Help Groups, Community Based Organization, *etc.*). The information content is developed in close interaction with the local people in a bottom-up manner. VKCs are connected with hub centres through a hybrid wired and wireless network. The VKCs provide information on agriculture, health, employment, weather, education, government entitlements, microenterprises training, *etc.* In each case, the community or local partner provides an accessible rent-free building, electricity and volunteers. In turn, the project provides all the needed equipment, training and helps in collecting data. Gender concerns are central to the project. More than half of the volunteers are women. This has positively reflected in the increase of the number of women users.

6.2. ISRO -Village Resource Centres (VRCs)

Indian Satellite Research Organization (ISRO) has initiated the VRC concept *i.e.,* setting up of the Village Resource Centres (VRCs) with a view to integrating its capabilities in satellite communications and satellite based earth observations to disseminate a variety of services emanating from the space systems and other IT tools to address the changing and critical needs of rural communities (VRC-WP, 2006). Village Resource Centres will become single window delivery mechanism for tele-medicine, tele-education, natural resources data, agriculture advisories, land & water resources advisories, interactive farmers' advisories, e-governance services and weather advisories. It will also be involved in capacity building of the community to enhance its awareness and knowledge levels. The VRCs will provide information on health, education, nutrition, gender issues, legal services and women empowerment.

Village Resource Centres (VRCs) was inaugurated in 2004. Under this VRC programme, most of the MSSRF -VKCs are connected through Indian Space Research Organization's (ISRO) uplink and downlink satellite facilities. The satellite based ISRO-VRCs, aims at digital connectivity to remote villages for providing multiple services such as tele-medicine, tele-education, interactive farm and fishery advisories, government schemes and entitlements, weather services and remote sensing applications through a single window. Users located at one node of this network can fully interact with others located at another node through video and audio links. Each node can further be expanded using different technologies such as notice boards, pamphlets, public address system, community newspaper, press releases, cable TV, audio/ video conferencing through wireless, telephone, meetings, mobile phone and intranet web site for dissemination of the useful and necessary information. To strengthen the management among VRC, VKCs and strategic partners (data providers, generators, capacity building providers, *etc.*) the Jamsetji Tata National Virtual Academy (NVA) developed a three-tier knowledge network. The main aim is to develop a user-controlled, user-owned and user-managed network, which will help to reach the unreached and include the excluded in terms of information, knowledge and skill empowerment. As on September, 2009, 18 VRCs and 101 VKCs have been set up by National Virtual Academy (NVA), MSSRF in India. For setting up and running of VKCs the MSSRF is working with 127 partners. For providing locale-specific demand driven content and capacity building of rural communities the VRCs and VKCs are working with 315 partners consisting of Government Departments-Agriculture, Animal Husbandry, Fisheries, Education, Rural Development, Health, *etc.*), NGOs, Academic and Research Institutions, Hospitals, *etc.*, (Senthilkumar, 2009).

6.3. Community Information Centres (CICs)

The eight North-Eastern (NE) states of India have traditionally been less developed than the other states due to their geographic remoteness and difficult hilly terrain. The Government of India in 2002, set up 487 Community Information Centres (CICs) at the block level in the eight NE states. They are equipped with computer communication equipments and are internet-enabled. The CICs, besides offering basic services like internet browsing, e-mail and training in computer fundamentals, also provide citizen-centric or government-to-citizen (G2C) services. *e-Suvidha* is a single-window front end for such services and is offered by many CICs. The CIC operators teach the villagers how the internet can help them obtain necessary information regarding farming, agriculture, health, education, *etc.*

6.4. Mission 2007

It was way back in 2003 that the idea was seeded to have a Mission targeting the 60th year of Indian Independence named as Mission 2007 which would successfully let the concept of the ICT enabling of all the villages of India evolve into a movement. Today after 15th August of 2007 the Grameen Gyaan Abhiyan proudly holds hands with its 400 odd partners across the country with one intention of creating a Knowledge revolution in rural India (Senthilkumar, 2009). The Grameen Gyaan Abhiyan is Rural Knowledge Movement in India which has been working with an aim to achieve ICT enabling of 637 000 villages of India has built a multi stake holder partnership with the different ICT4D models present in India. They include the community based models, the entrepreneurial models, government models, the business models or corporate models, the cooperative models and the combinations of all these models in pairs or more (http://www.mission2007.in).

6.5. Common Service Centres (CSCs) Scheme

Common Service Centres (CSCs) Scheme is the nationwide initiative of Government of India to provide support for establishing 100 000 Common Service Cantres in 600 000 villages of India. CSCs scheme has been started in 2004 with the vision to develop these centres as a front-end delivery points for Government, private and social sector services to rural citizens of India in an integrated manner.

7. WARANA WIRED VILLAGE PROJECT

The project was initiated in 1988 to increase the efficiency and productivity of sugarcane sector and provide a wide range of information and services to 70 villages around Warana. It comprises 25 co-operative societies and a membership of over 20000 farmers with a total turnover of over US $ 130 Million. There are 54 functional village information kiosks which are open from 10 am to 6 pm and frequented by 30 to 100 farmers daily (Patil *et al.*, 2009). An evaluative study after the seven years of implementation indicated that the personalised sugarcane processing information that allowed farmers to obtain information from the co-operative records at their local kiosks was mostly used by the farmers (Veeraraghavan *et al.*, 2009).

8. WEB PORTALS

8.1. AGRISNET

AGRISNET (Agricultural Resources Information System and Networking), a project funded by the Department of Agriculture and Cooperation, Ministry of Agriculture, Govt. of India. Under this scheme most of the State Governments are established information rich agricultural websites. For example, Sikkim AGRISNET (http://www.sikkimagrisnet.org),

Andhra Pradesh agri-portal, http://www.apagrisnet.gov.in, Uttar Pradesh (UP) Agrisnet Knowledge Portal (http://agriculture.up.nic.in), AGRISNET - Himachal Pradesh (http://203.193.179.168/default.aspx) - Expert Advisory Services (http://www.hp.gov.in/expertadvisory/SignUp.aspx).

AGRISNET projects, pilots for 17 states, have been approved to provide following services; market prices, soil information, crop diseases and management, good practices for horticulture, sericulture, *etc.* Twenty three priority services and processes for reengineering have been identified. Several initiatives have been taken by central government and states such as ASHA in Assam, KISSAN and e-Krishi in Kerala, Krishi Maratha Vahini in Karnataka, *etc.*, aimed at meeting challenges facing the agriculture sector in the country and have met with varying degree of success. As such, to consolidate learning from the past, integrate present diverse and disparate efforts and upscale them to cover the entire country, Agriculture has been included as a Mission Mode Projects (MMP) in National e-Goverence Plan (NeGP) and is to be operationalised by Department of Agriculture and Cooperation (DoA&C). The typical services envisaged in Agriculture as an MMP include: information to farmers on seeds, fertilizers, pesticides, government schemes, soil recommendations, crop management, weather and marketing of agriculture produce. The DoA&C has adopted twin strategy to spearhead implementation of MMP in Agriculture through AGRISNET & two portals AGMARKNET & DACNET. Through the AGMARKNET portal, information on about 300 commodities from 2000 mandis (markets) is made available. The information is updated regularly, directly from the markets. DACNET portal provides subject/ crop specific information from various Directorates under the Department of Agriculture and Cooperation (http://www.mit.gov.in/default.aspx?id=863).

8.2. DACNET – An e-Governance Project of Government of India

DACNET (Department of Agriculture and Co-operation Network) is an e-governance project sanctioned by the Department of Agriculture and Co-operation, Ministry of Agriculture, GoI and executed by the National Informatics Centre (NIC) to facilitate Indian agriculture on-line. It is built using the key criteria such as ease of use, speed of delivery, simplicity of procedure, single window access and affordable services, *etc.*, DACNET has four major components: Network together directorates, regional directorates and field units for Internet and intranet access with the central DACNET resources, establish a local-area network throughout Department of Agriculture field locations, empower employees through specialized system training programs, web-based applications at the directorate, regional directorate, and field units. Applications under DACNET are Plant Quarantine Information System, Crop Weather Watch, Market Prices Analysis, Bio-fertilizers Informatics Online, Integrated Pest Management Information System, Computerized Registration of Pesticides, Knowledge Management System 'e-Granthalaya' and Farm Machinery Informatics online (www.dacnet.nic.in).

Joint Secretary (IT), DoA&C, GoI, 2009 indicated that under the DACNET scheme, 46 web sites and 39 applications are developed (75 were developed and functional), which include web portals on complete information on 9 crop directorates, extension services, Integrated Nutrient Management, Marketing, Mechanisation and Technology, Economics and statistics (www.agricoop.nic.in/Kharif2009/JS(IT).ppt). Some of the notable applications are: CSMS (Content Schedule Management Program), the Mass Media Portal (http://dacnet.nic.in/csms) for reporting and dissemination of Agriculture Programmes produced by National, regional and Narrowcast Centres of Doordarshan and FM stations (http://navkrishi.nic.in/), SEEDNET (www.seednet.gov.in); Seed management, certification, variety management system, monitoring watershed details (www.dacnet.nic.in/rfs/), district wise crop production (www.dacnet.nic.in/pulses/report.htm) and land use online data entry, pest and diseases monitoring system, online data reporting and monitoring of technology mission on integrated development of horticulture in North-Eastern States of India (http://dacnet.nic.in/techmissionscheme).

8.3. InDG- India Development Gateway Portal

India Development Gateway (InDG) is a nation wide initiative of the Department of Information Technology, Government of India and implemented by Centre for Development of Advanced Computing (C-DAC), that targets specific country needs in the domain of rural and social development. InDG aims to provide credible information products and services in local languages that respond to the real and strategic needs of the rural people especially the farming communities. The use of Content Management System (CMS) enables shared ownership of the portal with various stakeholders. Any authenticated user can upload content to a specified topic in the portal. The content includes various forms such as words, audio and video so as to reach out effectively to rural masses and first level service providers. The information requirements of the rural people are first identified through various mechanisms. Once the information needs are identified, they are mapped to the existing content available with various development stakeholders such as Government, NGOs, Community based organizations and private institutions. The required content is collected, validated by the respective domain experts, edited for better presentation and published on the site for the end users. If the required content is unavailable, it is generated by an identified agency with expertise in the relevant field. Direct content uploading feature through portal is one of the important features of www.indg.in portal. The Content Management System available in the portal helps the user to directly upload the content into the portal by following simple steps. It saves the time in content flow from content providers and user end. All the registered members will be able to access this facility. Special privileges like 'edit/modify the content' will be given to such partners who validate the content online. The guiding strategies of InDG are three fold. First is to leverage the unique partnership structure to generate information products and services for the unreached in their local languages. InDG banks on consortiums of agencies for content contribution, validation and translation. Building dense networks of partners having expertise in the identified domains and leveraging C-DAC's technology expertise lays InDG's foundation. InDG is following a decentralized content management system for content on the gateway. The content publishing has been distributed and the role of the gateway team is of quality assurance and monitoring. In each of the focus areas the strategic partners of the InDG have managed the content. They are responsible for content generation, development, editing and final publishing on the gateway.

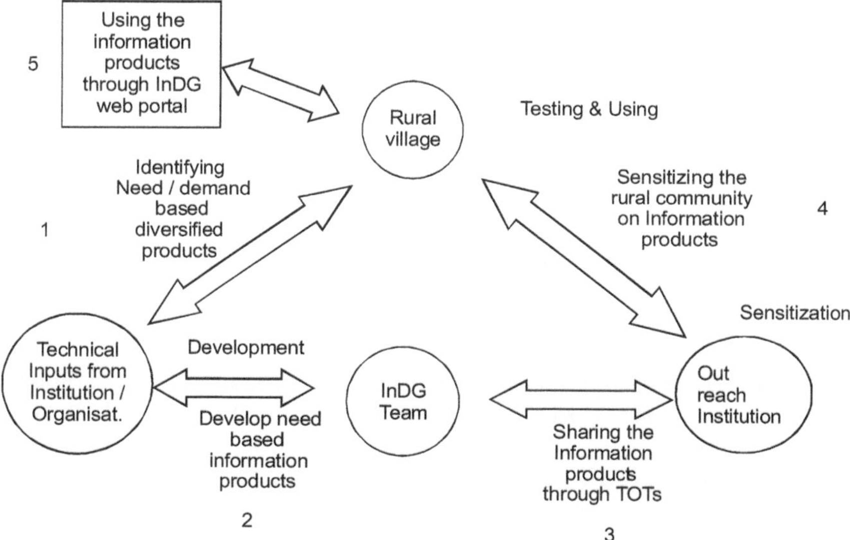

Figure 6 : Information Productization and Implementation Process of InDG

The second is to deliver value added content and services with a set of community building and collaborative tools. Keeping in mind the developmental needs of the target communities the gateway has both online as well as offline presence. The online mode provides knowledge management services on the thematic areas, collaborative tools and a platform for learning and networking to its target customers. The offline mode will complement the online mode wherever connectivity and other delivery issues are present. In the offline mode the gateway would operate through the network of its partners to provide consulting, research, training and knowledge management services to its target customers.

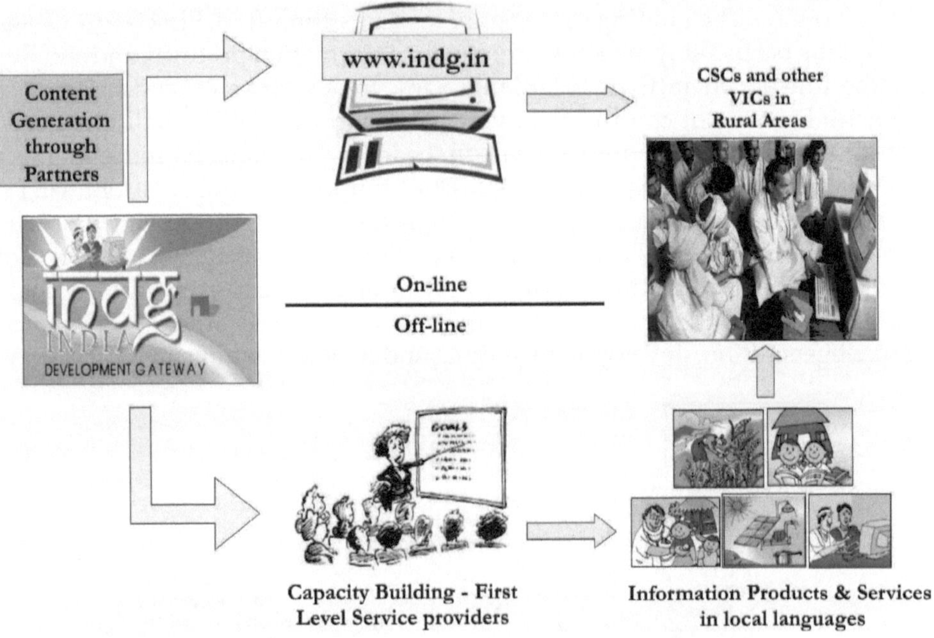

Figure 7 : Online and Offline Initiatives by India Development Gateway

The third is to develop a revenue model that ensures the financial and social sustainability of the project beyond the stipulated project period. The business model hopes to generate revenue from those users who can afford to pay for the services of the InDG portal. The services include priced information products, consultancies, membership fee and advertisements, paid trainings/ workshops and fund raising from donors (Kathiresan, 2007; Vijayalakshmi, 2008; Kathiresan *et al.*, 2009).

8.4. DEAL - Digital Ecosystem for Agriculture and Livelihood

DEAL is an ontology driven digital info mediation architecture; different user groups and expert groups will co-create content through knowledge exchange and dynamic aggregation which will be enabled for delivery over multiple media. This will allow different e service in this domain to exchange knowledge objects according to a commonly agreed ontology (www.dealindia.org). DEAL is a research project which focuses on developing an Ontology Server (OS) in the Indian agriculture and rural livelihood (ARL) domain (IARL). The agenda here is to build an action oriented vehicle to provide consistent usable access to information for the rural Community searching for knowledge, as well as to assist those whose interaction (*e.g.* Farm Science Centres (Krishi Vigyan Kendra-KVK) scientists and farmers) to create new knowledge. Firstly, the project now attempts to build the OS for IARL as a reference tool that can structure and standardize agricultural terminology in multiple Indian languages for use by different systems. This ontology will contain terms, the definition of those terms and the specification of relationships among those terms. It will start with a Thesaurus, the conversion of the Agrovoc hosted by the Food and Agriculture Organization of the United Nations in Hindi and later in other Indian language. But the aim is to work in parallel on enhancement to the Thesaurus, so that the system can provide the basic relationships inherent in the thesaurus. It will aim to capture and structure the knowledge in the ARL domain (Chatterjee and Prabhakar, 2009).

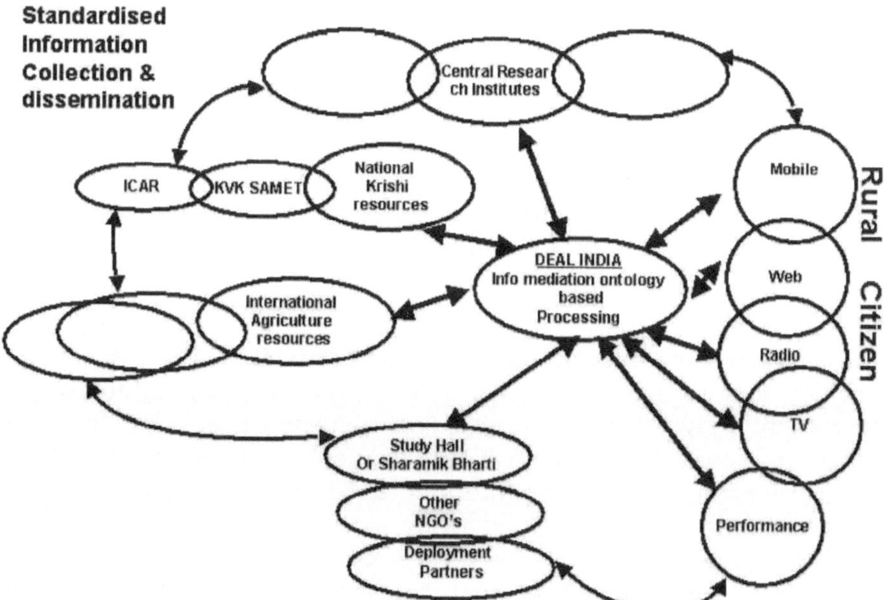

Figure 8 : Conceptual Architecture of Knowledge-Net of the DEAL (Chatterjee and Prabhakar, 2009)

The DEAL (Digital Ecosystem for Agriculture and Livelihood) project is a step towards addressing these issues by assembling a technology enhanced agricultural extension intervention in a Digital Ecosystem (DE) framework. Conceived by Indian Institute of Technology, Kanpur (IIT-K) and funded by Media Lab Asia. DEAL is an ICT enhanced network built on an existing framework of tele-centres in rural institutes, village schools, village level agriculture extension centres and other deployment partners. The project aims to create a digital knowledge base by involving the various actors in the existing system in the content creation process and making this knowledge accessible to farmers and other agricultural practitioners. The entire process of content creation and dissemination is capable of self generation, node independence and self-sustainability using an electronic medium. The moderating node in this system is IIT- Kanpur- providing the collaboration and collation technology platform, skills and resources to assist knowledge flows through the network. The presence of Government agencies helps to build trust in the network. The agricultural experts and educational institutions are responsible for verification of content generated.

Unlike conventional ICT interventions, which adopt either a top-down or bottom-up approach, DEAL focuses on increasing ties between members in the same layer, while also building links across the different layers. DEAL had provided a platform for different KVKs to share their extension experiences with each other through hosting a website for each of them. Horizontal ties between farmers from disperse geographical areas are enhanced through the use of the kissan blog for sharing agricultural experiences. This is a distinguishing feature of a digital ecosystem as creating these ties lead to shared norms and values which make ICT interventions successful. In the case of DEAL, this would fulfil the goal of sustained and voluntary content co-creation (Rajagopalan and Sarkar, 2007).

8.5. iKisan

iKisan is a comprehensive Agricultural Portal initiated by the Nagarjuna Fertilizers and Chemicals Ltd., in South India. iKisan is addressing the Information, Knowledge and Business requirements of various players in the Agriculture arena *viz.*, Farmers, Trade channel partners and Agri Input/ Output companies. Leveraging Information Technology and extensive field presence, Ikisan is positioned as an Information/ Knowledge exchange. Agriculture information; Crop Specific, Package of Practices of crops, Animal Husbandry, Aromatic and Medicinal Plants, Agricultural Machinery, Allied Agriculture, Sprayers, Rural Credit, Insurance iKisan crop solutions; farmers have a critical need to get timely solutions for protecting and nurturing their crops to get best yields. Addressing this key need, iKisan has developed easy-to-use diagnostic packages for different crops which will be provided on demand. Further, it also provides local agri news, weather and market information to the farmers (http://www.ikisan.com).

8.6. e-Krishi

The e-Krishi project, Market driven Agricultural Initiative through IT enabled Agri Business Centres in Kerala, addresses the existing gap in agriculture information flow and transaction management. The project envisages facilitating and enabling farmers and other Stakeholders through Agri Business Centres to interact with Agricultural Service Providers in the Private, Government and Non Government sectors. The project provides a web based solution enabling the small and medium farmers as well as owners of large landholdings. Piloting was done in Malappuram with the participation of the existing Akshaya e-Kendra Entrepreneurs. The facilities and resources of Akshaya e-Kendras in terms of computers, printers, scanners, cameras, *etc.* and intranet/internet connectivity, already established throughout Malappuram District of Kerala (about 550 e-Kendras), can be leveraged to reach the masses of farming community and other stakeholders in Agriculture sector (http://www.e-krishi.org). The e-Krishi portal also provides Farm Advisory Services (Crop Information, Fertiliser Recommendation, Planting Material Availability, Fertilizers and Pesticides, Weather Information), Agri-Market Information (Vegetable and Fruit Price, Daily Market Price, Market Analysis Report and Commodity Exchange), Resource Library: e-Krishi Eco-System, e-Krishi information centres, Photo Gallery, Model e-Krishi Project, KISSAN Video Channel. Further, e-Krishi news and online expert advisory, trade login and call centre support are the special features of the e-Krishi web portal (http://www.e-krishi.org).

8.7. ASHA

The ASHA project was launched by Assam Small Farmers Consortium (ASFC) to provide ICT based agribusiness services through 219 Community Information Centres using portal www.assamagribusiness.nic.in, since 2005. The project attempted to address the issues of farmers' access to technology, government endeavour (schemes), institutions, markets, resources and services to tackle poverty and poor growth through agriculture medium in Assam. Assam Small Farmers' Agribusiness Consortium (Assam SFAC) undertook the mission of providing access to technology, government endeavor, markets, institutions and services to farmers Media using ICT with relevant and need based Content under a brand name ASHA meaning Hope. ASHA services are made available through the Community Information Centres (CICs) located in 219 Community Rural Development Blocks of the state where farmers can access technology, Institutions, Government schemes, Markets and Services without barriers at his convenience either over telephone or through walk in and engage with them.

8.8. IFFCO- Agri-Portal

The Indian Farmers Fertiliser Cooperative (IFFCO) is one of the World's largest manufacturers of fertiliser. The IFFCO's agri-portal'-Sixteen states have been covered with information of relevance to farmers in local languages and can be accessed through IFFCO's website www.iffco.nic.in. User-friendly intuitive graphic based navigation is provided to facilitate viewing in touch screen environment. IFFCO has also installed about 100 Farmers Information Kiosks in 16 states. Training programmes and farmers meetings are conducted to encourage farmers to use the facilities provided in farmers' information kiosks (Patil *et al.*, 2009).

8.9. Agriwatch Portal

The agriwatch.com is the largest agribusiness portal in India and enables access to a large amount of agribusiness related information covering more than 15 sub sectors within the Agricultural and Food Industry. The daily, weekly and fortnightly Agriwatch Trade Research Reports are published (Patil *et al*, 2009).

8.10. iShakti

The iShakti information 'kiosk' is implemented by the Hindustan Unilever Limited, operated and hosted by the Shakti Entrepreneur. It provides a web-portal with content relevant to the rural audience. Local users can access information on categories including education, employment, agriculture, health, e-governance, personal care, and entertainment. They can post queries on these subjects to local experts who respond within 48 hours. Rural telephone connections are often unreliable, preventing routine Internet use. iShakti uses highly efficient and robust synchronisation to allow new content to be delivered to an autonomous village-based PC. This enables IT-based information and services to be delivered to regions where this was previously difficult or impossible (http://www.stockholmchallenge.se/data/ishakti_bridging_digital_)

9. ICTs FOR MARKET INFORMATION AND AGRI-BUSINESS

9.1. AGMARKNET

The project is about empowering farming community with the knowledge of latest commodity prices and arrivals information through innovative usage of ICT by networking agricultural produce wholesale markets in the country. It was initiated with the objectives of Networking 2800 major agricultural produce wholesale markets; imparting computer awareness and application usage training to about 5000 market personnel; dissemination of daily commodity prices and arrivals information in major Indian languages. In order to bring the farmers in a better bargaining position and to promote a culture of good agricultural marketing practices in the country, Directorate of Marketing and Inspection (DMI), Ministry of Agriculture, Government of India has embarked upon an ICT Project: NICNET based Agricultural Marketing Information System Network (AGMARKNET) as part of the Central Sector Scheme : "Marketing Research and Information Network" . The project, launched during later part of Ninth Plan Period (1997-2002), involves progressively linking all the agricultural produce wholesale markets, State Agricultural Marketing Boards/ Directorates and DMI offices for effective exchange of market information. National Informatics Centre (NIC) is executing the project on turnkey basis.

Project components include installation of basic ICT infrastructure at markets and their networking, capacity building, application development to facilitate creation of market level database as well as a national database on daily commodity arrivals and prices and a portal to gradually serve as a single window facility for disseminating all information pertaining to agricultural marketing. Presently, more than 1000 markets from different parts of the country are reporting data regularly to the portal. Functional days of markets being different, more than 300 markets are sharing information on daily basis for the use of public at large (Source: http://www.stockholmchallenge.se/data/agmarknet).

9.2. e-KRISHI VIPANAN

e-Krishi Vipanan project is the e-Agriculture Marketing project of Government Madhya Pradesh, India, conceived and executed for the benefit of farmers and traders. The project is executed by Madhya Pradesh Agricultural Marketing Board, Government of Madhya Pradesh and Madhya Pradesh Agency for Promotion of Information Technology (MAP-IT), on build, own and operate basis with a Consortium of vendors on Public -Private - Participation model to make operations effective and transparent by collecting and disseminating real time information, on-line and help the concerned stake holders in effective decision making, which will eventually lead to grainless mandis.

9.3. ITC-e-Choupal

ITC's Agri Business Division, one of India's largest exporters of agricultural commodities, has conceived e-Choupal as a more efficient supply chain aimed at delivering value to its customers around the world on a sustainable basis. The e-Choupal model has been specifically designed to tackle the challenges posed by the unique features of Indian agriculture, characterised by fragmented farms, weak infrastructure and the involvement of numerous intermediaries, among others. 'e-Choupal' also unshackles the potential of Indian farmer who has been trapped in a vicious cycle of low risk taking ability - low investment - low productivity - weak market orientation - low value addition - low margin - low risk taking ability. This made him and Indian agribusiness sector globally uncompetitive, despite rich and abundant natural resources. Such a market-led business model can enhance the competitiveness of Indian agriculture and trigger a virtuous cycle of higher productivity, higher income, enlarged capacity for farmer risk management, larger investments, higher quality and productivity.

Appreciating the imperative of intermediaries in the Indian context, 'e-Choupal' leverages Information Technology to virtually cluster all the value chain participants, delivering the same benefits as vertical integration. 'e-Choupal' makes use of the physical transmission capabilities of current intermediaries – aggregation, logistics, counter-party risk and bridge financing -while disinter mediating them from the chain of information flow and market signals. With a judicious blend of click and mortar capabilities, village internet kiosks managed by farmers – called sanchalaks – themselves, enable the agricultural community access ready information in their local language on the weather and market prices, disseminate knowledge on scientific farm practices and risk management, facilitate the sale of farm inputs and purchase farm produce from the farmers' doorsteps.

Real-time information and customised knowledge provided by 'e-Choupal' enhance the ability of farmers to take decisions and align their farm output with market demand and secure quality and productivity. The aggregation of the demand for farm inputs from individual farmers gives them access to high quality inputs from established and reputed manufacturers at fair prices. As a direct marketing channel, virtually linked to the 'mandi' system for price discovery, 'e-Choupal' eliminates wasteful intermediation and multiple handling. Thereby it significantly reduces transaction costs.

'e-Choupal' ensures world-class quality in delivering all these goods & services through several product / service specific partnerships with the leaders in the respective fields, in addition to ITC's own expertise. While the farmers benefit through enhanced farm productivity and higher farm gate prices, ITC benefits from the lower net cost of procurement (despite offering better prices to the farmer) having eliminated costs in the supply chain that do not add value.

Launched in June 2000, "e-Choupal" has already become the largest initiative among all Internet-based interventions in rural India. "e-Choupal" services today reach out to more than 4 million farmers growing a range of crops - soyabean, coffee, wheat, rice, pulses, shrimp - in over 40,000 villages through 6450 kiosks across 8 states (Madhya Pradesh, Karnataka, Andhra Pradesh, Uttar Pradesh, Maharashtra, Rajasthan, Uttaranchal and Tamil Nadu). The problems encountered while setting up and managing these 'e-Choupals' are primarily of infrastructural inadequacies, including power supply, telecom connectivity and bandwidth, apart from the challenge of imparting skills to the first time internet users in remote and inaccessible areas of rural India (Sourse: http://www.echoupal.com).

9.4. EID Parry- Indiagriline

EID Parry- Indiagriline is an attempt to catalyse e-commerce in rural agricultural and non-farm products and provide economic well being to rural areas initiated by EID Parry. Indiagriline provides a network of partnerships which bring commerce, technology and value added information to help expand the choices of rural India and make farming and SME activities a profitable business (http://www.indiagriline.com/english/corp/index.html). Recognizing the significance of information and communication technology (ICT) as a powerful tool for bridging the infrastructure gaps in rural India, EID Parry regarded the Internet as the next logical medium for delivering its farm extension services. Indiagriline is an effort to provide an end-to-end solution addressing the needs of the farming community in South India. EID Parry launched this project in early 2001 by setting up Internet kiosks in 16 villages around its sugar factory in Nellikuppam, Tamilnadu State. These kiosks were called Parry's Corners, and were intended to be business hubs of their respective villages-a one-stop shop that acted as a storefront for buying farm inputs, a market for selling goods, and an Internet café for communication and information services. The strategic goal of the Murugappa Group was to ultimately develop the following capabilities: Distribution infrastructure – this infrastructure would be capable of supporting bidirectional distribution of products and services into and out of rural India. Therefore, developing a low-cost channel for rural distribution was a key goal; and Trading infrastructure – this would serve as the foundation to a platform for trading agricultural commodities and rural industry manufactured goods. Towards achieving these ends, EID Parry forged and facilitated partnerships among a wide range of organizations, including Tamil Nadu Agriculture University (TNAU) and its research stations, Tamil Nadu University for Veterinary and Animal Sciences (TANUVAS), National Horticulture Board, AMM Foundation, and Murugappa Chettiar Research Center, to create the agriportal Indiagriline.

www.indiagriline.com was developed by using in-house expertise. It fashioned a franchise-based business model to meet the demand for information and connectivity. These kiosks are owned and operated by franchisees trained to operate the system. Although EID Parry covers the cost of establishing the infrastructure for voice and data connectivity, the franchisee also contributes his portion and attends to the operational expenditure of the kiosk. This cooperative relationship builds a win-win situation for both parties involved. Each Kiosk has a corDECT wall set, a PC, printer, telephone, furniture and a power source with a back up. The franchisees can leverage the EID Parry brand name to attract customers to their location for selling products or services. They also benefit from a wealth of knowledge transferred to them by EID Parry on how to successfully manage and operate the Parry's Corners. EID Parry also offers assistance in financing the franchisees through arrangements with third-party lending institutions such as Indian Bank. In a rural setting where literacy rates are low, fear and resistance to technology are high, however, this facility is slowly changing the life of the farmers, who, instead of travelling to the company or other places to get their business done, now walks into the neighbourhood for information.

Farmers can gather information directly from the kiosk or communicate with an agronomist to get specific, customized advice *via* e-mail. The typical turnaround time is a day. Services such as crop diagnostics actually can be performed remotely. The franchisee can use the digital camera to take a picture of the crop to be inspected and e-mail the image to the agronomist. The agronomist then will be able to follow up with his diagnosis. All this can be done without the farmers leaving the village (http://www.eidparry.com/casestudy.asp).

10. TELEPHONE/ MOBILE TELEPHONY

10.1. Farmer Call Centre (Kissan Call Centre)

The Department of Agriculture & Cooperation (DoA&C), Ministry of Agriculture, Govt. of India launched Farmer Call Centres on January 21, 2004 across the country to deliver extension services to the farming community. The purpose of these call centres is to respond to issues raised by farmers, instantly, in the local language. There are call centres for every state which are expected to handle traffic from any part of the country. Queries related to agriculture and allied sectors are being addressed through these call centres. The Farmer Call Centre is a synthesis of two hitherto separate technologies namely, the Information and Communication Technology (ICT) and the Agricultural Technology. Both have their specialized domains and work cultures. To optimally utilize the strengths of both these systems, it was proposed to take full advantage of professionally managed Call Centre mechanism and dovetail it with the specialized Subject Matter Specialists knowledge of Agricultural Scientists and Extension Officers, so as to facilitate its reach to the farming community. It is accordingly proposed to make use of existing specialized infrastructure of Call Centres (which are normally industry-driven and serve to high-end and many a times, mission critical service sector) and make this communication backbone available to the Subject Matter Specialists of Agriculture, Horticulture, Animal Husbandry, Marketing and other related areas. The Farmer Call Centre consists of three levels - namely Level-I (the basic Call Centre interface, with high quality bandwidth and local language proficient Agriculture Graduate), Level-II (Subject Matter Specialists on concerned important crops and enterprises, connected through good bandwidth telecom and computer connectivity) and Level-III (the Management Group to ensure ultimate answering and resolution of all the farmers' queries which are not resolved at Level-II, connected on off line mode).

10.2. SMS Broadcast Service by KVK

The Farm Science Centre (Krishi Viigyan Kendra -KVK), Babhaleshwar, India has pioneered in the IT enabled service aiding instant messaging from Farm Science Centre to individual farmers for extending Agricultural information through SMS alerts. The service comprises sending Short Message Service alerts on cellular phones registered at Farm Science Centres by individual farmers. Weekly SMS alerts are issued on various agricultural developments like weather forecast, disease forecast and market information. The service is also being used as a medium to send information on important trainings and other programmes to the members of the Farmers Clubs and SHG network under the Farm Science Centre. The service is an important milestone in reaching out to millions of farmers at a stroke of a mouse click and enables the farmer to have information access and derive the fruits of technological prowess and face the challenge of an upcoming free market.

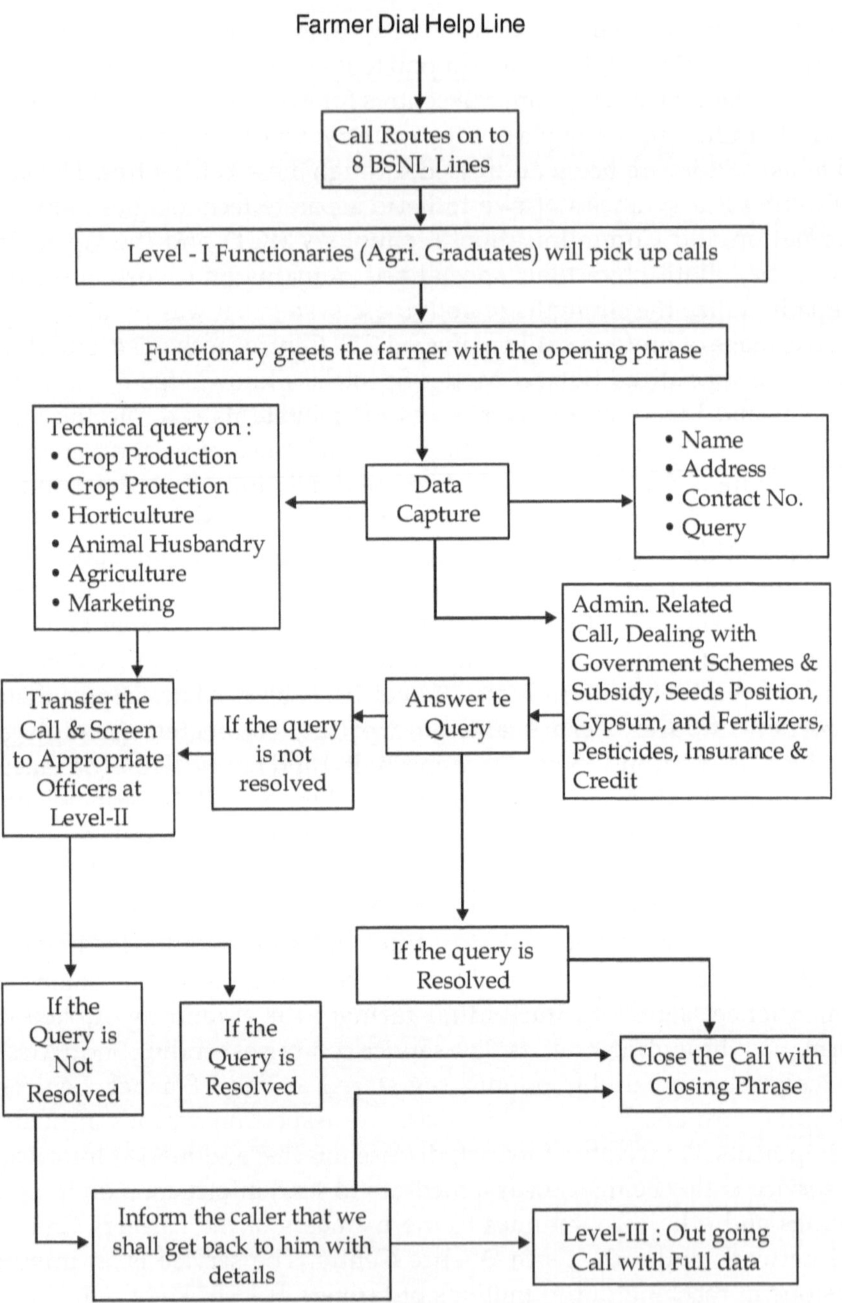

Figure 9 : Schematic Representation of Farmers Call Centre Operational Mechanism (Source: http://www.manage.gov.in/kisan/Operational_mechanism.htm)

The service is a boon for hundreds and thousands of farmers around the Farm Science Centre within Ahmednagar District. There are currently 241 registered farmers availing this service on payment basis from Farm Science Centre, Babhaleshwar. All registered farmers who are having cellular mobile handsets supporting all major GSM/CDMA networks compatible for Devnagari Unicode fonts within the country can receive the SMSs from Farm Science Centre without incurring any cost. The Farm Science Centre bears all costs of sending the SMS across all the registered farmers having cell phones. In continuation with the efforts of the Farm Science Centre in dissemination of agriculture technological information, the Farm Science Centre have further started the Marathi based SMS text delivery for the cellular handsets users within the District. The vernacular delivery module was commissioned during 2008. The service has been well appreciated by the farmers who have said such a service by the Farm Science Centre is just in the nick of time when farmers have their crops ready for market. The number of farmers coming to the Farm Science Centre is growing day by day and the Farm Science Centre expects to add another couple of hundred farmers by the end of this year (http://www.kvk.pravara.com).

11. ICT INITIATIVES OF CHAMBAL FERTILIZERS AND CHEMICALS LIMITED

Chambal Fertilisers and Chemicals Limited is one of the largest private sector fertiliser producers in India. A dedicated farmer website uttamkrishi.com, farmer helpline- Hello Uttam and mailers such as 'Chambal ki Chitthi' (Letter from Chambal) are important channels to share information with farmers. The web, phone and mobile based extension services such as; Uttamkrishi.com, Hello Uttam and *Via* SMS are offered to the farmers (http://chambalfertilisers.in).

11.1. Farmers' Website – uttamkrishi.com

Chambal set up uttamkrishi.com a website for farmers in the year 2001. It has a bottom-up approach to farming solutions as the website has been designed region wise to address the local issues. Broad information on crops and agronomy has been posted. Farmers can access it free and post queries that are answered by experts.

11.2. Farmers' Helpline – Hello Uttam

Telephonic help lines called "Hello Uttam" have been set-up. Farmers can raise issues over phone by calling one of the local numbers of "Hello Uttam" helpline. Information about the help line is widely circulated; and for areas where the help line is not operational, the farmers are urged to seek clarity from Department of Agriculture, Govt. of India's Farmers' Call Centre facility.

11.3. Mailers and AVs

Chambal is particular on proper dissemination of information. For that it has a quarterly mailer "Chambal ki Chitthi" that is hand-delivered to every Uttam Bandhan farmer. It contains valuable farm related information on activities pertaining to that quarter. Handouts, leaflets, farm calendar, farmer diary, *etc.*, are distributed extensively. It arranges radio talks, AVs and programmes on Doordarshan for the information of the farmers (http://chambalfertilisers.in/cfcl.asp).

12. ICT INITIATIVES OF NGO

12.1. DHAN Foundation

Development of Humane Action (DHAN) Foundation, a professional development organisation, was initiated on October 2, 1997. DHAN Foundation has taken up Information Technology for poor as a new theme with the following objectives: making information technology accessible to poor by developing relevant schemes through research and pilot activities and collaborating with research and academic institutions on e-governance and computer education at schools in rural areas. As part of the programme, internet kiosks in the rural areas as well as urban slums are set up through which services such as computer education, e-mail, ePost, agricultural market intelligence, *etc.*, are rendered (http://www.dhan.org/themes/itforpoor.php).

12.2. ISAP - Indian Society of Agribusiness Professionals

12.2.1. ISAP- Community Technology Learning Centres (CTLCs)

Indian Society of Agribusiness Professionals (ISAP), a non-governmental professional society- with support of Microsoft - Unlimited Potential Programme has established 'Community Technology Learning Centres (CTLCs)' in remote villages of Maharashtra to provide IT training to 450 00 farmers and unemployed youth. Under two-year programme, ISAP would be setting up 250 CTLCs at village level for imparting IT training to rural community and increase their income earning potential.

ISAP is working on online weekly price monitoring system of herbal and medicinal plants with the funding support of National Medicinal Plant Board. ISAP gathers and manages authentic data about the weekly price and demand for 101 medicinal plants from 50-marketing centers in different states of the country. These data are weekly upgraded on the basis of prices and quantity offered for different medicinal plants (ISAP, 2008).

12.2.2. ISAP- Query Redress Services (QRS)

Transfer of technical and scientific knowledge to the farmers is one of the primary functions of ISAP. QRS enables ISAP to provide solutions to farmer's queries pertaining to agricultural practices, problems, productivity improvement, scientific farming and improved technology for production to farming community. ISAP is running a query redress services (QRS) funded by One World South Asia (OWSA). The queries are received *via* electronic mail, by post and through telephone and responded within 24 hours. Currently ISAP receives more than 300 queries per day from farmers of Himachal Pradesh, Madhya Pradesh and Uttar Pradesh farmers, where it runs this service (ISAP, 2008).

12.2.3. ISAP- Community Radio Stations (CRS)

Timely availability of reliable information is the key to achieve sustainable food production and mitigate risks. Toward this community radio stations will act as an effective tool of communication and create platform to share experiences, perspectives and innovations to increase yield and reduce labour. ISAP has been identified as one of seven organizations in the country to establish community radio station (ISAP, 2008).

13. EXPERT SYSTEMS, DECISION SUPPORT SYSTEMS, CDs AND OTHERS

13.1. Agricultural Decision Support System by Agro-Climate Planning and Information Bank (APIB)

A pilot project to create "Single Window Agricultural Knowledge Base" was initiated at Regional Remote Sensing Centre, ISRO, Bangalore, India during mid 1990's. As a part of project, user needs assessment, spatial data and non-spatial data were collected. Based on the information collected, 17 agricultural decision support modules were developed and deployed in the selected three district level offices of the Department of Agriculture and Horticulture for field testing in three districts of Karnataka State of India. However, case study revealed that there was a lack of conceptual inadequacy and lack of skill to use modules among the extension officials, infrastructure inadequacy to deploy larger area and hence the modules were not used as expected (Saravanan, 2002).

13.2. Expert System on Pests and Diseases of Major Crops in Andhra Pradesh

Software has been developed with user-friendly menus using VB.net a relational database management system for information retrieval, and it has been used as a diagnostic tool to speed up decision making in identification and recommendation of control strategies of major pests and diseases on the spot. This is the most effective extension tool to take the technology from the scientists to the farmers directly without any dilution of the content which normally creeps in because of the number of the agencies in normal technology transfer systems. Using this system, farmers, scientists, extension workers and researchers can identify the pest or diseases within a short time and effective decision making can be taken up to manage the problem. This software has the unique provision to store the data by embedding the picture in the database itself. It can play a vital role in the maintenance and monitoring complete details of the pests and diseases (Ravishankar *et al.*, 2006).

13.3. Pesticide Advisor (Verson 2005.1.0): An Expert System for Judicious use of Pesticides for Management of Pests

The pesticide advisor software works as a simple interactive decision making tool providing a choice of recommended curative as well as preventive pesticides as per the recommendations of the central insecticide board and registration committee, Government of India to manage key pests, in major crops. The different modules, which are in the software, are as follows; pesticides complete details, pesticides recommendation, pesticide selection criteria, field use rating, pesticides quantity calculator, pesticide delivery system, data on bio-pesticides, safety measures, pesticide manufacturers, pesticides scenario and pesticides general information (Amerika Singh *et al.*, 2006).

13.4. Vasundhara: Software for Soil and Water Test Based Nutrient Recommendations by KVK Ahmednagar, Maharashtra

The KVK has successfully demonstrated the use of ICT in disseminating improved agricultural technologies in rural areas under its Cyber Extension Programme. The KVK pioneered the pest and disease forewarning through its SMS based alert system to the registered users. A number of technology CDs has been developed on crops, livestock and other agri-enterprises for use by the farmers. It also developed software called 'Vasundhara' for soil and water test based nutrient recommendations and also released CDs on statistical information and FAQs on farmer's needs and problems. The KVK suggested providing facilities for automatic weather station, infrastructure for market intelligence, quality assessment and certification (Sajeev, 2009).

13.5. TCS - mKrishi

mKrishi is a mobile-based crop advisory service. The application provides personalized information and expert advice to rural farmers, and runs on Tata Indicom's Code Division Multiple Access through network, and it uses various applications including sensors, camera phones, and Global Positioning System technologies. Through mKrishi, farmers get answers to queries related to agriculture, such as advice on use of fertilizers, pesticides and growth hormones. It also provides up-to-date weather and market information through text messages on cell phones. Services such as assistance with crop insurance, loan services, rural yellow pages and government policy information will be included in a phased manner. The mKrishi application can be customized in farmers' local languages, both text and voice. Currently the application runs as pilot project in four villages and few hundreds of farmers each holding an average of three to four hectors of land. The service includes weather and soil sensors for these villages (www.tcs.com; http://satellite.tmcnet.com/topics/satellite/articles/41495-tcs-mkrishi-earns-wall-street-journal-it-award.htm; Pande and Arve, 2009).

13.6. Digital Data Banks - Agricultural Planning and Information Bank (APIB)

APIB, an agricultural digital data bank was a joint project developed for the East Khasi Hills District of Meghalaya State of India by the North Eastern Space Applications Centre (NESAC) in collaboration with the Department of Agriculture, Government of Meghalaya, India. It is a single window access to information related to agriculture and allied sectors useful for the farmers, extension personnel and planners. It contains information modules on natural resources, Package of Practices for 25 crops, Soil Crop Suitability, Integrated Plant Protection (12 crops), Varietal Selection (16 crops), Post Harvest Information (8 crops), Agro-meteorological Services, Food Processing, Fertilizers, Farm Machinery & Hand tools, Allied sectors & occupations, Soil conservation and Forestry, Water resources, Micro Credit and Self Help Groups. Natural resources module provides block-wise details of the Land Resource Development Plan (LRDP), current Land Use Practices, Soil Type map, Ground Water Prospective map and Watershed map. Extension officials can also use this module. Package of practices are provided in both standard and comprehensive versions. This module provides the correct version for each of the major crops. Integrated plant protection provides vital and practical information necessary for pest and disease prevention, control and mitigation. Integrated approach encompassing cultural, mechanical, biological and chemical methods have been considered. Harvest and post harvest module provides crop-wise information on traditional storage devices, such as harvest tips, post harvest handling and processing. Allied sectors gives information on Animal Husbandry, Fisheries, Sericulture and other non farming activities essential for the farmers of this region (http://megapib.nic.in/about_apib.htm).

14. VALUE ADDED SERVICES

14.1. IFFCO Kisan Sanchar Ltd.

Bharti Airtel Limited, India's leading integrated telecommunications services provider, and Indian Farmers Fertiliser Cooperative limited (IFFCO) launched a joint venture company IFFCO Kisan Sanchar Limited (IKSL), that is set to provide a major boost to Indian agriculture and the rural economy at large during 2008. The joint venture company will harness the power of telecom to add value to the farm sector and empower the rural farmer by giving him access to vital information, which will enhance his livelihood and quality of life. Innovative telecom products and services, especially created for the farming community, will enhance their productivity and play a bigger role in India's growth story. The Indian farmer can now look forward to the benefits of mobile telephony and mobile internet through this initiative. Synergizing on IFFCO's presence in about 98 per cent of villages throughout the country, Airtel would use IKSL as a vehicle to make its mobile telecom services reach the rural masses (IFFCO, 2008). IFFCO Kisan Sanchar Ltd. is, therefore, a unique venture to strengthen the Cooperative movement in rural India, to empower the farmers through information, to boost the rural economy, and also to create rural employment. IKSL will offer products and services, specifically designed for farmers, through IFFCO societies in villages across the country. On offer will be affordable mobile handsets bundled with Airtel mobile connection. The farmers will also get access to a unique VAS platform that will broadcast 5 free voice messages on mandi prices, farming techniques, weather forecasts, diary farming, animal husbandry, rural health initiatives and fertiliser availability *etc.* on a daily basis. In addition, the farmer will be able to call a dedicated helpline, manned by experts from various fields, to get answers to their specific queries. This is expected to promote community building within the society and rural community at large (http://www.iffco.nic.in).

14.2. BSNL- Mandi on Mobile Service

Uttar Pradesh State farmers are able to know rates of agriculture commodities in any market in the State on their mobile phones, service was launched by the State-run telecom major Bharat Sanchar Nigam Ltd (BSNL) teamed up with the Uttar Pradesh Agricultural Marketing Board (Mandi Parishad) to launch the 'Mandi on Mobile' service for the farmers (Manoranjan, 2009). The service would be voice-based. To know the rates of over 100 commodities including crops, vegetables and other items, the farmers need to dial specific number from their BSNL cellular phones and follow the voice command subsequently. The service would immensely benefit the farmers, especially those who used to sell their agriculture produce to middlemen at low prices without knowing the market rates (IANS, 2008).

14.3. Nokia Life Tools

Nokia Life Tools is a range of services which includes Agriculture, Education and Entertainment services designed specially for the consumers in small towns and rural areas of the emerging markets. The service provides timely and relevant information customized to the user's location and personal preferences directly on their mobile phones. Nokia Life Tools Agriculture services aim to plug the information gaps and needs of farmers, by providing information on seeds, fertilizers, pesticides, market prices, and weather (temperature, rainfall, wind conditions) *via* their mobile phones. Information on weather, agriculture tips and techniques, as well as market prices are provided to improve farmers' productivity and earnings. Farmers are empowered with tailored and reliable information in synch with the cropping cycles delivered regularly to their mobile phone (http://www.nokia.co.in/explore-services/nokialifetools).

14.4. Fisher Friend Project

QUALCOMM, M. S. Swaminathan Research Foundation (MSSRF), Tata Teleservices and Astute System Technology were collaborated to develop a BREW-based mobile application, which empowers fishing communities in India. Fisher Friend is a mobile application, which is based on QUALCOMM's BREW solution and works on 3G CDMA and WCDMA handsets. The new application allows fishing communities to use ICT and wireless technology to earn their livelihood. Fishermen can send request form from menu-driven client software on the mobile phone. Fishermen can access vital updates on opportunities, risk and market information in their local language. Presently, the application is implemented in the coastal areas of Tamil Nadu, India.

Fisher Friend project aims to be extended in other communities in the coastal belts of India (News-i4d, 2009).

14.5. Rubber Board, India- Market Price by SMS

The Rubber Board provides the update of both national and international rates of natural rubber through SMS throughout the country by the rubber farmers and dealers in India (especially Kerala State of South India) are tracking the prices of the commodity real time by SMS. The growers are helped by a service by the Rubber Board which through SMS updates the farmers with rates in the global as well as domestic markets, which is also displayed in the Rubber board's web portal (www.rubberboard.org.in).

14.6. SMS Service to Farmers by the Department of Agriculture, Haryana State

Farmers of Chandigarh (India) are using mobile phones to sort out agricultural related problems. The Haryana Agriculture Department has introduced short messaging service (SMS) for farmers. The service was available by the agriculture department's Kissan (Farmer) call centre. More than 800 farmers had sent their queries through SMS service and replies had been sent by officials' concerned and agricultural scientists. The free of cost SMS service is available to farmers on mobile number (News-i4d, 2007).

15. ICT INITIATIVES BY THE NATIONAL AGRICULTURAL INNOVATION PROJECT (NAIP), INDIAN COUNCIL OF AGRICULTURAL RESEARCH (ICAR)

The World Bank aided National Agricultural Innovation Project (NAIP) has been conceived to pilot this innovation in conducting agricultural research. The projects which have policy relevance to the ICTs for Agricultural extension are mentioned below and these projects are under progress. Through these pilot projects, the policy recommendations, well-tested models for digital knowledge repository and dissemination of information to the farm stakeholders are expected during the year 2011-12.

15.1. Re-designing the Farmer-Extension-Agricultural Research-Education Continuum in India with ICT-mediated Knowledge Management

Through this project, a consortium of partners from the agricultural research and education sector and ICT for development area will work together to organise a national pilot project. They will develop new and revived linkages between research and education sectors with agricultural extension, through the use of ICT mediation and contemporary practices in knowledge and information management. The engagement with farmers will be through online as well as offline methods. The emphasis is on providing information support services to the extension personnel and farmers in a seamless fashion. This integration of back-end digital services to generate an access and delivery system for different stakeholders in online and offline modes is at the core of this action-research effort (source: http://www.akmindia.in).

As a part of this pilot project, developing "agropedia" aims to develop a comprehensive digital content framework, platform, and tools in support of agricultural extension and outreach. In other words, it aspires to be a one stop shop for any information, pedagogic or practical knowledge related to extension services in Indian agriculture - an audiovisual encyclopedia, to enchant, educate and transform the process of digital content creation and organization completely (Source: http://agropedia.iitk.ac.in/?q=content/welcome-agropedia-2).

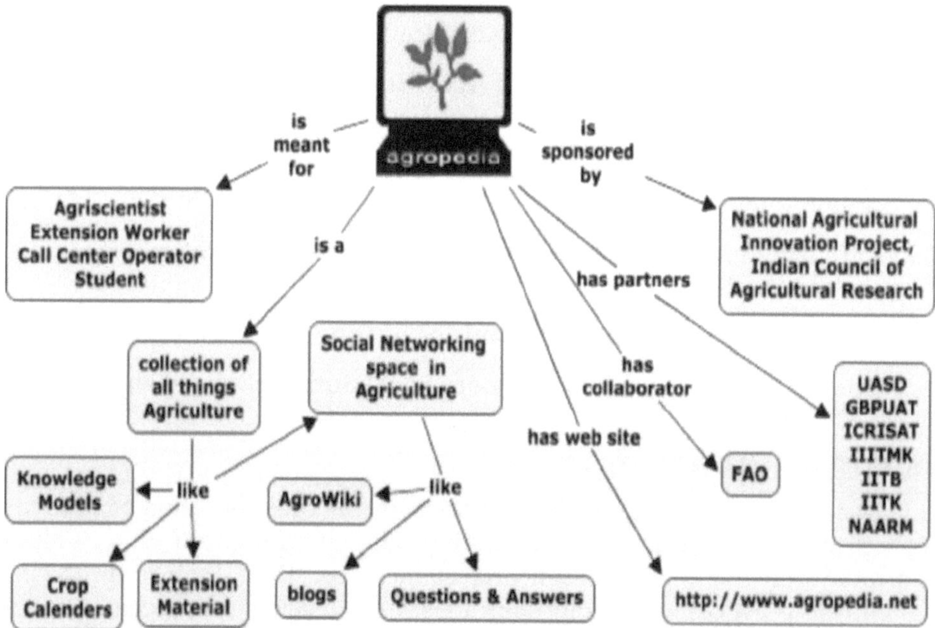

Figure 10 : Schematic Representation of Agropedia- A Comprehensive Digital Content Framework

15.2. Development of a Set of Alternative ICT Models Based on a Study and Analysis of the Major ICT Initiatives in Agriculture in India to Meet the Information Need of the Indian Farmers

The project will study the effectiveness and impacts of the various existing ICT based knowledge dissemination systems for the farmers in India and propose ways to make the systems more effective, with deeper and more extensive impact to fulfil the need of various stakeholders. The project proposes to undertake assessment of the major ICT initiatives in agriculture vis-à-vis the need of the farmer in various agro socio-economic situations and suggest a set of alternative ICT models based on the assessment for the major situations (http://www.naip.icar.org.in/downloads/C30012.pdf). The NAIP's other ICT's for Agriculture intiatives are;

15.3. AGROWEB - Digital Dissemination System for Indian Agricultural Research (ADDSIAR)

15.4. e-Publishing and Knowledge System in Agricultural Research

15.5. Developing a Decision Support System for Agricultural Commodity Market Outlook

15.6. Development and Maintenance of Rice Knowledge Management Portal

15.7. Establishing and Networking of Agricultural Market Intelligence Centres in India

15.8. Decision Support System for Enhancing Productivity in Irrigated Saline Environment using Remote Sensing, Modelling and GIS

15.9. Strengthening of Digital Library and Information Management under NARS (e-GRANTH)

15.10. Mobilizing Mass Media Support for Sharing Agro- Information

(Sourse: http://www.naip.icar.org.in; Yaduraju *et al.*, 2009)

16. IMPACT OF ICT PROJECTS

Systematic and comprehensive impact studies on application of ICTs for agricultural extension are not available. However, Gandhi *et al.*, (2008) indicated that the Digital Green project increased the adoption of certain agriculture practices seven-fold over a classic extension approaches. Digital Green project was shown to be ten times more effective per dollar spent. Further, 85 per cent of adoption of improved technologies achieved as against 11 per cent of adoption by traditional extension methods. Similarly

Krishnareddy and Ankaiah, (2005) reported that deploying e-Sagu prototype increased income of the farmers for the tune of INR. 3075 (63 USD) per ha and also reduced the pesticide usage. Further, their rudimentary estimate of economic advantage indicated that if the e-Sagu prototype used for 1000 farmers, overall net benefit with the proposed ICT based system is INR 100 Million (USD 204800). Saravanan (2008) reported the cost and time indicators comparing traditional extension system and e-Arik (e-agriculture) project, sixteen fold and three fold less time were required to the clientele availing and extension system delivering extension services, respectively. He further reported that 3.4 fold economic benefit as compared to the expenditure of deploying e-agriculture prototype. Interestingly, Ganghi *et al.*, (2009) reported positive social side effects and other qualitative results of Digital Green project on participatory video for agricultural extension.

17. LESSONS FROM ICTs FOR AGRICULTURAL EXTENSION PROJECTS

Keniston (2002) stated that "At least fifty grassroots projects are currently using modern ICT for development in India. Surprisingly, these projects have rarely been studied. No comparisons have been made between them. They are seldom in touch with each other. Lessons learned in one project are not transmitted to others. Appropriate technologies are rarely evaluated. Central questions of financial sustainability, scalability and cost recovery are hardly ever addressed. So, opportunities to learn from the diverse, creative Indian experience so far remain almost entirely wasted". Even after experimenting hundreds of ICT projects for rural development in the last one decade, observations mentioned above are still very much relevant; and also preliminary hypotheses on Grassroots ICT projects in India by Keniston (2002) are yet to be tested fully. However, with the available literature and author's personal experience in implementing e-Arik (e-Agriculture), e-village and e-Kiosk (Touch screen kiosk for farm technology transfer) projects and literature review, the following lessons were drawn.

1. *Pilot Project Syndrome*: Most of the ICT based Agricultural extension projects were implemented as "Pilot Projects"; and after the pilot period, most of the projects are never implemented in larger scale. Efforts for continuance of pilot projects are not taken sincerely by the implementing and also funding (Donor) agencies.

2. *Users Unwilling to Pay:* Most of the ICT for Agricultural extension projects beneficiaries (generally farmers) are not willing to pay for the services they receive. Similar to most developing countries farmers, in India also

most farmers feel that "agricultural advisory services" are welfare activity of the State and National Governments. And hence, they are unwilling to pay for the services.

3. *Small Scale of Operation:* The ICTs for agricultural extension projects were implemented in very limited geographical area and covering few hundreds or at maximum thousands of farmers. Exceptionally, few projects like farmers call centres and e-Soil Health Card Programme covers entire country and Gujarat State, respectively. And few web portals are developed for larger farm stakeholders (AGMARKNET, InDG, e-Krishi). However, continuous updating and maintaining web portals require sufficient resources, which are lacking after few years.

4. *Islands of Learning:* In almost all the projects, the participation of agricultural education, and research institutions appears to be marginal (Balaji *et al.*, 2007). Most of the projects do not have collaboration with other farm research and extension stakeholders. Practical challenges or constraints in implementing the ICT projects are seldom disclosed and shared with others. Learning experience of one project to another project is seldom shared.

5. *Lack of Systematic Evaluation:* Most of the projects never revealed actual evaluation results, Generally they report 'positive' results, and most common difficulties such as; inadequate rural ICT infrastructure (especially frequent power-cuts) and difficulty in content localisation and customisation were indicated. Systematic and objective evaluation or impact of the projects was seldom done. Similar type of projects, with little modification, was implemented in isolated manner. Except few projects, large number of projects evaluation results were never published or communicated.

6. *Knowledge Middle Men with Less Permanency:* Most published projects are from educational/ research institutions, which generally, ignored traditional extension system and extension personnel, those who are serving over a long period in rural India. They implemented time bound ICT projects and hired "facilitators"/ "intermediaries". Once, project completes stated objectives and targets, facilitators also disappear along with the project. In this regard, Digital Green used the services of the public extension personnel. Even, if project winds-up the learning took place among extension personnel will be remain for a longer time and more useful to the farmers. In e-Arik case, public extension personnel are unwilling to collaborate with the ICT project, because of most of the field level extension personnel never used internet and lack of skill in

using other ICTs. However, Subject Matter Specialists from Farm Science Centre (KVK- Krishi Vigyan Kendra) was willingly collaborated with the e-Arik project (Saravanan, 2008).

7. *Information is Not Ultimate* : Along with ICT based advisory services, input supply and testing need to be integrated for the greater impact (Balaji *et al.*, 2007). In e-Arik project of North-East India, farmers demanded inputs as per recommendations of the project research fellows.
8. *Virtual Seeing is Not Believing* : The e-Arik (e-Agriculture: ICTs for Agricultural Extension) project was implemented in most backward area of North-East India; experiences indicated that advisory services and information alone were unable to convince the farmers and they demanded field level demonstrations.
9. *Difficulty in Localisation of Content* : Content need to be aggregated from different sources but it needs to be sorted in granular format for rapid adaptation for local use. Localisation and customizability of content are still are not practiced on a significant scale (Balaji *et al.*, 2007). If sufficient scientific information is not available, content need to be generated, tested, refined and used for further advisory services through ICTs. Most of the web portals lack relevant content in local language.
10. *Lack of Co-ordination* : In the absence of collective and coordinated efforts by the public-private agricultural research and extension institutions, ICTs have not penetrated satisfactorily in rural India despite time, money and efforts invested so far (Patil *et al.*, 2009).
11. *Limited ICT Infrastructure* : Planning Commission of Government of India report (2005) indicated that still 10 per cent of villages yet to get electricity. Further, State wise village electricity coverage varies from 100 per cent (in Punjab State) to 26 per cent (in Jharkhand State). Household wise electricity varies from 94.80 (in Himachal Pradesh State) to 10.30 per cent (in Bihar State). Rural tele density is 9.03 (telephone subscribers per 100 individual). Only 1.2 per cent people have internet access in rural India and rural mobile penetration remains 4.92 per cent. Computer and internet access is difficult task for rural farmers, even though rural tele-centres (Village knowledge centres, village resource centres, common service centres *etc.*,) are taken-up on a massive scale by the national government, developmental institutions and private sector.

18. THE WAY FORWARD

1. *National and State Governments e-Agriculture Policy* : National and state e-Agriculture policy need to be formulated. It should explore and outline the possibilities of leveraging ICT for the Agricultural Extension.

2. *Human Resource Development*: Creating awareness on ICT potentials, ICT using skill and capacity development among the extension personnel of the public and private extension systems and also among farmers and other stakeholders in the extension systems.
3. *Strengthening ICT Infrastructure*: Extension organisations need to be equipped with ICTs.
4. *Localisation and Customisation of Content*: Research, educational institutions and extension systems should continuously strive for the appropriate content localisation and customisation as per the demand of the farmers and other stakeholders.
5. *Integration of ICTs with Public-Private Extension System*: Appropriate ICTs to be identified and deployed in the extension system to complement ongoing extension efforts.
6. *Research and Farm Developmental Institutions Collaboration*: Establishing strong working collaboration among the ICT initiatives of the research and developmental institutions such as; Indian Council of Research (ICAR), National Informatics Centre (NIC), Centre for Development of Advanced Computing (C-DAC) and Ministries of Information Technology & Communications, Agriculture and Co-operation, Rural Development. The leading research and educational institutions in agricultural and information technology should join together to leverage ICT penetration for agricultural extension and they should also guide the other ICT initiatives in rural India.
7. *Convergence of Communication Methods*: For effective agricultural extension service delivery, the convergence of traditional extension communication methods (personal contact methods, print media, radio and TV) and new ICTs are to be appropriately used to reach farm stake holders.

In India, during the last one decade, hundreds of Grassroots ICT projects are implemented. Invariably, agriculture becomes one of the indispensable parts of the project service menu. However, we yet to get substantial results in increase of agricultural production because of deployment of ICTs. Most projects are implemented in smaller geographical area and covering few hundred farmers and hence, drawing generalisations may not appropriate. Much hyped ICT projects are yet to make any break through in agricultural information dissemination. Even though, ICTs are promising to make difference and also accelerating information access by some farmers, but, most of the ICT practitioners in agriculture are over excited. Author's personal experience shows that "ICTs are not going to supplant traditional extension system and it can play only complementary role". ICTs for agricultural extension projects need to be compared and evaluated objectively. At the same time, experiences are indicating that ICT are going to play greater role in private sector agribusiness, market information and market intelligence. Further, certain type of farm information (*e.g.* Informing government schemes) and online monitoring of the progress of the governmental schemes are proved successful. Hence, it is high time to find out appropriate information to provide through ICTs. As indicated earlier, formulating National and State level e-Agriculture policy, human resource development, strengthening ICT infrastructure, localisation and customisation of appropriate content are to be taken-up to harvest the benefits of ICTs for agricultural extension services provision and agricultural development.

References

Amerika Singh, Sharma, O.P., Dhandapani, A., Niranjan Singh, Nagarajan, S., Sathya Kumar, S., and Neelam Mehta, 2006. Pesticide advisor (Version 2005.1.0): An expert system for judicious use of pesticides for management of pests. *ICAR News,12*(4).

Balaji, V., Meera, S.N., and Dixit, Sreenath., 2007. ICT enabled knowledge sharing in support of extension: addressing the agrarian challenges of the developing world threatened by climate change, with a case study from India. *SAT eJournal*, eJournal.icrisat.org, December, 2007, 4(1).

Chatterjee, J and Prabhakar, T.V., 2009. On to action- Building digital ecosystem for knowledge diffusion in Rural India. http://www.cse.iitk.ac.in/users/tvp/papers/vof_pasadena.pdf (Accessed on 5th July, 2009).

DoA&C., 2000. Policy framework for agricultural extension (draft). Extension division, Department of Agriculture and Co-operation, Ministry of Agriculture, Government of India, New Delhi.

Gandhi, Rikin, Veeraraghavan, Rajesh., Toyama, Kentaro., and Ramprasad, Vanaja., 2008. Digital Green: Participatory video for agricultural extension. *I4d Magazine*, June, 2008 Issue, www.i4donline.net

Gandhi, Rikin, Veeraraghavan, Rajesh., Toyama, Kentaro., and Ramprasad, Vanaja., 2009. Digital Green: Partcipatory Video and Mediated instruction for Agricultural Extension. *Information Technologies and International Development*, 5 (1), Spring 2009.1-15. http://itidjournal.org/itid/article/view/327/150 (Accessed on 15th July, 2009).

IANS, 2008. Farmers to get market rates on BSNL mobiles. http://www.thaindian.com/newsportal/business/farmers-to-get-market-rates-on-bsnl-mobiles_100116668.html (Accessed on 5th July, 2009).

ICAR-FFTDDSA, 2008. ICAR framework for technology development and delivery system in agriculture. www.icar.org.in/miscell/tdd_final.pdf (Accessed on 14th December, 2008).

IFFCO, 2008. IFFCO and Airtel join hands to usher in the second green revolution. Joint venture company IFFCO Kisan Sanchar Ltd. to provide big boost to Indian agriculture and rural economy.http://www.iffco.nic.in/applications/iffcowebr5.nsf/0/89b6f7f9c1950e3465257442003 ae142?OpenDocument (Accessed on 5th July, 2009).

ISAP, 2008. Indian Society of Agribusiness Professionals - A Profile, ***Information Bulletin***, New Delhi.

Joint Secretary (IT), DAC, GoI, 2009. Presentation on use of information technology in agriculture in India, National conference on agriculture for Kharif campaign 2009 (20th-21st March, 2009), http://agricoop.nic.in/kharif09.html (Accessed on 25th June, 2009).

Kathiresan, C, Vijayalakshmi, B and Sarat Chandra Babu, N., 2009. India Development Gateway Initiative - A National Level ICT Platform to Reach the Unreached, In: Saravanan, R., (Ed.). *Agricultural Extension*. New India Publishing Agency, New Delhi (In Press).

Kathiresan, C, 2007. India Development Gateway (InDG) - Reaching the Unreached with Rural Development Perspective using ICT. In: Sasikumar, M., Balaji, R. and Gangaprasad, G.L. (Eds), *Proceedings of the International Conference on ICT Solutions for Socio-Economic Development (ISED 2007)*, Bangalore.

Keniston, Kenneth., 2002. IT for the common man: Lessons from India. *NIAS Special Publication*, sp7-02. Bangalore. National Institute of Advanced Studies, Indian Institute of Science.

Krishnareddy, P., and Ankaiah, R., 2005. A framework of information technology- based agriculture information dessemination system to improve crop productivity. *Current Science*, 88(12).

Kuttayan Annamalai, Sachin Rao and C. K. Prahalad, (n.d). Known Problems and System wide Reforms, Section IV, The EID Parry Story, www.eidparry.com (Accessed on 5th June, 2008).

Lall, Anusha and Sahi, Swati., 2009. Taking ICTs to the Grassroots- A case study of the Lifelines India initiative, *IFIP news letter*, 19 (1), www.iimahd.ernet.in/egov/ifip/feb2009/anusha-lall.htm (Accessed on 3rd July, 2009).

Manoranjan, Mohapatra., 2009. Value added services in rural India, *Connect-World*, No. 26, India issue 2009. http://www.comviva.com/media/Moa's_Article_VAS%20in%20rural%20India.pdf (Accessed on 5th July, 2009).

Mathur, Dhrupad, Piyush Gupta and Sridevi, A., 2009. e-Governance approach in India- The National e-Governance Plan (NeGP), In: R K Bagga and Piyush Gupta (Eds.) *Transforming Government: e-Government initiatives in India*, The ICFAI University Press, Hyderabad.

NSAE, 2009. Backround Note. National Seminar on Agriculture Extension, 27- 28th February, 2009, NAS Complex, PUSA, New Delhi. http://www.syngenta foundation.org/db/1/657.pdf (Accessed on 2nd May, 2009).

NSSO, 2005, Access to modern technology for farming, situation assessment survey of farmers, 59th Round, *Report No. 499*, National Sample Survey Organisation (NSSO), Ministry of Statistics and Programme Implementation, Government of India, New Delhi.

Navarro, Rex and V. Balaji., 2003. A Virtual University for Semi-Arid Tropics: A concept note. *Electronic Journal Information Systems in Developing Countries*, An open access journal, 14. http://www.ejisdc.org/ojs2/index.php/ejisdc/article/view/90/90

News-i4d, 2007. Haryana farmers use SMS service for agricultural problems. http://www.i4donline.net/news/news-details.asp?catid=1&newsid=8910 (Accessed on 5th July, 2009).

News-i4d, 2009. QUALCOMM launches fisher friend to empower fishing communities in India. http://www.i4donline.net/articles/current-article.asp?articleid=1392&typ=Features (Accessed on 5th July, 2009).

NPFF, 2007. National policy for farmers 2007, Department of Agriculture and Co-operation, Ministry of Agriculture , Government of India, New Delhi.

Pande, Arun K., and Arve, Subhash., 2009. mKRISHI based grape farming. *I4d Magazine*, 7 (2), February, 2009. www.i4donline.net

PC, GoI, 2007. Recommendations of working group on agricultural extension for formulation of eleventh five year plan (2007-12), Planning Commission, Government of India, New Delhi. http://planning commission.nic.in/plans/planrel/11thf.htm

Patil, V.C., Gelb, Ehud, Yaduraju, N.T., Moni, M. and Patil, Roopa. S., 2009. Web based agriculture in India, http://www.fao.org/docs/eims/upload//257364/Patil_presentation.pdf

Rajagopalan, Radhika and Sarkar, Runa., 2007. The impact of DEAL on community networks: A case review, *OPAALS Conference 2007*. http://opaals.iitk.ac.in/deal/embed.jsp?url=other/whtp.htm (Accessed on 2nd July, 2009).

Ramamritham, Krithi., Bahuman, Anil., Duttagupta, Subhasri., Bahuman, Chaitra., Balasundaram, Srividya., (2005). Innovative ICT Tools for Information Provision in Agricultural Extension. www.cse.iitb.ac.in/krithipapersICD2006.pdf (Accessed on 21st February, 2008)

Ramamritham, Krithi., Bahuman, Anil., RuchiKumar, Aditya Chand, Duttagupta, Subhasri., Raja Kumar, G.V. , and Chaitra Rao, 2009. aAQUA - A multilingual, multimedia forum for the community. www.dil.iitb.ac.in/docs/aAQUA.pdf (Accessed on 2nd July, 2009).

Ramaraju, G. V., and Krishna Reddy, P., 2009. eSagu: An IT- based personalized agro-advisory system for augmenting livelihoods of farmers. www.solutionexchange-un.net.in/emp/e-discuss/res11070719.p (Accessed on 1st July, 2009)

Ramkumar, S., and Rao, S.V.N., 2004. Knowledge dissemination on cattle health through information kiosks in veterinary centres - Designing, delivering and evaluation of farmer –usable touch screen information kiosk. *Information Bulletin*, Rajiv Gandhi College of Veterinary and Animal Sciences, Pondicherry, India.

Ravishankar, H., Raju C.A., and Jagadeesh Chandra, 2006. Expert system on pests and diseases of major crops in Andhra Pradesh. *ICAR News*, 12 (2).

Sajeev, M.V., 2009. Innovations in farm technology assessment, refinement and demonstration: Role of Krishi Vigyan Kendras, In: Saravanan, R., (Ed.) *Agricultural Extension*, New India Publishing Agency, New Delhi.

Saravanan, R., 2002. Agricultural decision support system by Agro-climatic Planning and Information Bank (APIB), *Seminar Report*, Department of Agricultural Extension, GKVK, University of Agricultural Sciences, Bangalore.

Saravanan, R., 2008. e-Arik. In: Bagga R. K., and Piyush Gupta, (Eds.) *Transforming Government: e-Government Initiatives in India*, The ICFAI University Press, Hyderabad.

Saravanan, R., and Indra Devi, T., 2008. e-Arik: Using ICTs for Agricultural Extension. *I4d Magazine (Information for Development (i4d) Magazine)*, November issue. www.i4donline.net, Pp: 38-40.

Senthilkumar, 2009. Birth and Growth of Village Knowledge Centres and Village Resource Centres and the process of setting up / running the VRCs and VKCs (Activities of Jamsetji Tata National Virtual Academy). *Personal Communication*, in relation with ICT4ARD-2009 Seminar at Pasighat, Arunachal Pradesh, India, during September 9-11, 2009.

SOOCHNA SE SAMADHAN, 2009. SOOCHNA SE SAMADHAN -Knowledge Services just on the dial of phone. www.solutionexchange-un.net.in/decn/cr/res05110601.doc (Accessed on 2nd July, 2009).

TARI, 2009. Information note to the press (Press Release No. 56/2009), Telecom Regulatory Authority of India (TARI). www.trai.gov.in (Accessed on 25th June, 2009).

Veeraraghavan, Rajesh., Yasodhar, Naga., and Toyama, Kentaro., 2009. Warana Unwired: Replacing PCs with Mobile phones in a Rural Sugarcane Cooperative. *Information Technologies and International Development*, .5 (1), Spring 2009. 81-95. http://itidjournal.org/itid/article/view/327/150 (Accessed on 15th July, 2009).

Vijayalakshmi, B., 2008. India Development Gateway Initiative. *Proceedings of the eINDIA 2008*, New Delhi, 29-31 July, 2008.

VRC-WP, 2006. Village Resource Centres (VRC) - Working Paper. http://egovindia.wordpress.com/2006/06/26/village-resource-centres-vrc-%E2%80%93-working-paper/ (Posted by egovindia on June 26, 2006; Accessed on 10th July, 2009).

Yaduraju, N.T., Agrawal, R.C., and Himanshu, 2009. NAIP's e-Agriculture initiatives. *I4d Magazine*, 7(2), February, 2009. www.i4donline.net (Accessed on 30th July, 2009).

Author Information

Dr. R. Saravanan

Associate Professor &
In-Charge, School of Social Sciences

College of Post Graduate Studies
Central Agricultural University (CAU)
Umiam, Barapani- 793 103
Meghalaya, INDIA
Phone No. (O): +91-3642570031; (M): +91-9436054939;
Fax No.: +91-3642570030
e-mail: saravananraj@hotmail.com; saravanancau@gmail.com
URL: www.saravananraj.net

Dr. R. Saravanan is an Associate Professor (Communication), School of Social Sciences, College of Post Graduate Studies, Central Agricultural University (CAU), Barapani, Meghalaya, India, since 2009. He is undertaking teaching, research and extension in agriculture. His publications include a book on "Agricultural Extension: Worldwide Innovations", 30 articles in the referred scientific international and national journals, books and proceedings. He prepared three manuals, three multimedia CDs on Experiential Learning Programme and another multimedia CD on Participatory Rural Appraisal. He is currently undertaking three innovative e-Agriculture research projects; e-Arik (e-Agriculture), e-Village and e-Kiosk sponsored by the DSIR, Ministry of Science and Technology, Government of India (GoI), C-DAC, DIT, Ministry of Communications & Information Technology, GoI and Rural Innovation Fund, NABARD, respectively. One of the e-Agriculture based research projects was awarded National level "Best e-Governed Project- Special Jury Award 2008" by the Computer Society of India (CSI) - Nihilent e-Governance Awards 2007-08.

His First Book : Agriculture Extension : Worldwide Innovations, 2008, ISBN : 81-89422-96-0, Published by New India Publishing Agency, (www.bookfactoryindia.com) accepted worldwide by the readers.

❑❑❑

Chapter 5

IRAN

H. Shabanali Fami, Malihe Falaki and
Javad Ghasemi

1. INTRODUCTION

The agricultural paradigm in the developing world including Iran will have to be recast to take advantage of knowledge availability to achieve multiple goals of income, food, jobs, *etc.* Although, many farming communities in developing countries need timely expert advices to make them more productive and competitive, they are still facing a multitude of problems to maximize agricultural productivity. In spite of successful and promising researches conducted on new agricultural practices, majority of the farmers are not getting upper-bound yield due to several reasons. One of them is that scientific agricultural advice is not reaching the farming community in a timely manner. A wide information gap exists between research and practice.

To cope with this problem different countries have taken various measures. What is common in the extension systems of developing countries is that in a backdrop of rapid organizational and social changes in agriculture, new hardware and software devices have to be applied for better responsiveness to increasing farmers demand for timely and accurate information. At the pace of recent progress in ICT, within a decade it will be possible to provide instantaneous connectivity to millions of people, enabling mass customization and personalized services (Reddy and Ankaiah, 2005). Data mining technology can be used to extract useful knowledge from huge database and simulation technology can be used to predict in to the future (Wiederhold, 2002).

In the 1990s, the key actors (UN international agencies, bilateral agencies such as USAID and so on) were the first to address the issue of ICTs in the debates on development (Helvetas, 2003). One can observe that during the same decade, progress in information technology is affecting all spheres of our life. Due to progress in hardware technologies, people are able to procure high-speed reliable computers with huge storage capacities at affordable cost. Also, database and data warehousing technologies can be used to store and retrieve large amounts of information, which can be coupled with internet technology to deliver information, instantaneously to the needy people. Recent ICTs developments enable the maintenance of huge information (text, image, audio and video) repositories with negligible down-time. The stored information can be quickly extracted by millions of users, simultaneously. Currently, Internet speed is doubling every nine months (Reddy, 2001). Such ICT-based developments provide new opportunities to improve the utilization and performance of livelihood technologies such as agriculture, education, library, health and medical services, and artesian technologies. The research challenge here is to identify the areas where progress in ICTs could be used to improve the performance of these services and technologies, and build cost-effective ICT-based systems that improve the living standards of the rural population. It is now a responsibility for us to improve the utilization and performance of agricultural technology by exploiting recent progress in ICTs (Reddy and Ankaiah, 2005). ICT is a generic technology and its application to agriculture and rural development are considered extensive and pervasive. ICTs can be used as a broad tool for providing local farming communities with scientific knowledge and establishing knowledge societies in the rural areas of the developing world. There is a vast literature on the benefits and potentials of ICTs as tools for enhancing people's lives whether by increasing access to information, relevant to their economic livelihood or better access to other information sources. GoI (2001) emphasized the necessity of developing the capacity to generate, absorb, disseminate and protect knowledge and exploit ICT as a powerful tool to derive societal transformation. Information technology is seen as an important means of achieving such a transformation. When used as a broad tool for providing local farming communities with scientific knowledge, ICT heralds the formation of knowledge societies in the rural areas of the developing world. ICT can give a new impetus to the social organizations and productive activity of agriculture which, if nurtured effectively, could become transformational factors (Meera *et al.*, 2004).

ICT has also a significant role to perform in evolving such a paradigm, as was evident from the interdisciplinary dialogue on Information Technology (IT): Reaching the Unreached (Swaminathan, 1993). ICT can also play an important role in bringing about sustainable agricultural development when used to document both organic and traditional cultivation practices. At present, ICTs include, but are not limited to computers, software, high capacity storage, networks, telecommunications, databases, data warehouses, multimedia, training, the internet and its world wide web, Geographic Information System (GIS), Computer-Aided Design (CAD), online services, video-conferencing, Executive Information Systems (EIS), electronic-mail and expert systems (Zazueta and Vergot, 2003).

In the early 1980s, only less than 0.5 per cent of the United States farmers used computer in agricultural production, while in 2001, 55 per cent of the 2,165,000 farmers had computer access and about 43 per cent had internet access. In addition, the per cent of ownership, including owned or leased computers, increased from 31 per cent in 1997 to 50 per cent in 2001 (ECOP, 1980; NASS, 2001; Meera *et al.*, 2004).

During 1994 and 1995, University of Florida/ IFAS held fifty half-day workshops each year, directed at farmers, on using the internet for business. Today Florida ranks at the top in internet purchases of agricultural inputs and second in marketing over the internet (Zazueta and Vergot, 2003).

However, ICT can bring new information services to rural areas where farmers, as users, will have much greater control than before over current information channels. Access to such new information sources is a crucial requirement for the sustainable development of the farming systems. ICT can give a new impetus to the social organizations and productive activity of agriculture which, if nurtured effectively, could become transformational factors (Meera *et al.*, 2004). In addition, IT is a strategic technology in the international competitiveness of agricultural producers. Information and Communication Technologies (ICTs) can be cost-effectively and practically employed to facilitate information delivery and knowledge sharing among farmers, extension agents and other stakeholders of agricultural development paradigm (Annor-Frempong *et al.*, 2006).

Applications of ICT in support of agricultural and rural development fall into five main areas, as outlined by Don Richardson. These are: economic development of agricultural producers; community development; research and education; small and medium enterprises development and media networks.

From other viewpoint, some agricultural development services that can be provided in the developing world, using ICT, are:

- Online services for information, education and training, monitoring and consultation, diagnosis and monitoring, and transaction and processing;
- e-commerce for direct linkages between local producers, traders, retailers and suppliers;
- The facilitation of interaction among researchers, extension (knowledge) workers, and farmers;
- Question-and-answer services where experts respond to queries on specialized subjects ICT services to block- and district-level developmental officials for greater efficiency in delivering services for overall agricultural development;
- Up-to-date information, supplied to farmers as early as possible, about subjects such as packages of practices, market information, weather forecasting, input supplies, credit availability, *etc.*;

- Creation of databases with details of the resources of local villages and villagers, site-specific information systems, expert systems, *etc.*;
- Provision of early warning systems about disease/ pest problems, information regarding rural development programmes and crop insurances, post harvest technology, *etc.*;
- Facilitation of land records and online registration services;
- Improved marketing of milk and milk products;
- Services providing information to farmers regarding farm business and management;
- increased efficiency and productivity of cooperative societies through the computer communication network and the latest database technology;
- Tele-education for farmers;
- Websites established by agricultural research institutes, making the latest information available to extension (knowledge) workers and obtaining their feedback (Meera *et al.*, 2004).

Keeping the capacities of ICTs in view, this new fast emerging technology seems to have a very fascinating power to tackle the problems faced by different agricultural systems throughout the world.

2. AGRICULTURE IN IRAN

Despite the obvious dominance of oil, agriculture is one of the most important economic sectors in Iran. Rapid agricultural growth continues to be the key to poverty alleviation, food security and overall economic development in the country. Iran is situated between 25° to 40 ° N latitudes and between 44 ° to 63 ° E longitudes. Roughly one-third of the country's total surface area is suited for farmland, but because of poor soil and lack of adequate water distribution in many areas, most of it is not under cultivation. In fact, out of a total area of 165 million hectares, at present, only 20 million hectares are under cultivation. From the total arable land, 12.5 million hectare is under cultivation each year (arable land, orchards and vineyards). The western and northwestern portions of the country have the most fertile soils.

Iran has three distinct agro-ecological zones: humid, semi-arid and arid regions. The country has enormous potential to meet the food and nutritional security challenges in the new millennium. Agriculture accounts for about 20 per cent of GDP, 25 per cent of employment opportunities, 80 per cent of food supply and 35 per cent of non-oil exports as well as in the provision of considerable raw materials of industrial use (Earthtrends, 2003; Mobasser, 2003). The major constraint to further development of the agricultural sector is not the availability of land, rather the availability of water. Only 10 per cent of the country receives adequate rainfall for agriculture and irrigation techniques are heavily relied upon limited water supply. About 54 per cent of the country's arable land is rainfed. The main food grains produced by Iranian farmers are wheat, rice and barley. These are staple foods and are the main sources of protein in an Iranian diet. Wheat alone represents 50 per cent of land usage, while fruit accounts for 15 per cent of irrigated land. Two farming systems are primarily used to cultivate these lands. Man-made irrigation systems are used in certain areas, while rain-fed and dry farming systems are used in the western and north-western parts of Iran. All agricultural lands are owned by the people, except a very small portion which belongs to government's bodies (Badripour, 2004; CET, 2004).

The 1979 Revolution sought self-sufficiency in foodstuffs as part of its overall goal of decreased economic dependence on the west so that in the constitution, strong emphasis has been given on some of the key preconditions for national food security such as agricultural development, science and technology, environmental protection, social security, poverty alleviation, protection of women's rights and human dignity (Wikipedia,2007; Shabanali Fami, 2000).

Iran has become more or less self-sufficient in the production of certain fruits, vegetables and livestock. There are different livestock raising systems in Iran, *viz.*, grazing system, mixed farming system and landless or industrial system. The first system is common in rural and nomad-based areas, while the last one is practiced mostly around the highly-populated areas like small towns and cities. Nomadic tribesmen perform animal husbandry as a sideline to crop farming. Mixed farming system seems to be the most dominant livestock production system in Iran. However, some major agricultural outputs are nuts, cotton, dairy products, wool and caviar. Iran has recently undertaken various initiatives to move away from subsistence agriculture and achieve modernization. These include dam building, subsidization of irrigation, more governmental transparency and application of high technologies such as nanotechnology and information technology. The key issue that how rapidly the country should open itself to the modernizing is influenced by the outside world. The government's last Five-Year Development Plan ended in 2005, helped in liberalizing production and marketing in the sector, allocated foreign exchange at a preferential rate for the import of agricultural machinery and continued to pay subsidies on domestic production (CET, 2003; CET, 2004;

Badripour, 2004). In this line, the Ministry of Agriculture has been urging peasants to consolidate their small, scattered farming plots and to form Rural Production Cooperatives (RPC's). Optimally, each RPC will cover 2,000 to 92,500 hectares of land and employ 200 to 250 farmers, who will maintain their individual ownership. While the movement towards RPCs has been slow, it is estimated that by 2022 almost all of Iran's farms will consist of RPCs or large commercial farms (CET, 2004).

Although, a significant attention has been paid to agriculture, major problems plaguing the sector's development include shortages of water, inefficient irrigation, outdated farm machineries, lack of raw materials and the common practice of subsistence farming. The government has made some progress in developing the sector including the vast improvement of rural infrastructure, such as bringing electricity to rural communities and road building, but the task is far from over.

However, in food security terms, the Iranian government strongly supports agricultural development, increased food production, rural development and self-sufficiency in strategic food security. Over recent years, the government has been calling for even more attention to be paid to the issues related to agriculture such as access to land, better incentives to farmers, easy and affordable access to production inputs, improved knowledge, enhanced participation in rural development and sustainable agriculture in a broader sense (CAPES, 1996; CET, 2004).

Although the per capita output of the agricultural sector is far below that of the developed or even developing countries, it is still comparably higher than other economic units in Iran. Despite mammoth investments made in the industry sector, its per capita production is below that of the agricultural sector. Although the agricultural sector is facing extra manpower, its productivity rate is 4.2 per cent, compared with 3.1 per cent for other economic sectors. With the removal of the current problems of the sector and further mechanization of farms, the productivity rate would grow higher (ICCIM, 2007).

Iranian agriculture is still facing a multitude of challenges to maximize productivity and the national extension system serves as the Iranian main source of technological/information support for national agricultural development and facilitating the development processes in agriculture and rural development. Agricultural extension is known as the application of scientific research and new knowledge to agricultural practices through farmer education (Wikipedia, 2007). According to many policy-makers and planners, the main agricultural development strategies will not be fulfilled unless the capacities of extension system are taken into consideration and enhanced. Therefore, for sustainable agriculture to succeed, agricultural extension must arise in a new way. It should be enabling, participatory and ICT-based, creating the conditions for agricultural development based more on locally available resources and on local skills and knowledge.

3. AGRICULTURAL EXTENSION IN IRAN

In Iran, to reach the objectives of agricultural development, enhancement of extension system is an important strategy. Agricultural extension work was started before 1958 in Iran (Iravani, 1992). Historically, extension system of the country has mainly involved technology transfer expanding its role to new areas. It has been playing an important role in agricultural and rural development in Iran. During its more than 50 years of activities and experiences, agricultural extension has undergone many changes. However, one of the primary goal of extension services in Iran is to develop competences of farmers so that they may have better control over their farming systems. The approach of extension system towards agricultural and rural development is to help and empower farmers to discuss, recognize and define their own needs and encourage them to involve in diverse programs to satisfy their needs.

At present, agricultural extension system in Iran is facing many challenges to extend its services to all the three millions farming systems throughout the country. The most important of them are: low accountability to the needs of farmers, financial constraints, dealing with diverse agro-ecological zones, lack of sufficient numbers of extension workers, lack of responsiveness of government facilities to deliver full-scale extension services and unsatisfactory level of farmer's participation in technology development and transfer. Accordingly, the financial crunches of public extension system as well as the pressure on the government to maintain the vast extension network are the two main factors which stimulate the emergence of infusing ICTs in the agricultural extension system of Iran (Ministry of Agriculture, 2005).

4. APPLYING ICTs IN AGRICULTURAL EXTENSION IN IRAN

Keeping in view the advances in the information technology that have triggered deep changes in agricultural extension, agricultural extension system in Iran is going to take advantage of ICTs developments to enhance its ability to expand its services coverage. ICTs can be used by extension staff to provide farmers with better and timely information. Recent years have been witnessing a surge in extension services delivery, offering online information to the Iranian farmers. Some new applied researches of some academics recognized the potential of computer in agricultural extension activities in Iran (Movahed Mohammadi, 2002; Rezaee, 2004; Rasouli Azar, 2004). A number of measures have also been taken in Iran to develop and apply ICTs in different areas of national agricultural development during the past decade. However, It is believed in the extension system of Iran that the use of ICT is a strategic move to overcome many of the shortcomings of the earlier extension programs and approaches. In Iran, extension is a public service provided at no charge to Iranian farmers with a broad range of interests and needs for agricultural information and services. Recent years are witnessing a surge in extension services delivery, offerings, online information to farmers provided under governmental sponsorship. The provision of personalized advice online, where users determine the kind of information they need and interact directly with advisors, is a promising emergent application in Iran.

5. AREAS OF APPLYING ICTs IN AGRICULTURAL EXTENSION IN IRAN

It is an increasingly known fact that application of ICTs by agricultural extension systems can be effectively harvested for overall agricultural and rural development, if they are equipped with the full potential of ICTs. Accordingly, the country needs a primary and scalable computer-based system which can be incrementally developed and extended to cover all areas of extension activities. The research challenge here is to identify the areas where progress in ICTs could be used to improve the effectiveness of extension services and technologies that improve both the productivity of farming systems and enhancing the living standards of farmers. A national study conducted in Iran identified some main areas in which ICTs can play significant roles as depicted in Figure 1.

6. CHALLENGES AND DRIVING FORCES OF APPLYING ICTs IN AGRICULTURAL EXTENSION IN IRAN

Recent years have seen an increase in the use of ICTs in almost all spheres of rural lives in different countries, despite persisting problems of access, connectivity, literacy, content and costs. In order to properly infuse of ICTs into agricultural extension, many challenges need to be addressed such as developing the capacity of agricultural producers and extension workers to enable them to use ICTs effectively as part of their routine activities. Although ICTs can play a crucial role for improving effectiveness of agricultural extension services, there are still many limitations to harness the full potential of this high technology for agricultural development. Major hindrances of utilizing IT in agricultural extension activities can be classified as follows:

1. Preference on traditional information media by some farmers.
2. Lack of appropriate and timely scientific advice about modern farming to be rapidly transferred to farmers.
3. Limited access to handheld computers for field workers.
4. Limited access to high-speed and broadband internet services.
5. Increasing digital divide between urban and rural areas.
6. Difficulties in the design of an appropriate information system (Hautzer *et al.*, 1997).

Figure 1 : The Main Areas of Applying ICT in Agricultural Extension

Other challenges being faced by extension systems may include: provision of relevant contents in local language and useful applications, development of telecommunication infrastructure in farming areas and communities (Haddad and Draxler, 2002), reduction of the allied costs (Gell *et al.*, 1999; Maru, 2003), eliminating legal barriers (Haddad and Draxler, 2002), access to appropriate content as well as relevant software (Haddad and Draxler,

2002; Maru, 2003), lack of dynamic organizational culture (Sepehry, 2001), lack of appropriate inter-sectoral relationships and collaboration (Raab, 2005),low level of extension professional knowledge on computer and internet use (Barani and Qodsi Rasi, 2003; Gell *et al.*, 1999), lack of access to personal computer or laptop at home, high level of computer anxiety among extension personnel (Martin *et al.*, 2001; Fletcher and Deeds, 1991; Smith and Kotrlik, 1990), lack of comprehensive information plan(Jalali Shahkoohi, 2003); lack of reliable and related databases; resistance to change and lack of sufficient Persian-based internet contents.

In order to ensure that ICTs are effective in supporting agricultural extension activities, the essential preconditions should be provided. From the perspective of related studies, the following driving forces of integrating IT into agricultural extension need to be adequately taken into account: expansion of computer availability and usage, development of information networks, professional development of extension staff, establishment of necessary databases, provision of Persian-based internet contents for the conditions of Iran, financial support of ICT-based extension activities, conducting training courses on IT for extension agents, motivating extension professionals to use IT, computerization and automation of organizational affairs of extension offices, provision of broadband internet and forming favorable attitude towards IT application in extension (Sepehry, 2001; Haji Ketabi and Noor Mohammadi, 2001, Sharma, 2002; Anonymous, 2003; Maru, 2003; Maningas and Mancebo, 2004; Rezaee, 2004; Rasouli Azar, 2004).

Therefore, within the context of the revolution in information technology, it is a challenging decade for Iranian extension system to enhance its old mandates and operations so that it is more responsive to different categories of farmers. The introduction of new information technologies and concepts can change the ways through which agricultural information is gathered, analyzed, transferred and managed. Agricultural extension system in Iran will continue to take advantage of ICTs developments to enhance its ability to expand its services in remote rural areas. The main focus of the most recent national 5-years development plan has also been on the application of ICTs in agricultural extension. In order to achieve this objective, it is necessary to identify driving forces and hindrances of utilizing ICTs in agricultural extension system of country. There are a number of challenges related to applying ICTs in agricultural extension that must be overcome before extension organizations in different provinces can successfully implement their projects on the basis of farmers' daily needs and requests. The overall result of a national study revealed (Figure 2) the following hindrances limits extension system of Iran to achieve the goals of a real cyber extension system in the country: financial limitations, lack of skills for information management and acquisition, lack of organizational culture compatible with application of IT in extension system, psychological obstacles, infrastructural obstacles and personal obstacles (Falaki, 2005).It is quite clear that many of these challenges can be removed by taking the related driving forces into account as depicted in Figure 2.

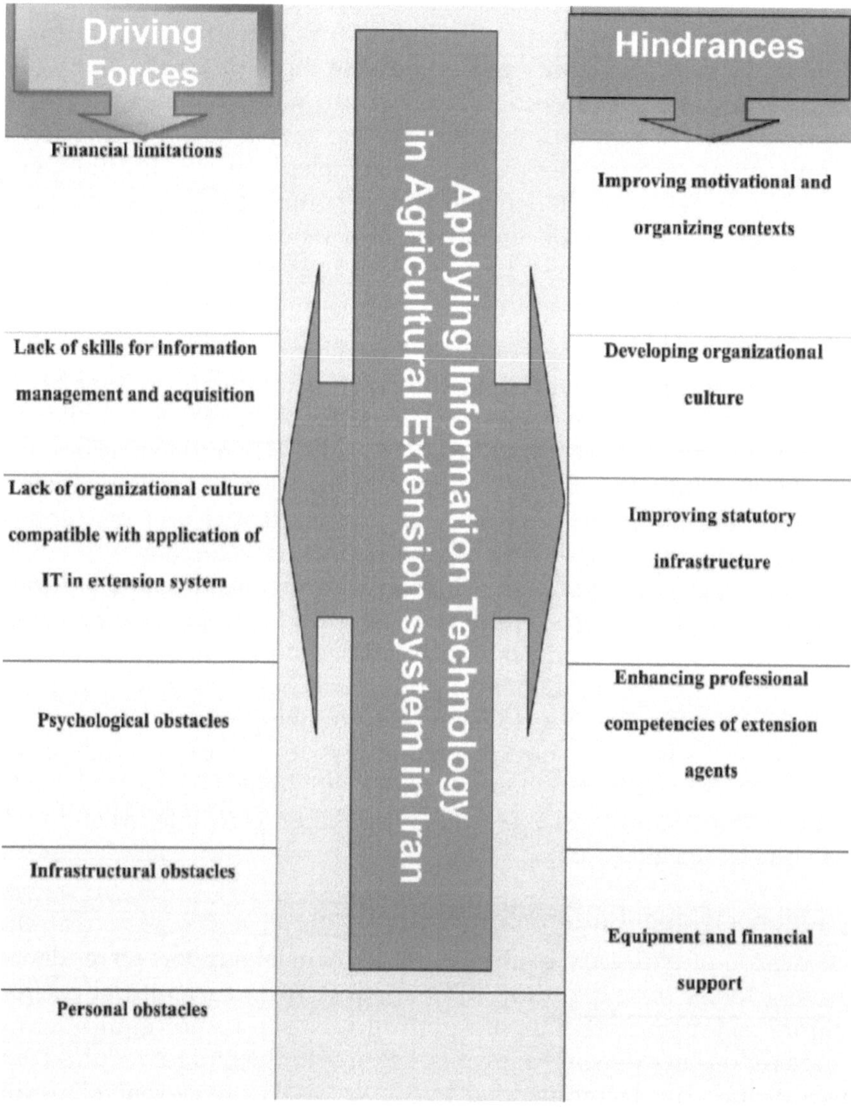

Figure 2 : The Main Driving Forces and Hindrances for Applying IT in Agricultural Extension in Iran

The main hindrances of applying ICTs in agricultural extension are discussed here along with the proposed driving forces which can in turn be of significant importance to any innovative extension system.

6.1. Infrastructural Obstacles

The call for demand-driven extension opens the door for examination of how ICTs can be cost-effective and practical tools for facilitating and channeling farmers' demands, and addressing those demands. But when so many rural areas of developing countries are without access to the basic telecommunication services that support so many key Information and Communication Technologies (ICTs), like the telephone and the internet, this examination is severely challenged. Any resulting programs and projects remain totally dependent on the strength of national universal access and telecommunication policy rises to the surface as a primary enabler or obstacle (Richardson, 2003).

According to the study conducted at national level (Falaki, 2005) to facilitate information delivery and to bridge up the gap existing between different areas of extension system, the government has to try to build up the necessary infrastructure required for integrating ICTs into extension system. As a rule of thumb, the efforts to apply information technology should be started at locations which have necessary infrastructure and pre-requisites. A number of measures have been taken in Iran to develop and apply ICTs in the different areas of national development for the past decade, but there is still the need to develop hardware and software to support extensionists in the use of ICTs for delivering extension services to farming communities. In Iran, there are significant numbers of private and public extension institutions providing agri-based services to farmers by utilizing various ICTs tools and applications. Thus, they should have access to a network and personal computer. Moreover, every extension center should also be equipped with an established minimum standard for ICTs capability to be able to conduct ICT-based courses (Falaki, 2005).

6.2. Financial and Equipment Limitations

Provision of necessary equipment and financial support were identified as driving forces for expanding ICTs application in agricultural extension. Therefore, extension officials are getting furnished with equipment and software for use in extension activities. More over the budget required for the growth of ICTs are recommended in the successive five- year plans and a certain per cent of total budget is planned to be allowed for ICTs application in the state and central sponsored extension projects.

6.3. Personal and Psychological Obstacles and Lack of Skills for Information Management

Along with reliable infrastructure, competent human resources are fundamental tools and means of assessing, planning, managing and adopting ICTs for achieving a cost-effective and efficient extension system. In fact, success of extension organizations depends on the ability of their members to communicate by applying advanced methods and media such as information technologies. Extension workers play a significant role in the process of disseminating new agricultural information to farming communities in Iran. As the results of a study showed, enhancing professional competencies of extension officials was a key factor for application of ICTs (Falaki, 2005). Therefore, appropriate training courses such as ICDL (International Computer Driving License) package, internet skills and computer maintenance skills are conducted for extension personnel at different levels. It is obvious that paying adequate attention to the use of ICTs is a strategic move to overcome many of the shortcomings of the earlier programs.

6.4. Lack of Organizational Culture

As noted earlier, any extension organization subsists just when its culture is changed and updated by new advanced information and communication technologies. In other words, empowerment of extension organizations relates to create and manage their knowledge asset which is not possible unless the system is prepared to use new information technologies. According to the findings of a national study, organizational culture of extension systems can be enhanced by networking extension centers and related stakeholders (Falaki, 2005). It is because knowledge can only become a means of power if communication of networks of relevant actors evolves. Networking improves coordination among different departments which otherwise poses several problems and reduces organizational productivity. However, as extension organizations should serve farmers in very vast geographical areas as well as in highly varied agro-ecological zones they need to capture and store complex and diverse information by the means of ICTs.

The identified driving forces which could prove useful to cope with the problems of infusing ICTs into extension activities can be categorized as follows: improving motivational and organizing contexts, developing organizational culture, improving statutory infrastructure, enhancing professional competencies of extension agents and equipment and financial support. These identified factors deserve more attention than they have been afforded previously in the process of extension services delivery by grasping the potential of new technologies. In addition, despite the need, the obstacles outlined above have hindered successful deployment of information technologies in agricultural extension system. The hindrances were very varied from personal limitations to organizational and even cultural barriers. However, to overcome these obstacles the study investigates the driving forces. A more complete list of strategies or solutions for alleviating the identified obstacles is suggested as below:

- To provide a reliable, efficient, cost-effective and secured gateway to the national and global information infrastructure and networks by establishing and developing national databases and ICTs infrastructure in rural and farming areas.
- Providing regional and local extension centers with high–speed internet connectivity to the national and global information infrastructure.
- To establish and encourage significant public- private partnership and collaborative investment in ICT-based extension activities particularly in software development and e-advisory services.
- Re-conceptualizing and reshaping agricultural extension and communication by taking ICTs potentials into account leading to cyber-extension and e-communication for agricultural development.
- Motivate and build confidence as well as competence among extension workers to apply ICTs in doing their professional tasks.
- A variety of attractive extension and training materials should be prepared with the help of computer, using creative and interactive techniques.
- Develop and apply information technology tools to facilitate the work of extension workers at field level but with adequate human backstopping.
- Establishing substantial financial commitment to the advancement of IT capacity and training for its staff (Falaki, 2005).
- Lack of sufficient extension staff competency.

In a changing environment created by ICTs, staff development is crucial. In such a circumstance, extension staff needs to get in-depth knowledge on ICTs potential and capacities. In fact, according to the national tentative plan for infusing ICTs into agricultural extension in Iran, generating awareness among extension agents about the capabilities of ICTs services is the first step which has to be considered in ICTs initiatives. Extension personnel needs also to be provided with new communication skills and tools so that they can better share their knowledge and experiences with other extension agents across large spatial distances, languages and cultures. Accordingly, they are brought into the chain of ICT networks for agricultural information exchange. An important point is that the extension agents are encouraged to make initiative for the acquisition and dissemination of knowledge on ICT-related subjects, such as setting up the device, using e-mail, word processing, use of spread sheets, presentation software, web browsing, document readers, downloading and installing new applications, tasks, contacts, notes, calendar, developing database and internet services. Some courses are being designed to give the agricultural agents an introduction to the internet, equipping them with the knowledge and confidence to explore what the internet can do for them to help farmers and to advance agricultural development projects. Keeping the needs of Iranian farmers in mind, the country has also planned to develop and deploy innovative ICTs tools for dissemination of agricultural information in the national extension system, through professional development of the extension staff working at different organizational levels. Up to now, a number of training courses has been conducted to empower the extension staff capacity on ICTs-related issues and it is planned to improve the common pool of knowledge and expertise of extension agents in the area of ICTs. In order to enhance professional competencies of extension agents the following measures are taken into consideration: paying more attention to ICTs training for extension agents particularly about English and ICDL skills, equipping extension centres/offices with ICTs facilities, provision of extension agents with PC, PDA, portable computer and broad band internet connections, promoting extension agents to make use of ICT potentials and establishing affordable telecommunication infrastructure for extension offices.

7. INTERNET USE IN AGRICULTURAL EXTENSION IN IRAN

Among ICTs, internet has been specifically designed to be open and decentralized tools for people to share information and communicate more freely (Chapman and Slaymaker, 2002). Information can be received through internet in the form of both text and image and quick dissemination of research findings to the needy farmers *via* internet services such as e-mail or messenger. Although, narrow bandwidth is a basic barrier in the developing countries to make efficient use of internet, currently, internet speed is doubling every nine months (Reddy, 2001).

Moreover, knowledge can become a means of power if coalitions of networks of relevant actors evolve. Hence, internet can solve the problem of lack of coordination between different agricultural departments. Internet has also a huge potential in automating the process of delivering extension services to farmers and in collecting and sharing a variety of information so that they will be having better access to information and knowledge. For agricultural sections such as extension to be deemed relevant and important to society in the future, it must immerse itself in the internet - using it, promoting it, and making the use of information technology as a way of life (Anonymous, 2003).

Software application in agriculture is also diverse and included expert system (ES), computer-aided design (CAD), computer-aided management (CAM), information delivering systems (IDS), simulations and automated control. In particular, IDS evolved into rich content applications from desktop and mainframe database into CD-ROM and eventually, web-based systems.

Given the importance of internet, different agricultural extension bodies at national, provincial and local levels attempt to make use of this technology through various measures such as designing specific websites and enhancing access of extension agents to different internet services. In fact, adequate attention is paid to encourage organizational initiatives for the acquisition and dissemination of knowledge on internet by agricultural extension workers through successive national development five-year plans. In other words, to facilitate information delivery and to bridge up the gap existing between different actors of agricultural knowledge and information system, the government makes more attempts to build up the necessary infrastructure required for integrating ICTs into agricultural sector, out of which professional development on internet skills is highly important. In addition, equipment and training related to internet use is being provided to the extension agents. Agricultural extension are also planned to be equipped with PDA, lap top computers and broad band internet connections. Rural ICTs centre in Iran are also gaining significant role in delivering extension services to the villages along with their other general services such as e-banking, e-commerce, e-learning and expanding interpersonal communications *via* e-mail.

8. RURAL ICT STRATEGIC PLAN IN IRAN

Based on the official census in the year 1996, there are 68122 villages in Iran in which approximately 23 million people live. According to this statistic the rural population was about 38 per cent of the total country population. From this, about 70 per cent were living in only 18 per cent of the villages. This means that due to the scattering of villages and their low population, approximately 82 per cent of the country's villages includes only 30 per cent of the rural population. In 1996, 12223 villages had a population of more than 500 people in each village, and the rest of 55899 villages had less than

500 people per village. It may be concluded that a huge amount of the country's rural population is condensed in one-fifth of the country's villages.

Earlier to the fourth national development plan, ICT development in the country's villages had not yet been extensive, and was confined to the development of telephony and post services. Hence, selecting the villages with more than 500 inhabitants was seen to be suitable for rural ICT development in the scope of the national strategic plan for ICT development in rural areas. It was because those villages had basic requirement for the development of ICTs and played role of central village for their neighboring villages in term of providing different services including ICT-related services. In most of these villages, educational and health care facilities, as well as facilities like electricity and telephone were existed. The government in order to enhance village's access to ICTs services formulated a Rural ICT development strategic plan keeping in view the visions of national 20-year vision document, the fourth development plan, the national ICT documents and the ICT section of the fourth development plan. This plan considered as one of the subdivision documents of the national ICT development strategic plan took place with the support and assistance of the national ICT plan. From the hierarchical point of view, the visions of the rural ICT should be compatible to those of the national ICT.

In the World Summit on the Information Society (WSIS) of Geneva 2003, Iran had agreed on the basis of moving towards the Information Society to expand ICTs services to rural areas. In the national third 5-year development plan, expansion of telecommunications and telephony infrastructure in rural areas was seriously taken into consideration. It is while in the Fourth 5-year development plan equipping the rural areas to ICT was directly noticed. Accordingly, the country planned to establish rural ICTs centers in 10000 villages with provision of various services to the villages. This project was started in 2004 under the management of the Ministry of Information and Communication. The main objectives of these centers were defined as:

- Providing villagers with integrated information based services.
- Increasing the Telephone penetration factor in rural areas and the access of rural families to Telephone services.
- Providing the basic requirements of developing the idea of e-government and e-services.
- Providing village residents with means of connecting to the internet as the main information gateway.

- Providing other governmental organizations such agricultural extension centers based in rural districts the sufficient means of providing adequate services to villagers at a lower cost (Sadri, 2005).

9. CASE STUDIES

A study was conducted by Falaki *et al.*, (2008) to assess the attitude of extension professionals to examine its effect on the adoption of IT by extension system. The population of the study consisted of 1145 extension professionals working throughout the country in different provinces. Using a stratified random sampling technique, 241 extension experts were selected as the sample for the study. The reliability and validity of the tools of study were confirmed by a pilot study and expert judgment. The results of the study indicated that the respondents had higher knowledge on word processing software, windows and file management in the area of computer application. In contrast, in the area of internet application they were mainly interested in getting information from specialized sites, searching by search engines of Yahoo and Google and exchanging e-mails. Correlation analysis showed that there was a significant relationship between the attitude of extension professionals towards ICTs application in agricultural extension and their age, educational attainments, knowledge of IT, its application and English language proficiency. Meanwhile, there was a significant difference between attitudes of male and female extension experts towards IT.

Papzan *et al.*, (2008) carried out a study to investigate the expansion of ICTs in rural areas of Kermanshah County in Iran. They used Delphi technique and SWOT model. To draw an outlook of the ICT's expansion, in each part of the model, questions were presented and based on the received responses the required analysis were carried out in three rounds to draw an outlook. A pretest was carried out in a statistical population which included all the ICTs and rural development experts in Kermanshah county. Based on the received responses those who were not well familiar with ICTs omitted. Simultaneously the rural and urban expert views were compared. The existence of electricity, telecommunication office, a young educated class and tourist attractions were counted as points of strength, and poor telecommunication facilities, high overhead, little attraction for private sector to invest, the lack of local experts and low educational level were viewed as points of weakness. Access to world markets, reducing the middlemen's role, positive attitudes of those responsible, inclusion of ICTs in fourth five year plan and enhancement of rural tourism were mentioned as points of opportunities, and increase in consumerism, abnormal behavior were counted as points of threats. In addition, the results indicated that the two groups of rural and urban experts shared a common view that the following completions would result from the expansion of ICTs in rural areas: access to world market, enhancement of social participation, environmental conservation. However, the two groups did not hold a unified view concerning points such as a rise in unemployment and poverty, and the existence of local knowledge. In sum, the environment is optimistic towards the expansion of ICT in the area under study.

Mokhtarnia *et al.*, (2008) conducted a study to investigate the relationship between attitudes and skills in application of information technology; and knowledge Management of Agricultural Extension Experts in Iran. The study was a descriptive-correlational survey research. The statistical population included all Agricultural Extension Experts (N=2214) dealing with extension activities in eight regions as distinguished by the Ministry of Jihad-e-Agriculture based on their climate similarities. A sample of 160 extension personnel was selected by use of "multilevel stratified random sampling" method. A questionnaire was designed and administered to gather the required data. A panel of experts from the Department of Agricultural Extension and Education as well as from the college of Management in the University of Tehran verified the content validity of the questionnaire. To determine the reliability of the components of Knowledge Management, Cronbach alpha was calculated to prove the reliability. The findings revealed that, although the attitude of 93 per cent of the studied people towards Knowledge Management was either at medium or medium to high level, the range of skills of the majority of extension experts about application of Information Technology and components of Knowledge Management was from very low to low levels. It was just in application of one component of Knowledge Management *i.e.* filing management (Professional Information), that the experts had a medium level of skill. The results also demonstrated that there was a positive significant relationship between the dependent variable of the attitude of extension experts towards Knowledge Management; and their skill in application of Information Technology as well as the sum of six components of Knowledge Management (knowledge acquisition; knowledge sharing; knowledge evaluation; creation/ storing and omission of knowledge; application and distribution of Knowledge Management; and file management). Furthermore, a positive significant relationship between the dependent variable (attitude of extension experts towards Knowledge Management) and their skill on application of the two components of file management, and knowledge evaluation (at 5 per cent level of significance) was observed.

Lashgarara *et al.*, (2008) in their study found that knowledge is a primary resource, no less important than land, labor or capital for improving the lives of people. In the recent years, one of the best tools for delivering information in Iran is Information and Communication Technology (ICT). Hardware and software used to generate, prepare, transmit and store data, such as computers, radio, TV, computer programmers/tools; *etc.,* has been defined as ICT. ICT have this potential to offer new opportunities for a creation of innovative solutions to move towards the sustainable development goals. However, ICT might be considered as luxury and is often given low priority in Iran, but it can accelerate development by facilitating access to information. ICT can serve as effective tool to promote sustainable development in Iran by raising awareness and sharing knowledge, environmental monitoring, resources management, improving preparedness, enabling reliable early warnings and increasing situational awareness. The methodology of this research is descriptive- analytical and its main purpose is determination role and place of ICT in improving of sustainable development in Iran.

Sharifzadeh *et al.,* (2008) in a study conducted in Iran found that Iranian agricultural research system needs institutional supporting based on designing and implementing strengthening mechanisms. Systemically, integrating the information and communication technologies in agricultural research system of Iran seems as an appropriate strengthening mechanism. However, appropriate application of information and communication technologies and related systems such as national agricultural research information system require systemic and strategic planning based on identification and recognition of associated system-wide organizational implications. Regarding to this comment, this study indicated an implications of integrating information and communication technologies in the Iranian agricultural research system.

Omidi Najafabadi *et al.,* (2008) in their study examined the ways ICTs can be applied in agricultural insurance system. The Agricultural Insurance Fund (AIF) in Iran has hired private sectors named "brokers" to implement an agricultural insurance scheme. AIF has spent considerable time, effort, and money to train brokers. Unfortunately, many of their investments are met with disappointing results and traditional education is no longer effective. The study revealed that to overcome this challenge, an information and communication technology (ICT) based training system seems to be the best solution.

Ommani and Chizari (2008) conducted a study to design an Information Dissemination System (IDS) based e-learning in agricultural of Iran. A questionnaire was developed to designing Information Dissemination System. The questionnaire was distributed to 96 extension agents who work for Management of Extension and Farming System of Khuzestan province of Iran. According to the result there was a significant relationship between the age , IT skill and knowledge, years of extension work, the extend of information seeking motivation, level of job satisfaction and level of education with use of information technology by extension agent. According to extension agents five factors were ranked respectively as five top essential items to designing Information Dissemination System (IDS) based E-learning in agricultural of Iran. These factors include: 1) Establish communication between farmers, coordinators (extension agents), agricultural experts, research centers, and community by information technology. 2) The communication between all should be mutual. 3) The information must be based on farmers needs. 4) Internet used as a facility to transfer the advanced agricultural information to the farming community. 5) Farmers can be illiterate and speak a local and they are not expected to use the system directly. Knowledge produced by the agricultural scientist must be transformed in to computer understandable presentation. To designing Information Dissemination System, electronic communication, in the agricultural society and rural areas must be developed. This communication must be mutual between all factors.

10. CONCLUSION

Extension system in Iran increasingly acknowledges the importance of new information and communication technologies so that technologies are being considered as a basic component of any extension programs. In fact, ICTs modify the practice of extension services and open interesting avenues for the collection and dissemination of local knowledge and its integration with modern and global knowledge, which will require redefinition of knowledge management. Keeping the present situation in view, adequate attention is paid in Iran to encourage organizational initiatives for the acquisition and dissemination of knowledge on internet by agricultural extension experts through successive five-year plans. In other words, to facilitate information delivery and to bridge up the gap existing between different actors of agricultural development system, the government is to make more attempts to build up the necessary infrastructure required for integrating ICTs into agricultural sector, out of which professional development of agricultural staff is highly important. In addition, equipment and training are being provided to the agents in the use of internet. Agricultural extension workers are being equipped with PDA, lap top computers and broad band internet connections. Attending the agents in English courses is encouraged as a lot of resources on the net are in English and ICDL skills. In addition, further attempts are made to infuse ICTs into agricultural extension system in Iran as follows:

- Increasing efforts are made to incorporate ICTs in all endeavors related to agricultural development.
- The organizations and departments concerned with agricultural development have realized the potential of ICTs for the speedy dissemination of information to farmers.
- Government at national and provincial levels is to reorient agricultural policies so that a fully-fledged strategy is formed to harness ICT's potential for assisting overall agricultural development.
- When recruiting personnel for ICT initiatives, graduates - preferably in agriculture are being selected to work as the interface between ICTs and farmers, since this ensures effective delivery of services.
- Information technology is also introduced into agricultural technology development in order to improve efficiency of agricultural R&D.
- Some databases based on the research results of the research institutes for efficient agricultural technology dissemination are being established.
- Different training courses for extension agents on ICT-related subjects are vastly conducted throughout the country.
- Every extension centers are being equipped with an established minimum standard for ICTs capability to be able to conduct ICT-based courses and improve communication skills of extension staff.

References

Annor-Frempong, F., Kwarteng, J., Agunga, R., and Zinnah, M.M., 2006. Challenges and prospects of infusing information communication technologies (ICTs) in extension for agricultural and rural development in Ghana. *Proceedings of the AIAEE 22nd Annual Conference,* Florida, USA.

Anonymous, 2003. Information technology in extension: Goals and strategies for the future. *Extension Information Technology Texas Cooperative Extension.*

Badripour, H., 2004. Country Pasture, Forage Resources Profiles, Islamic Republic of Iran. *http://www.fao.org/ag/agp/agpc/doc/counprof/Iran/Iran.htm* (Accessed on 12th March, 2004).

Barani, H., and Qosdi Rasi, H., 2003. Hopes and fears in developing ICTs for rural areas. *Proceedings of the National Forum on ICTs application in Rural Areas,* Elmo Sanat University of Technology, Electronic Research Center, Iran.

CET., 2003. Agriculture sector-Market brief-Iran. Canadian Embassy in Tehran. *http://atn-raiae.agr.ca/africa/3888_e.htm* (Accessed on 4th April, 2003).

CET., 2004. Agri-Food country profile, Statistical Overview-Iran. Canadian Embassy in Tehran. *http://atn-raiae.agr.ca/africa/3860_e.htm* (Accessed on 1st August, 2004).

CAPES., 1996. Food Security in the Islamic Republic of Iran: A Country Report to the World Summit. *Technical Committee of World Food Summit National Secretariat*, Center for Agricultural Planning and Economic Studies (CAPES), Tehran.

Chapman, R., and Slaymaker, T., 2002. ICTs and Rural Development: Review of the Literature, Current Interventions and Opportunities for Action. *Working Paper 192*, Overseas Development Institute, UK.

Earthtrends., 2003. Agriculture and Food-Islamic Republic of Iran, Country Profile. *http://earthtrends.wri.org* (Accessed on 12th December, 2004).

ECOP, 1980. Computers in Agriculture. Extension Committee on Organization and Policy (ECOP), Washington, DC: United States Department of Agriculture.

Falaki, M., 2005. A study to recognize main areas, driving forces and obstacles of applying information technology (IT) in Agricultural extension system. *M.Sc. (Agri.) thesis (Unpublished)*, Department of Agricultural Extension and Education, University of Tehran, Iran.

Falaki, M., Shabanali Fami, H., Iravani, H., and Movahed Mohammadi, H., 2008. Attitude of Extension Experts towards Application of Information Technology in Agricultural Extension System. *Journal of Science and Technology of Agriculture & Natural Resources*, 12(43) (A).

Fletcher, W.E., and Deeds, J.P., 1991. Computer anxiety and other factors preventing computer use among United States Secondary Agricultural educators. *Journal of Agricultural Education and Extension*, 35(2). Pp: 16-21.

Gell, E. Schiefer, G. Goschiefer, Parker, C. and Rosskopf, K., 1999. Why is the adoption rate by farmers so slow? Bonn, EFITA, *http://www.efita.dk/papers/ep4/Efitapaper-4.asp* (Accessed on 9th June, 2006).

GoI., 2001. *Report of Prime Minister's Task Force on India as Knowledge Superpower.* Planning Commission, Government of India, New Delhi.

Haddad,W., and Draxler, A., 2002. *Technology for Education*, UNESCO, Rome.

Haji Ketabi, A., and Noor Mohammadi, M., 2001. Role of It as instrument, component or context of development in the areas of employing university's graduates, Tehran. *Jihade-Daneshgahi*, Organization for Employment Collaboration among Graduates, Pp: 639-646.

Hautzer, H.J., Helbig, R. and Schiefer, G.W., 1997. Computer based information and report system for the support of agricultural extension presentation of a prototype. *Proceedings of the First European Conference for Information Technology in Agriculture*, Copenhagen, Denmark.

Helvetas, 2003. Helvetas and information and communication technologies (ICTs) for participation. *Swiss Yearbook of Development Policy,* Swiss Association for International Cooperation, Geneva.

ICCIM (Iran Chamber of Commerce, Industries and Mines)., 2007. *http://www.iccim.org/english/Magazine/iran_commerce/no23_2000/20.htm* (Accessed on 12th March, 2007).

Iravani, H, 1992, (Ed.), *History of Agricultural Extension Education in Iran,* Mashhad Publishing, Mashhad, Iran.

Jalali Shahkoohi, A.A., 2003. Role and application of ICTs in the villages. *Proceedings of First Symposium on ICT application in Agricultural Extension and Development,* Tehran, Iran.

Lashgarara, F., Farjjolah Hosseini, S.J., and Mirdamadi. S.M., 2008. Does Information and Communication Technology Have a Role in Improving of Sustainable Development in Iran? *Proceedings of the world Academy of science, engineering and technology,* 34, October 2008.

Maningas, R.V., and Mancebo, S.T., 2004. Utilization of IT-based Services and Communication technology Media by End-users of Agricultural Research and Extension Networks in the Philippines. *Proceedings of the AFITA/WCCA Joint Congress on IT in Agriculture.*

Martin, B.L., Stewart, D.L., and Hillison, J., 2001. Computer anxiety level of Virginia extension personnel, *Journal of Extension,* 39(1).

Maru, A., 2003. Information and Communications Technology Use in agricultural extension in India, CTA Observatory, ICT and Agricultural Extension. ISNAR (International Service for National Agricultural Research), *www.isnar.cgiar.org* (Accessed on 9th August, 2004).

Meera, S. N., Jhamtani, A., and Rao, D.U.M., 2004. Information and communication technology in agricultural development: a comparative analysis of three projects from India, AGREN, *Network Paper,* No. 135. ODI, UK.

Ministry of Agriculture, 2005. Strategic Document of Extension and Farming System in the Forth development Plan, *Committee of Programs' Coordination and Integration,* Deputy of Extension and Farming System, Ministry of Agriculture (Jihade-Keshavarsi), Tehran, Iran.

Mobasser, S., 2003. Focus on seed programs; the second industry in Iran (17), *Seed Control and Certification Department,* SPII, Karaj, Iran.

Mokhtarnia, M., Rezvanfar, A., Shaabanali Fami, H., and Jafarnezhad, A., 2008. An Analysis Of The Relationship Between Attitudes And Skills In Application Of Information Technology And Knowledge Management of Agricultural Extension Experts, *Agricultural Economics,* 1(2008). Pp: 215-225.

Movahed Mohammadi, H., 2002. Role of internet information network and web in the educational and research activities of postgraduate students of selected agricultural colleges in Iran, *Ph.D.(Agri.) thesis (Unpublished)*, Department of Agricultural Extension and Education, University of Tehran, Iran.

NASS., 2001. Farm Computer Usage and Ownership, *Washington, DC: Agricultural Statistics Board, US Department of Agriculture.* National Agricultural Statistics Service.

Omidi Najafabadi, M., Farajollah Hosseini, S.J., Mirdamadi, M., and Moghadasi, R., 2008. Designing an Efficient Information and Communication Technology (ICT) System to Train Private Agricultural Insurance Brokers in Iran, *Australian Journal of Basic and Applied Sciences*, 2(4). Pp: 1041-1051.

Ommani, A. R., and Chizari, M., 2008. Information Dissemination System (Ids) Based E-Learning In Agricultural Of Iran (Perception Of Iranian Extension Agents)", *Proceedings of the World Academy of Science, Engineering And Technology*, Volume 28, April 2008.

Papzan, A.H., Zarafshani, K., and Rasekhi, B., 2008. Outlook of the Expansion of ICT in Rural Areas of Kermanshah County in Iran, *Roosta Va Towse' E*, 10(4). Pp: 199-229.

Raab, R.T., 2005. Present status of ICT application in agriculture in Asia and the Pacific", *Proceedings of Seminar on Information Productivity and Competitiveness*, Yogyakarata, Indonesia.

Rasouli Azar, S., 2004. A a study on the extent of applying IT in the Iranian agricultural system, *M.Sc. (Agri.) thesis (Unpublished)*, Faculty of Agriculture, University of Tarbiat Modares, Tehran, Iran.

Reddy, P.K., and Ankaiah, R., 2005. A framework of information technology-based agriculture information dissemination system to improve crop productivity, *Current Science*, 88(12).

Reddy, R., 2001. Computing - The next 10 years. Talk presented at Georgia Tech 10th Anniversary Convocation. *http://www.rr.cs.edu/rrlong.html* (Accessed on 6th April, 2004).

Rezaee, R., 2004. Investigating factors influencing ICTs use by staff extension personnel at the headquarter of extension in the Ministry of Agriculture. *Research report (Unpublished)*, Department of Agricultural Extension and Education, University of Tehran, Iran.

Richardson, D., 2003. Agricultural Extension Transforming ICTs? Championing Universal Access. *Background paper for CTA's ICT Observatory 2003.* TeleCommons Development Group, *www.telecommons.com* (Accessed on 9th September, 2007).

Sadri, M.R., 2005. Rural ICT Strategic Plan. Vice Minister for ICT, Ministry of Information and Communication Technology, Iran *http://www.scict.ir/Portal/* (Accessed on 7th September, 2007).

Sepehry, M. R., 2001. Role of IT in human resources development and increasing job productivity. *Proceedings of the National Forum on the Role of ICTs in Employment,* Tehran, Jihade-Daneshgahi, Organization for Employment Collaboration among Gatuates, Iran.

Shabanali Fami, H., 2000. Participation of rural women in mixed farming in Iran. *Ph.D. (Agri.) thesis (Unpublished),* University of Agricultural Sciences, Bangalore, India.

Sharifzadeh, A., Abdollahzadeh, G.H., and Sharifi, M., 2008. Integrating information and communication technologies in the Iranian agricultural research system, *International Journal of Education and Development using ICT,* 4(4).

Sharma, V.P., 2002. Cyber extension: Changing the face of India agriculture. *http://www.manage.gov* (Accessed on 9th May, 2004).

Smith, M.L., and Kortrlik, J. W., 1990. Computer anxiety among extension agents. *Journal of Extension,* 28 (4). Pp: 14-16.

Swaminathan, M.S., 1993. (Ed.), *Information Technology: Reaching the Unreached.* Chennai, Macmillan India.

Wiederhold, G., 2002. Information systems that also predict into future. *Lecture Notes in Computer Science,* Springer-Verlag, pp. 1–14.

Wikipedia., 2007. Iran. Encyclopedia of the Wikipedia, *www.wikipedia.org* (Accessed on 7th July, 2007).

Zazueta, F.S., and Vergot, P.V., 2003. Use of handheld computers in agricultural extension programs. *Proceedings of the EFITA Conference,* Debrecen, Hungary.

Authors Information

Dr Hossein Shabanali Fami

Associate Professor of Agricultural Extension
Department of Agricultural Extension and Education
Faculty of Agricultural Economics and Development
University of Tehran
Karaj, Postal code: 31587-11167, P.O. Box: 4111
IRAN.
Phone No. (O) ++98-261-2238293
(M) ++98-912-3006115
e-mail: hfami@ut.ac.ir; hfami2001@yahoo.com
hfami@hotmail.com

Dr. Hossein Shabanali Fami is an Associate Professor of Agricultural Extension, holds Ph.D degree in Agricultural Extension from the University of Agricultural Sciences, Bangalore, India. He is a faculty member of the Faculty of Agricultural Economics and Development of the University of Tehran. He has published over 100 papers and some books. His main research areas are as follows: Role of ICTs in agricultural and rural development, Designing and management of farming systems (land and water utilization systems), Gender studies in agriculture and rural development and participatory approaches to agricultural and rural development. He has more than 20 years of experience in agricultural development and agricultural extension services in Iran.

Mrs. Malihe Falaki

Ph.D. Scholar

Department of Agricultural Extension and Education

Faculty of Agricultural Economics and Development

University of Tehran

Karaj, Postal code: 31587-11167, P.O. Box: 4111

IRAN.

Phone No. (O): ++98-261-2238293

(M): ++98-915-5612184

e-mail: Malihe_falaki@yahoo.com

mfalaki@ut.ac.ir

Mrs. Malihe Falaki is a Ph.D. scholar of Agricultural education. She holds M.Sc. in Agricultural extension from University of Tehran, Iran. Her current research interests include the following main areas, ICT application in agricultural extension system, designing knowledge management system (KMS) in agriculture, agricultural entrepreneurship and sustainable agriculture. For the past eight years, she has published some papers.

Mr. Javad Ghasemi

M.Sc. (Agricultural Education) Graduate

Department of Agricultural Extension and Education

Faculty of Agricultural Economics and Development

University of Tehran

Karaj, Postal code: 31587-11167, P.O. Box: 4111

IRAN.

Phone No. (O) ++98-261-2238293

(M) ++98-919-2934807

e-mails: javadghasemi710@gmail.com
javad_ghasemi710@yahoo.com

Mr. Javad Ghasemi is a M.Sc. graduate of agricultural education, holds M.Sc. degree from the University of Tehran, Iran. He has published over 10 papers and one book. He has conducted some research projects in different areas related to agricultural education and development. His main research areas are as follows: Agricultural education, e-learning in agriculture and farming system management.

❏❏❏

Chapter 6

IRELAND

Padraig Wims

1. INTRODUCTION

There is increasing recognition that information and knowledge are now at the heart of socio-economic development. Information and Communication Technology (ICT) and broadband are enabling tools and infrastructures for accessing, developing, using and sharing information and knowledge. Accordingly, Irish Government policy has focused on promoting the adoption and use of ICT in the home, in education and in the workplace.

There have been significant increases in levels of engagement with ICT in Ireland in recent years. Almost two-thirds of all adults now have access to the internet, compared with only 1 in 20 in 1996, while four out of every five adults own a mobile phone. This chapter critiques the level of adoption of ICTs among the farming community and compares this with levels of adoption by the general population. It analyses the use of ICTs by the government and by public and private extension agencies in Ireland and identifies constraints to their adoption.

The Ministry of Agriculture has been among the most enthusiastic supporters of the Government of Ireland's policy by promoting the adoption and use of ICTs among its clients. It sought to facilitate easier transmission of information to clients, to minimise the collection of new data by using existing electronic information and to integrate with clients' own systems. These projects are described in this chapter.

The use of ICTs by the public extension system is also critiqued; it has been innovative in using ICTs to disseminate information to its clients and the use of mobile telephony has been particularly effective. Internet based services to deliver online training and advisory information have also been developed and are critiqued.

Ireland also enjoys agricultural extension type services provided by the private sector such as agribusiness, particularly agricultural input suppliers and agricultural commodity processing companies. These have been innovative in the use of ICTs to facilitate dissemination of information among their clients. A novel extranet service developed by a large agribusiness is evaluated in this chapter; this service allows farmer clients to access all information relevant to their account with this agribusiness through a secure website. The lessons learned from this evaluation are also presented.

2. AGRICULTURE IN THE IRISH ECONOMY

Ireland is a small, trade-dependent economy in the extreme West of the European Union. It experienced rapid economic expansion from 1995 to 2007, when annual GDP growth averaged 6 per cent, but economic activity dropped sharply in 2008 and Ireland entered into a recession for the first time in more than a decade with the onset of the world financial crisis and subsequent severe slowdown in property and construction markets. Although the export sector, dominated by foreign multinationals, remains a key component of Ireland's economy, construction most recently fuelled economic growth along with strong consumer spending and business investment. Property prices rose more rapidly in Ireland in the decade up to 2006 than in any other developed world economy. Per capita GDP also surged during Ireland's high-growth years. Agriculture, once the most important sector, is now dwarfed by industry and services. Nonetheless, agriculture remains very important to the Irish economy; the agri-food sector is one of Ireland's most important indigenous industries and in 2007 accounted for almost 7 per cent of GDP, 8 per cent of employment, and 10 per cent of exports (Department of Agriculture, Fisheries and Food, 2008a). While agriculture is of major importance to economic welfare and development in Ireland it is particularly central to the economic and social vitality of rural communities. Employing about 155,000 people (out of a labour force of approximately two million people), it accounts for over half of the country's indigenous exports and the industry as a whole accounts for at least 35 per cent of the value added in Irish manufacturing. The dispersed nature and composition of the agri-food sector ensures that it continues to play an integral part in the viability of rural and coastal areas.

The land area of Ireland is 6.9 million hectares of which 4.3 million hectares (62.3 per cent) is used for agriculture and a further 724,000 hectares (10.5 per cent) for forestry (Department of Agriculture, Fisheries and Food, 2008b). About 80 per cent of the agricultural area in Ireland is devoted to grassland (3.4 million hectares), 11 per cent to rough grazing (mainly hill and mountain pastures, 0.5 million hectares) and 9 per cent to crop production (0.4 million hectares) (ibid). Beef and milk production currently account for 56 per cent of agricultural output at producer prices (CSO, 2008), both of these are grassland based farm enterprises. Data from CSO's Farm Structures Survey show that in 2005 (the most recent year for which data are available) there were 132,700 individual farm holdings, with an average farm size of 31.8 hectares, but 60 per cent of all farm holdings were under 30 hectares in size (Department of Agriculture, Fisheries and Food, 2008a).

3. AGRICULTURAL EXTENSION IN IRELAND

Ireland has a long history of agricultural extension. After the Great Famine of 1845-48 (approximately one million of the population died from starvation during this time, caused by failure of the potato crop due to Late Blight) the

British government established instructors that travelled Ireland encouraging small farmers to diversify (Jones, 1979). These instructors subsequently became known as "Itinerant Instructors" *i.e.* travelling from place to place imparting knowledge and information to farmers. Agricultural extension in Ireland was established more formally with the establishment, in 1900, of the Department of Agriculture and Technical Instruction. Agricultural advisors were established in each county through County Committees of Agriculture. Under the County Committees, the system consisted mainly of general advisors under the direction of a Chief Agricultural Officer. Advisors worked with farmers, largely transferring knowledge on agricultural matters. They were also involved in education through "Winter Farm Schools", although the formal agricultural education system was through the agricultural colleges, which were under the Department of Agriculture or were private. With the establishment of ACOT (An Chomhairle Oiliuna Talmhaiochta, Irish words meaning "Council for Development in Agriculture") in 1980 agricultural advisory services and education were brought together and the triangle was completed with the establishment of Teagasc (Irish word denoting "Agriculture and Food Development Authority") in 1988, when agricultural research, training and advice were brought together within one national organisation (Phelan, 2009).

Currently Teagasc offers four different services to farmers:

1. Business and Technology Programme: This service is geared towards the development of farm enterprises and operates on a 50 per cent cost recovery basis (National exchequer pays 50 per cent and farmers pay the remaining 50 per cent). The programme covers the main enterprises of dairy, drystock and tillage.

2. Good Farm Practice Programme: This programme covers REPS (Rural Environmental Protection Scheme, an agri-environmental programme), fertiliser and nutrient advice, water quality, emissions reduction and countryside management. REPS is full cost recovery while other services are based on a 50 per cent recovery rate.

3. Rural Development Programme: This service looks at optimising all farm profits and other options for the future. Advisers deal with succession, inheritance, and lifestyle issues and draw up a "Way Forward Action Plan" for the farm household. This service is fully funded from the National Exchequer.

4. Adult Training: This programme deals mainly with education and is 62 per cent funded from the National exchequer (Teagasc, 2008a).

Traditionally the extension system was public; however, with tightening budgets in the 1980s it engaged in cost recovery. This was helped significantly by the fact that farming became increasingly regulated and bureaucratic with numbers of different schemes providing income supports for farm households. Teagasc had 45,000 farmer clients in 2007 (Boyle, 2008). However, 40,000 farmers (not necessarily clients) were supported with their area aid applications while an additional 15,500 farmers were assisted with other schemes (such as Farm Waste Management Scheme and the Farm Investment Scheme). In addition the REPS planning service supported 24,300 farmers to avail of agri-environmental payments.

Teagasc employs over 1,500 staff at over 100 locations throughout Ireland. Research services are provided by 200 research scientists and 300 technicians at nine dedicated centres. There are 550 advisors and regional specialists located at regional, county and local offices. There are eight colleges and local training/research centres staffed by college lecturers, education officers and technicians.

Phelan (2009) maintained that significant changes have occurred in agricultural extension in Ireland. In the 1970s and 1980s extension workers were "Whole Farm Advisors" dealing with not only farm activities but also with livelihood issues that impacted on the farm household. Even sensitive topics like succession and inheritance were discussed. With cuts in budgets and the advent of cost recovery extension workers adopted the role of the "expert", providing high level technical information to farmers or through helping farmers to comply with and maximise their receipts from the many government schemes. This in the context of multifunctional agricultural households left information gaps, which began to be filled by the private sector such as agribusiness (particularly agricultural input suppliers and agricultural commodity processing companies) and indeed by other rural development groups. As a consequence, Ireland now enjoys agricultural extension type services provided by both the public sector and the private sector.

4. INFORMATION AND COMMUNICATION TECHNOLOGY POLICY IN IRELAND

There is increasing recognition that information and knowledge are now at the very heart of socio-economic development (O'Donnell *et al.*, 2003). ICT and broadband are enabling tools and infrastructures for accessing, developing, using and sharing information and knowledge (ibid). A study conducted by Chambers Ireland (2007), involving Irish businesses, showed that Information and Communication Technology (ICT) tools and e-business solutions had become crucial for increasing efficiencies across all sectors of business in Ireland.

Both the EU and Irish Government policy have focused on promoting adoption and use of ICT in the home, in education and in the workplace. In March 2000, in Lisbon, the European Union (EU) set itself the ambitious target of becoming the world's *"most competitive and dynamic knowledge based economy"* within ten years (European Commission, 2007). It recognised that attaining this goal depended on making the best possible use of ICT. The Lisbon Strategy placed greater emphasis on the knowledge-based society within existing policy processes and launched the eEurope 2002 Action Plan as a roadmap to modernise the European economy (ibid). The i2010 strategy was launched by the European Commission in June 2005 and will be in place until 2010. It builds on the eEurope initiative, which came to an end in 2005 (ibid).

Irish Government policy on ICT was built around two Action Plans. In January 1999, the Government launched the first plan for implementing the information society in Ireland. This set out a series of actions and initiatives for the period 1999-2001, including development of telecommunications infrastructure which included national availability of broadband, new legislation and other measures to enable both public and private businesses to operate on-line and a range of e-Government initiatives and projects (Government of Ireland, 1999).

A study by O'Donnell *et al.,* (2003) showed there had been significant increases in levels of engagement with ICT in recent years. Nearly 1 in every 2 adults now had access to the internet, compared with only 1 in 20 in 1996. Four out of every five adults owned a mobile phone. The dramatic adoption of ICT was further supported by the second Action Plan of Government (New Connections, 2002) which accepted that it was timely to assess how we had progressed, to review our priorities, and to put in place a new strategic framework to take the country forward.

The Government Ministry of Agriculture and Food (Department of Agriculture and Food, 1999) recognised that the country was embracing ICT at a rapid rate and this had major implications for rural areas in particular. However, it did caution that without the necessary infrastructural investment, rural areas could be disadvantaged. Since then much emphasis has been placed on engagement with business and the individual in relation to ICT. However, it was recognised from an early stage that some groups were not engaging with ICT and as a result were being excluded from the benefits accruing. These included agricultural workers (O'Donnell *et al.,* 2003). The application of e-Government to agriculture had the potential to totally transform the way farmers engage with the Department of Agriculture. The Comptroller and Auditor General (2008) concluded that:

"There has been mixed progress in e-Government projects that had a high transactional objective. There were a number of outstandingly successful projects including motor tax, ROS, On-line Area Aid, and animal disease eradication." (p.38)

However, the Department of Agriculture was held up as an example of how successful e-Government could be. Today Department of Agriculture and Food, together with Teagasc have a number of excellent projects operating under the e-Government initiative. These will be presented and discussed later in this chapter. These initiatives indicate a very definite commitment on behalf of the Department of Agriculture and Teagasc to ICT and e-Government initiatives.

5. ADOPTION OF ICTs BY THE GENERAL POPULATION AND BY THE FARMING COMMUNITY IN IRELAND

The adoption of ICT by the Irish population is a multifaceted area made up of a number of different subsections which will be presented and discussed:

- The level of computer availability in Irish households;
- The level of access to the internet;
- The level of broadband availability; and
- Mobile phone penetration.

5.1. Level of Computer Availability

When computer access in households was measured during the last three months of 2008 it was found that 67 per cent of Irish households had either a personal computer or a laptop (Millward Brown IMS, 2009). However, further analysis revealed that only 58 per cent of rural households had either a personal computer or a laptop indicating that rural Ireland is lagging behind in computer use (ibid).

5.2. Level of Access to the Internet

Internet use among the Irish population is increasing. At the end of 2007, 54 per cent of the population used the internet on a regular basis either from home or place of work (Millward Brown IMS, 2008a), while this increased significantly to 64 per cent by the end of 2008 (Millward Brown IMS, 2009).

Internet use among businesses was considerably higher; research by Millward Brown IMS (2008b) indicated that internet access among Small and Medium size Enterprises (SMEs, defined as companies with less than 100 employees) has remained steady at 88 per cent, while corporate access (*i.e.* companies employing 100 or more employees) was almost universal. Millward Brown IMS went on to report that smaller companies (*i.e.* employing less than 10 staff) were least likely to have internet access with 82 per cent connected to the internet. These findings correlate well with those of Chambers Ireland (2007) who reported that internet use among businesses was at 80 per cent. However, farmers were among the lowest internet users at 34 per cent (ibid.). This figure would indicate, once again, that rural areas continue to trail urban centres when embracing ICT.

Comreg (2008) reported that there were 1.38 million active internet subscriptions at the end of 2008. This represented a 3.2 per cent growth on the previous three months and a very significant 17.9 per cent increase on the same period the previous year. While the total number of subscribers is increasing rapidly the data in Table 1 below show that dial-up internet access is declining rapidly while broadband is increasing.

Table 1 : Total number of active Internet subscribers in Ireland

Subscription Type	Quarter (Q) 4, 2008 Subscriptions	Quarterly Growth Q 2 2008 - Q 3 2008	Year-on-Year Growth Q 3 2007 - Q 3 2008
Metered Narrowband	224,722	-9.4 per cent	-31.1 per cent
Flat Rate Narrowband	36,856	-10.5 per cent	-34.9 per cent
ADSL Broadband	633,536	+3.6 per cent	+24.9 per cent
Other Broadband	491,544	+10.9 per cent	+71.5 per cent
Total Internet Subscriptions	**1,386,658**	**+3 per cent**	**+17.9 per cent**

(Source: Comreg, 2008)

According to Millward Brown IMS (2008a) the main uses for the internet included sending and receiving e-mail, doing research and getting information, online banking, shopping and downloading music while the younger age groups used networking sites and played online games. It was also noted that online banking was becoming popular among adults over 25 which would suggest that this group is comfortable with using ICT and do not have any concerns with internet security.

5.3. Level of Broadband Availability

Broadband subscriptions are increasing at a rapid rate. Comreg (2008) reported that by the end of September, 2008, there were 1,125,080 broadband subscriptions in Ireland. This represented an increase of 6.7 per cent in the number of subscriptions for this quarter. Fixed Wireless Access subscriptions continued to decline in the third quarter of 2008 by 2.1 per cent while mobile broadband was the fastest growing platform, growing by 20.9 per cent during this period. Between the third quarters of 2007 and 2008, mobile broadband subscriptions increased by a very significant 204 per cent. Comreg (2008) calculated that the broadband per capita penetration rate in the third quarter of 2008 was 25.9 per cent. However, without mobile broadband the penetration rate was only 19.7 per cent. In contrast, Ireland's household broadband penetration rate was 58.9 per cent as of June 2008, compared to 46.8 per cent the previous year.

Comreg (ibid) compared broadband penetration rates in Ireland with other EU countries. It reported that Ireland ranked 12th highest penetration among the EU 27 countries in terms of household penetration rate; Denmark and the Netherlands had the highest household broadband penetration rates at 86 per cent and 83 per cent respectively. This situation was confirmed when fixed broadband per capita penetration rates for EU countries were compared. The EU Commission calculated Ireland's broadband penetration at 19.5 per cent in July 2008. This places Ireland below the EU average of 21.7 per cent and in 14th place in the EU 27.

Amarach Consulting (2007) surveyed subscribers' attitude to broadband. This study found that internet users outside of Dublin had far more problems with the poor service they were getting relative to Dublin both in terms of availability and competing choice. Meanwhile a national survey conducted by the Irish Farmers' Association (Smith, 2008) calculated that 75 per cent of rural households did not have access to broadband in 2007. It was also stated that infrastructure to provide broadband to rural areas was totally inadequate (*ibid*).

5.4. Mobile Phone Penetration

According to Comreg (2008), at the end of September 2008 there were 5.27 million mobile phone subscriptions (2G and 3G) in Ireland. This gives a penetration of mobile phones in Ireland of 121.5 per cent (calculated as the number of active SIM cards per 100 population). Mobile phone ownership was almost universal among people under 50 years while on the other hand farmers were among the least likely, at 82 per cent, to own a mobile phone (Millward Brown IMS, 2008a).

Comreg (2008) compared the Irish mobile penetration rate with that of EU countries and reported that the Irish penetration is marginally behind that of the EU average of 122.1 per cent.

Ireland has a reputation for unusually high usage of SMS (Short Message Service, or text messaging). Analysis by Comreg (2008) revealed that the total number of SMS messages sent by mobile users in Ireland totalled 2.52 billion from July to September, 2008, representing an average of 160 SMS messages per subscriber per month during this period. Among EU countries, Ireland had the second highest number of SMS messages sent per subscription, narrowly surpassed by Denmark.

5.5. Market Segmentation Analysis

It is possible to categorise the Irish consumer market based on their usage of and attitudes towards a suite of ICTs. Millward Brown IMS (2008c) surveyed by personal interview 1007 adults between 9th April and 16th May, 2008 and questioned them on their general attitudes to new technologies, mobile phone ownership, fixed line ownership, home PC ownership, internet usage and home internet access. Four distinct segments emerged based on their usage and attitudes: "Technology Enthusiasts", "Technology Savvy", "Mobile Converts" and "Technology Rejecters".

The largest group was identified as Technology Enthusiasts, these comprised 33 per cent of the sample and were characterised as being enthusiastic adopters of new technologies because they make life easier. Millward Brown IMS (2008c) found that the Technology Enthusiasts had a female bias (59 per cent), and the majority were employed (68 per cent). High home and mobile phone usage was typical of this group, as well as the highest incidence of home internet connections. A large proportion of this group was over the age of 35 years and had an interest in and appreciation of technology.

The second largest group, the Technology Savvy, comprised 30 per cent of the sample and were identified as wanting to be the first to adopt the new technologies and other people looking to them for advice about new technologies. These were predominantly a youth segment with 69 per cent aged less than 35 years and 25 per cent currently in 3rd or 2nd level education. The Technology Savvy segment was very comfortable with new technologies and they were likely to be important influencers of others' choices. There was quite a strong male bias, as well as a bias in favour of the higher socio-economic groupings.

The third category, Mobile Converts, comprised 28 per cent of the sample and were characterised as recognising that they needed to learn more about computers and new technologies. The main distinguishing feature of this segment was the high level of mobile phone usage compared to other technologies.

The fourth category, the Technology Rejecters, were categorised as being those who did not understand computers and new technologies. The majority of this group were over the age of 60 years and a high proportion (36 per cent) was retired. They had near universal home phone ownership (97 per cent) and did not tend to use mobile phones. Many admitted to not understanding new technologies and showed little interest in learning about them. Their use of the internet was very low.

6. ADOPTION OF ICTs BY THE FARMING COMMUNITY IN IRELAND

There has been a long-standing conviction in Ireland that farmers and farm families are among the slowest adopters of ICT. Morrow *et al.,* (2003) for instance reported that Irish farm families were falling behind other sectors of Irish Society with regard to home internet access and cited evidence which showed that 21 per cent of farmers had internet in their homes compared with the national average of 35 per cent at that time. They also commented that this compared unfavourably with the farming communities of other developed economies. In fact, most other countries of similar development status to Ireland conduct formal research in the area of ICT adoption by different socio-economic groups and occupations annually but this is not the situation in Ireland. This has made it difficult to state with confidence how trends have been moving and to accurately compare ICT adoption among Irish farmers with those in other countries. In addition to the issue of ICT adoption, there has been a dearth of data on the actual purpose to which PCs and the internet have been used by Irish farmers. In an effort to correct this information deficit, the author conducted research to measure the level of computer and internet usage among Irish farmers, to analyse the purpose to which these were put and to examine the factors which influenced the adoption of ICTs. The results were compared with the level of adoption among the general population in Ireland and with farm families in other comparable countries (Wims, 2007). Structured interview schedules were completed with farmer respondents who were selected at random from the Teagasc client lists. Questionnaires were personally administered face-to-face in the respondents' own homes or on their farm holdings during 2004. Analysis revealed that these respondents tended to be somewhat younger, with larger families, owning slightly larger areas of land, and engaging more in dairy enterprises (more profitable) than the national average. Nonetheless, it was considered that these respondents represented an excellent cross section of Irish farmers for whom PC and internet access and use should be a reasonable expectation.

In relation to access to a PC, it was found that 56 per cent of respondents owned a home PC, 11 per cent had access to a PC from some other source while 33 per cent did not have access to a PC. In terms of internet access, 48 per cent of respondents were found to have internet connections in their own homes while a further 6 per cent had internet access from elsewhere, thus 46 per cent of respondents did not have any internet access. Of the respondents with PC access, some 64 per cent indicated that they used the PC for farming matters while the remaining 36 per cent said they did not. Only 15 per cent of all respondents (27 per cent of those with a home PC) had a farm management software package on their home PC. Of the respondents who owned or had access to a PC, 34 per cent reported that they were the only members of their households who could operate the PC; in fact, the respondent was among those with the ability to operate the PC in only 69 per cent of cases. Respondents with access to the internet were asked if they used the internet for farming purposes; 61 per cent indicated that they did while the remaining 39 per cent reported that they did not use the internet for farming purposes. When probed about the purpose of internet access, a variety of responses were received; however the most common responses were general information retrieval, communicating with friends *via* e-mail and entertainment (music, games, movies). Significantly, any farm related business was reported by only a minority of those who accessed the internet. When the levels of ICT adoption by farmers was compared with that of the Irish population, it was found that 51 per cent of the Irish population owned a home PC and that 37 per cent had a home internet connection in 2004, the period corresponding to conducting the above research. Thus, the findings from the present study indicate that a slightly higher proportion of farmers than the general population had home PCs and internet connections. When international comparisons were made it was found that the Scandinavian countries were forging ahead when it came to adoption of ICT, with Finland, Norway and Sweden having 76 per cent, 75 per cent, and 74 per cent of farmers respectively having a home computer (Ross and Waksman, 2004). They were also at the top of the leader board with regard to internet access with all having in excess of 60 per cent of their farmers with internet access. Other countries with high levels of adoption were Switzerland, New Zealand, UK and USA. From the figures of ICT adoption established in this study, it emerged that Irish farmers were approaching a level of adoption that was similar to the UK, USA and New Zealand. However, the level of adoption of ICT by farming communities in the Scandinavian countries was quite a bit ahead of the figures for Irish farmers.

When the personal and farm situational factors associated with adoption of ICTs were examined the results were most revealing. A curvilinear relationship was found when examining the association between levels of ICT adoption and respondents' ages; 52.2 per cent of respondents under 35 years of age owned a home PC, while the corresponding figure for those between 35 and 50 years of age was 64.6 per cent and for those over 50 years it was 44.8 per cent. A similar trend appeared when examining the relationship between age and having a home internet connection: 43.5 per cent of respondents under 35 years of age were connected while 54.2 per cent of those between 35 and 50 years of age and 41.4 per cent of those over the age of 50 years had a home internet connection.

When the relationship between marital status and ICT adoption was examined it was found that married respondents were statistically more likely to own a home PC (chi sq = 12.54, DF=2, Sig = 0.002) and to have a home internet connection (chi sq = 12.12, DF=2, Sig = 0.002) than other (mainly single) respondents. An explanation for this trend becomes apparent when examining the relationship between ICT adoption and the presence of dependents - the highest level of PC ownership, at 75.9 per cent, was among respondents with dependent children, compared with 34.1 per cent of respondents who did not have dependents and 20 per cent of those with adult dependents; this relationship was statistically significant (chi sq = 19.278, DF=2, Sig = 0.0001). A similar association was found when analysing the relationship between having a home internet connection and dependent children. Further analysis revealed that of those respondents with dependent children, 76 per cent had a home PC and the age of their eldest child was 16.5 years while the age of the eldest child of those without a home PC was significantly younger at 11 years (t=2.86, DF=52, Sig=0.006). A similar relationship was found for home internet connections; dependent children of respondents with home internet were significantly older than those who did not have home internet (16.5 Vs 12.6 years, t=2.10, DF=52, Sig=0.04).

Some 83.3 per cent of respondents who had full-time off-farm work had home PCs while 57.9 per cent of those with part-time off-farm work and 50.7 per cent of those who did not have off-farm work had home PCs. Furthermore, 75 per cent of households where the respondent's spouse had a part-time job and a home PC, while 65 per cent of households where the spouse had a full-time job had a home PC and where no off-farm work was held by the spouse the household was least likely to have a home PC (50 per cent).

The relationship between farm size and ICT adoption was examined. It was found that respondents farming between 40 and 80 ha (*i.e.* medium-size farms) were most likely to have a home PC (72.1 per cent), while 50 per cent of those who cultivated more than 80 ha (large farms) owned a home PC and respondents farming less than 40 ha (small farms) were the least likely to have a home PC at 41.9 per cent (chi sq = 8.213, DF=2, Sig = 0.016). A similar trend was found when the association between home internet connection and farm size was examined; again respondents farming between 40 and 80 ha were most likely to have home internet access at 62.8 per cent, while 50 per cent of those with more than 80 ha and only 32 per cent of those farming less than 40 ha had home internet access (chi sq = 7.899, DF=2, Sig = 0.019).

In relation to farm enterprises practiced and ICT adoption, respondents most likely to have home PCs were engaged in specialist suckling (63 per cent with PC), dairying (58 per cent) and beef production (57 per cent). The majority of respondents with mixed enterprises (either mixed grazing or mixed crop and livestock) did not have home PCs. Some 44 per cent of respondents engaged in suckling as their main farming enterprise had an alternative source of income, compared with 23 per cent of those engaged in dairying. Teagasc (2002) supported this finding in its national farm survey and reported that cattle and sheep systems had the highest incidence of the farmer and or spouse having off-farm employment while dairy farmers had the lowest. When it is recalled that respondents and or spouses with alternative income sources were more likely to have a home computer than those farming full-time, it becomes evident that respondents engaged in suckling enterprises were more likely to have a home PC because they had an alternative source of income. This is confirmed when the farm enterprise of those with home internet connections is examined; only those with specialist suckling had significant home internet connections; 52 per cent of these had home internet connections whereas, a minority of respondents with all other farm enterprises had home internet connections.

It was concluded from this research that ICT adoption by Irish farmers compares favourably with the national average and with farmers in continental European countries. However, it is insufficient to merely examine levels of adoption as this can mask data on who is using ICT. This study found that the farmers themselves were using ICT in the majority of households. However, it is alarming that in almost one-third of farming households, children and or spouses were the only family members able to use the PC. The likelihood of ICT being used for farm business purposes in these households is remote.

It is a cause of concern that less than two-thirds of the farmers with access to ICT used this for farming purposes. In fact, ICT was used in more farming households for entertainment purposes than for farm business purposes. The majority of farm families applied their PC to their farm businesses in very basic form; mainly through record keeping, herd recording and accounts functions; this was mainly achieved using the most popular computer applications of MS Word™ and MS Excel™. Only 15 per cent of all respondents had a farm management software package employing more detailed farm record keeping and analysis.

Irish farmers who are married and with dependent children, particularly older dependent children, are much more likely to have adopted ICT within their homes; it is evident that these children have greater demand for ICT for educational purposes and for recreational purposes and are a considerable influence in encouraging adoption of ICT at household level. In addition, there were trends whereby farmers and or their spouses who have off-farm employment are more likely to have adopted ICT than full-time farmers. This is likely to occur for two reasons, one being the increased level of disposable income available in these households and the other being exposure to sources of information about ICT external to their immediate social system. Farmers engaged in specialist enterprises were more likely to have adopted ICT; however, the proportions engaged in less profitable beef and suckling enterprises that have adopted ICT was surprisingly high but it is suggested that this can be attributed to the fact that these farmers or their spouses were more likely to have off-farm employment than those engaged in dairying. Finally, it emerged that those with medium-sizes farms were most likely to have adopted ICT; *a priori* expectations were that those farming larger areas would have greater requirements for ICT within their farming systems, it was discovered that this category were least likely to have off-farm employment, thus reducing their disposable income and their exposure to ICT in the workplace.

7. DEPLOYMENT OF ICT BY IRISH MINISTRY OF AGRICULTURE

The Government of Ireland, as already highlighted, is strongly promoting the adoption and use of ICTs among its clients and the Ministry of Agriculture is among the most enthusiastic supporters of this. According to Hanly (2008), the Ministry of Agriculture's policy in relation to the use of ICT is to "... make the maximum use of electronic services, to facilitate easier transmission of information to clients, to minimise the collection of new data by using existing electronic information and to integrate with clients' own systems." The online services developed by the Ministry of Agriculture are mainly concerned with the administration of schemes and regulations. The most significant of these include a SPS (Single Payment Scheme) and AIM (Animal Identification and Movement) scheme.

The SPS is a direct subsidy payment to landowners under the Common Agricultural Policy of the European Union, agreed in 2003, and designed to remove the link that formerly existed between production of produce and subsidy received. The SPS online service for farmers allows them to submit their Single Farm Payment application electronically. In addition, they can check their application status online, view their application details, view scanned images of all correspondence with the Ministry and view colour mapping and aerial photographs of their farmland. This service was first launched in 2007 and during that year some 7,500 applications (6 per cent of all applications) were received electronically. This increased significantly to 20,000 applications (16 per cent) in 2008. According to Hanly (2008), the Ministry of Agriculture has recorded a 25 per cent reduction in errors in online applications compared with paper transactions, illustrating that the online facility simplifies the procedures for farmers to claim this annual payment.

AIM is an integrated system which traces all movements of farm animals; it is designed to reassure consumers that the food they eat can be traced back to the farm where it was sourced. The AIM (Animal Identification and Movement) online service allows registered farmers to register the births of their calves online; apply for a Movement Compliance Certificate; view, print and download their bovine herd profiles for any date; and view, print and download details of the movement of bovine animals into and out of their herds. In 2008, according to Hanly (2008), 10 per cent of all herds and 16 per cent of all calf births were registered electronically, instead of the more traditional paper-based method.

8. DEPLOYMENT OF ONLINE SERVICES BY TEAGASC, THE PUBLIC EXTENSION SERVICE

Information distribution is being supported by the use of different types of website in the case of Teagasc, the Irish public extension service. Here internet, intranet and extranet sites all play an important role in the delivery channels for improved information delivery. Teagasc uses a range of online systems for the purpose of disseminating information both within its own organisation and to its customers; the most significant of these include:

1. The Teagasc website (www.teagasc.ie) which is open to all on the internet;
2. The Teagasc intranet site (Client Information Management System, CIMS) which holds a large database of information for advisory staff and is updated on a daily basis; and
3. The client.net website, an extranet website, which is available to farmers who use the Teagasc Advisory Services and are paid-up clients of Teagasc.

The general Teagasc website (www.teagasc.ie), open to all web browsers, is a good promotional tool for the services offered by the organisation: it advertises the advisory services offered and explains the various types of research carried out. It also advertises the educational services offered, including the Teagasc e-College, which coordinates the development and delivery of online courses within Teagasc. The most significant of these courses are the Certificate in Animal and Crop Production (100 hours) and Certificate in Farm Business (80 hours). In 2007, a total of 450 students participated in these courses (Teagasc, 2008b). A "blended learning" approach is used, consisting of both traditional and online methods of instruction. Students use resource material posted to them by regular mail and attend practical training sessions at Teagasc centres for skills development. The online component is used to facilitate interactions and discussions with tutors. Students also submit assignments by e-mail and answer questions online to complete the course. The tutors are Teagasc college teaching staff, local education officers and advisory staff who have been trained in the use of this system.

The CIMS is a vital component of Teagasc's management strategy. This holds all client information such as farm details, services availed of, schemes applied for, status of contract with Teagasc, contact details and details of which Teagasc staff the client deals with. This intranet site can be accessed from any Teagasc office.

The client.net extranet service offers Teagasc clients up to date advice on topics such as beef husbandry, dairying, buildings and machinery maintenance, farm management, tillage husbandry, *etc.* Any Teagasc client can avail of this service and there is no additional charge for accessing this service. A major benefit of the extranet service to Teagasc clients is that they can avail of Teagasc's e-Profit monitor, a financial analysis programme that deals with farm finances by giving clients an understanding of the profit their farm business is generating. For instance, the e-Profit monitor assesses a dairy farm's profitability on a "cent per litre of milk produced" basis and compares input costs (fixed and variable costs) incurred with benchmark figures to enable the dairy farmer to compare the efficiency of his enterprise with other similar farms. Individual farmer reports are also generated through this service; including physical farm details such as cow numbers, milk production and feed and fertiliser used, together with financial details and milk prices attained, and compares all these figures to those of the previous year to generate reports to monitor progress. Teagasc (2008b) reported that a total of 1,200 farmers (of the 45,000 farmer clients of Teagasc) completed an analysis of their physical and financial performance during 2007. More recently, Connolly (2009) reported that by the end of February, 2009, over 800 dairy farmers (of the 22,000 approximately in Ireland) had used this eProfit Monitor programme to analyse their 2008 costs and budget for the 2009 season. These rates of uptake appear low (at 2.7 per cent and 3.6 per cent of the population, respectively) but it must be considered in the overall context that only 5 per cent of Irish farmers use farm management software to assist in the management of their farms (Department of Agriculture, Fisheries and Food, 2008c).

9. DEPLOYMENT OF MOBILE TELEPHONY IN AGRICULTURAL EXTENSION

The demographic structure of Irish farming has changed dramatically since accession to the European Union in 1973. Farming is no longer a family-centred occupation but it is now operated more as a business. One unintended consequence of this transition is the isolation experienced by members of the farming community as they manage their farm businesses. Because of this isolation, Irish farmers have enthusiastically embraced mobile telephony as a way of enhancing their communications with family, neighbours and business contacts. The Teagasc organisation has recognised that mobile telephones provide an opportunity to disseminate information rapidly to Irish farmers. Moore (2008) estimated that 90 per cent of Teagasc clients had mobile telephones. He reported that during 2007 Teagasc sent approximately 200,000 SMS messages and that this is likely to increase rapidly in the future as both farmers and extension workers experience the benefits of this medium. In terms of type of usage, the most common activity is to remind farmers of group meetings and local farm walks. SMS messages are sent the day before the event and are usually used to transmit additional information such as the exact location of the event. Extension workers typically set this up in advance so that the reminder is sent automatically the day before, night before or at the time agreed by the group. Moore (2008) also commented that an additional benefit of this medium is that once groups have been set up extension workers can easily send urgent messages to the group if there is a last minute change of venue or other detail. Teagasc disseminates SMS messages from either mobile telephones or the internet based at the extension workers' office. The cost of sending group SMS messages (c. 5-6 Euro cents per message) is significantly cheaper than post or voice messages. Moore also argued that there are sound principles supporting and underpinning the increased use of SMS messaging - receivers remember messages more accurately when they read the written message (as in SMS) compared to simply hearing the message (with voice messages).

In addition to disseminating routine messages, Teagasc extension workers are also using SMS messaging to disseminate "Just-in-Time" or time-sensitive information to farmer clients. One practical application of this is found in the horticulture industry. Horticultural extension workers use traps to identify when the second generation of Carrot Root Fly is on the move. This information is rapidly disseminated to growers so that spray applications can be based on need (identified by trapping) rather than routine spraying as was previously the norm.

SMS messaging is also increasingly used by Irish agribusiness to facilitate rapid communication with their clients. Dairy processors conduct regular tests on samples of milk supplied for processing and the majority of these companies use SMS messaging to inform their clients of the results of tests for fat content, protein content and milk hygiene as soon as these results are available. Farmers themselves comment positively on this method of information dissemination comparing it favourably with e-mail, as they constantly have their mobile telephone on their possession and this facilitates accessing messages delivered by this method, unlike e-mail which is typically not accessed as regularly.

10. DEPLOYMENT OF EXTRANET BY AGRIBUSINESS

In this section, a detailed case study of an extranet website is presented. The name of this extranet site is "AgriLink" and it was developed by Glanbia, an international cheese and nutrition company, whose headquarters are in Ireland. Glanbia was formed in 1997 from the merger of two previously long-established dairy food companies. The group is organised into three operating divisions of Agribusiness and Property, Consumer Foods and Food Ingredients and Nutritionals. Glanbia has 4,500 farmer milk suppliers and so is one of the larger milk processing companies in Ireland. Glanbia set up an extranet (www.agrilink.ie) which it named "AgriLink" in 1999 on a pilot basis and launched it among all its suppliers in 2000. It was set up as an extranet website independent of the Glanbia website in order to provide a link between the Milk Division and Agribusiness section of Glanbia and the farmer customers. It was a dedicated secure site whose main purpose was to allow farmers have access to all information relevant to their business with Glanbia. Milk suppliers have access to a range of information regarding their milk account. This includes the latest milk test results, individual milk collection details and copies of milk statements. A superlevy statement is updated after every collection and help pages are provided with advice on non-conforming test results. Trading account holders can view the current balance on their accounts. Grain suppliers can select by grain type or intake branch to confirm the latest test results, weight, price paid and value of grain supplied. Copies of grain statements are also provided. AgriLink also provides a comprehensive news section. This includes updates on the Glanbia Superlevy position, technical information, links to other agri-websites and the latest Glanbia promotions. Thus, AgriLink has an important extension element. Registering for access to the AgriLink website has always been free to customers of Glanbia. In order to access AgriLink milk suppliers require a Glanbia account number and access to the internet. This extranet site is not unique in Ireland; in fact almost all of the large dairy cooperatives have established similar websites in recent years.

The main benefits for clients of using AgriLink are:

1. 24 hour per day access to account information through AgriLink seven days a week.
2. AgriLink is updated daily, therefore the most current information regarding accounts is provided.
3. AgriLink also provides historical information since 1999; this allows the farmer to compare one month against the same month of another year.
4. Copy statements for milk, trading and grain accounts are provided.
5. Supplier's superlevy position updated after every milk collection.
6. Branch transactions updated daily after close of business.
7. Trading Statement download. This service allows a customer to select any month and a file containing a copy of the customer's trading or milk account statements will be forwarded by email to the customer. This file can then be downloaded into a farm management accounts package. The benefit is the time saved as there is no need to key-in branch dockets into the accounts package.

The home page of AgriLink is presented in Figure 1.

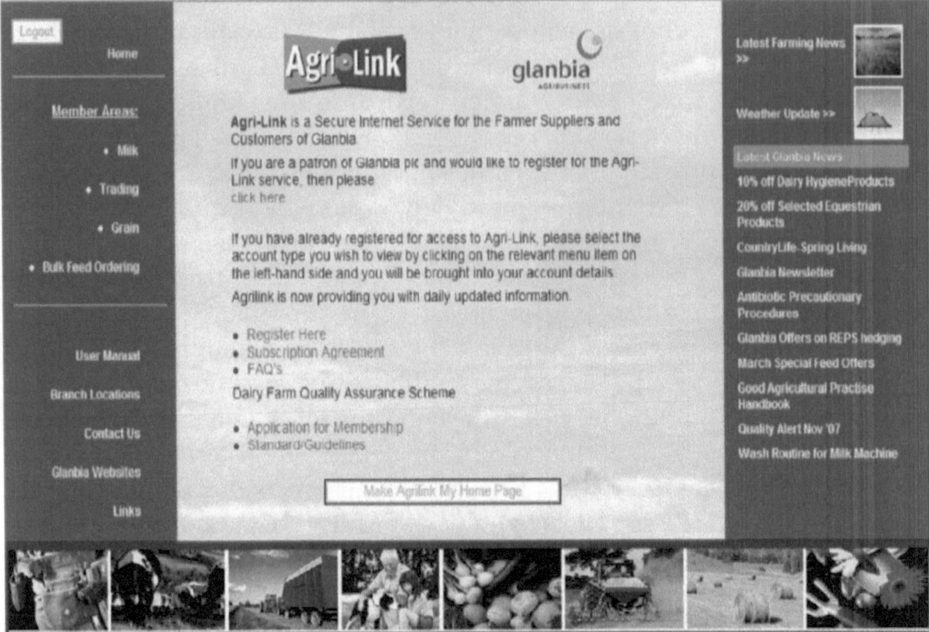

Figure 1 : AgriLink Interface, 2009

The features on the AgriLink website that were most accessed during 2007 are presented in Table 2.

Table 2 : Hits on AgriLink during 2007 (N=All 857 suppliers who used AgriLink)

Agrilink Feature Accessed	Number of Times	Per cent
Milk Tests	32,287	30
Milk supply Call	24,237	22
Superlevy situation	13,670	12
Milk monthly summary	9,306	9
Milk Statement	8,946	8
Quota Details	5,712	5
Trading Statement	5,503	5
Trading Balance	2,250	2
Financial Summary	1,998	2
Trading monthly sales	1,451	1
Milk Statement Interface	1,257	1
Other	2,782	3
Total	**109,399**	**100**

Research was recently conducted under the author's supervision to evaluate the AgriLink website (McEvoy, 2008). The overall objective of this research was to explore the effectiveness of the AgriLink project in providing information to its farmers and to determine how this could be improved. The specific objectives included, *inter alia*, to describe the history and evolution of the website; to describe the present content and how it is presented; to determine how many of the target users were actually using it; to analyse/ measure the success of the website; and to examine what constraints were impeding farmers from using it.

Data were obtained from the Glanbia website development team, Agribusiness Division and Farm Development Section together with dairy farmers who supplied milk to Glanbia. Quantitative research took the form of a postal questionnaire to a sample of Glanbia dairy farmers in order to establish farmer views on the adoption and use of ICT and AgriLink. Sample selection was based on the fact that of the 4,500 milk suppliers to Glanbia, 2,200 were registered to use AgriLink and 857 of these used the site during 2007 (Source: Glanbia IT section, personal communication). For this research, a sample of 165 dairy farmers in three categories was identified: 55 frequent users, 55 occasional users and 55 non users of AgriLink. The number of hits on the website by farmers ranged from 1 to 2,500 during 2007 among the suppliers registered to use it. The 55 frequent users for the sample were those farmers who recorded hits on the AgriLink site of more than 460 times and up to 2500 times during the year. The middle group of users had between 192-295 hits during this period. Some 113 fully completed questionnaires were returned, representing a response rate of 68 per cent. Of these 113, some 43 were frequent users, 41 were occasional users and 22 were non AgriLink users.

Analysis of the data revealed that the profile of the typical AgriLink user was male, engaged in full time farming, either young or middle aged (*i.e.* 35-50 years of age), married with dependent children, and owner or joint owner of the farm business. He was likely to be a Teagasc client, well educated and with a good level of agricultural education. In contrast, non users of AgriLink were also likely to be male, full-time farmers but were older (in age range 35 to 65 years), less likely to be a Teagasc client and also had a lower level of education, including agricultural education.

When home PC ownership was examined it was found 94 per cent of respondents had a home PC; this was expected as all AgriLink users had a home PC. However, 76 per cent of respondents who did not use AgriLink also had a home PC. This was higher than the national average of 67 per cent. Connectivity to the internet was examined and it was established that

49 per cent of respondents used broadband, 2 per cent used ISDN and the remaining 48 per cent used dial up. This half and half split between broadband and dial up was somewhat poorer than the rate of broadband penetration nationally. The distribution of respondents by their means of accessing the internet is presented in Table 3.

Table 3 : Type of Internet connection among Respondents by AgriLink User group (n = 113)

Internet connection type	Frequent Users (n=43)		Occasional Users (n=41)		Non Users (n=29)		Total
	No	Per cent	No	Per cent	No	Per cent	No
Dial-up	16	37 per cent	22	54 per cent	12	41 per cent	**50**
Broadband	25	58 per cent	18	44 per cent	7	24 per cent	**50**
ISDN	2	5 per cent	1	2 per cent	0	0	**3**
None	0	0	0	0	10	35 per cent	**10**
Total	**43**	**100 per cent**	**41**	**100 per cent**	**29**	**100 per cent**	**113**

When the three user groups were examined, the level of internet access was 100 per cent among users while access was at 65 per cent among non users. The use of broadband was also much lower among non users at 24 per cent whereas over half of the frequent users had broadband.

11. ADOPTION AND USE OF PC AND INTERNET AMONG RESPONDENTS

When PC usage was examined it was found that everybody in the respondents' households could operate the computer in 55 per cent of cases while the respondent alone operated the computer in 34 per cent of households. The remaining 11 per cent of respondents indicated that their spouses (4 per cent) or their children (7 per cent) could operate the PC. When the results were analysed by AgriLink user groups a different picture emerges and the results are presented in Table 4.

Table 4 indicates that respondents themselves were operating home PCs in almost all cases where they used AgriLink whereas in non-users households only 60 per cent of respondents could use a computer.

Computer use by respondents was examined to ascertain what the PC was used for and the findings are presented in Table 5.

Table 4 : Distribution of Respondent by who in household is capable of operating PC (n = 106)

	Frequent Users (n=43)		Occasional Users (n=41)		Non Users (n=22)		Total
PC operator	**No**	**Per cent**	**No**	**Per cent**	**No**	**Per cent**	**No**
Respondent	19	44 per cent	13	32 per cent	4	18 per cent	**36**
Respondent's spouse	1	2 per cent	2	5 per cent	2	9 per cent	**5**
Respondent's children	0	0 per cent	0	0 per cent	7	32 per cent	**7**
Everyone	23	54 per cent	26	63 per cent	9	41 per cent	**58**
Total	**43**	**100 per cent**	**41**	**100 per cent**	**22**	**100 per cent**	**106**

Table 5 : Distribution of Respondents by purpose to which PC is used (n=106)

Purpose *(Multiple Response allowed)*	**No. of Respondents**	**per cent of Respondents**
AgriLink	84	79 per cent
Recording the herd	79	75 per cent
Misc Farm Records	72	68 per cent
Entertainment	64	60 per cent
Word Processing	59	56 per cent
Information purposes	53	50 per cent
Spreadsheets	45	42 per cent
Accounts	44	41 per cent
School/College work	30	28 per cent
Off Farm employment	12	11 per cent
Other	11	10 per cent

From Table 5 it is clear that the respondents who used computers could apply them to farm use. The four options selected most often were AgriLink, Herd Records, Farm Records and Entertainment.

12. AGRILINK USAGE AMONG RESPONDENTS

Respondents who used AgriLink were asked how often they did so and their replies are summarised in Table 6.

Table 6 : Distribution of Respondents according to frequency of use of AgriLink (n=88)

Frequency of Use	No. of Respondents	per cent of Respondents
Daily or nearly everyday	28	32 per cent
Several times per week	5	6 per cent
About once a week	34	39 per cent
Once every two weeks	2	2 per cent
About once a month	18	20 per cent
Less than once a month	1	1 per cent
Total	**88**	**100 per cent**

From Table 6 it emerged that the majority of respondents used AgriLink at least once every month but a significant proportion also used the facility daily with 77 per cent of respondents using it at least weekly.

Respondents were also asked for what purpose they used AgriLink and their replied are summarised in Table 7.

Table 7 : Distribution of Respondents according to purpose of AgriLink use (n=84)

Feature *(Multiple responses allowed)*	No. of Respondents	per cent of Respondents
Milk monthly summary	80	95 per cent
Milk test results	84	100 per cent
Milk Statement	71	85 per cent
Milk supply details	84	100 per cent
Milk Superlevy statement	81	96 per cent
Trading statement	50	60 per cent
Total purchases	38	45 per cent
General news	31	37 per cent
Other	3	4 per cent

Respondents were asked to rate eleven features on AgriLink. Each feature was evaluated on the basis of being very useful, of some use or of no use to the respondents. All eleven features were easily accessed on AgriLink and were strongly promoted by the website developers. The distribution of respondents by their responses to these features is presented in Table 8.

Table 8 : Distribution of Respondents by how useful they rated AgriLink features (n=88)

AgriLink Feature	Very Useful	Of some Use	Of no Use	Have not tried it	Total Responses
Milk Monthly Summary	68	13	3	0	**84**
Milk Supply Details	77	6	1	0	**84**
Milk Test Results	79	4	1	0	**84**
Milk Statement	54	26	4	0	**84**
Super Levy Statement	73	6	3	2	**84**
Annual Summary	37	31	6	6	**80**
Trading Statement	39	22	8	12	**81**
Total Purchases	28	26	9	13	**76**
General News	10	29	26	13	**78**
Bulk Feed Ordering	5	4	17	47	**73**
Statement Download	19	16	10	30	**75**

From Table 8 it is clear that respondents who used AgriLink were very familiar with all milk related features on the website. However, Milk Statements and Annual Summaries were not as popular as all other milk related features; despite this most respondents still found them useful or of some use. On the other hand all agribusiness and online business features were much less popular.

The bulk feed ordering feature was used by 26 respondents but only nine of these reported it to be of any use. The General News section was also regarded as poor with only ten respondents finding it very useful. The Statement Download facility was only useful to those who had a farm software package to utilize it. While some 60 per cent of respondents did indicate that they had a farm management software package it is clear that not all of them were incorporating AgriLink into it and obtaining the full benefits of it.

13. FACTORS WHICH INFLUENCED PC AND INTERNET ADOPTION AND NON-USE OF AGRILINK

In this study PC and internet adoption were found to be universal for users of AgriLink. For non users it was necessary to examine the reasons for slower uptake of ICT. While 76 per cent of respondents who did not use AgriLink had access to computers at home, it was considered unusual that they had not engaged with this resource and other factors, other than access to a home PC, was clearly preventing them from realising the benefits of AgriLink.

14. COMPUTER TRAINING

It was found that non users of AgriLink were much less likely to have completed computer training than either frequent users or occasional users. The distribution of respondents by their level of computer training is presented in Table 9.

From Table 9 it is notable that 56 per cent of AgriLink users had engaged with computer training whereas only 33 per cent of non users had taken computer training. It may be concluded that this low level of training among

Table 9 : Level of computer training completed by respondents according to AgriLink user group (n=110)

Computer training	Frequent Users (n=42)		Occasional Users (n=41)		Non Users (n=27)		Total
	No	Per cent	No	Per cent	No	Per cent	No
Beginners course	16	38 per cent	18	44 per cent	9	33 per cent	**43**
Advanced ECDL	8	19 per cent	5	12 per cent	0	0	**13**
No Training	18	43 per cent	18	44 per cent	18	67 per cent	**54**
Total	**42**	**100 per cent**	**41**	**100 per cent**	**27**	**100 per cent**	**110**

non users has influenced the adoption of ICT and the use of AgriLink. Analysis revealed that 75 per cent of non AgriLink users had a home PC and 66 per cent were connected to the internet. Table 10 indicates what would encourage Non User Respondents to use AgriLink.

Table 10 : What would encourage Non User Respondents to use AgriLink (n=29)

Issue *(Multiple responses allowed)*	No. of Respondents	Per cent of Respondents
Better computer skills	14	48 per cent
Trading/ milk Account details	7	24 per cent
Improved Internet access	6	20 per cent
Basic AgriLink Training	6	20 per cent
Prices on bulk products	6	20 per cent
Products available to order	3	10 per cent
Phone/ Helpdesk access	1	3 per cent
Other	2	6 per cent

From Table 10 it is clear that a lack of computer training was a major impediment to increased computer use and increased AgriLink use.

Infrastructure was also an issue as 20 per cent of respondents felt that improved internet access would encourage them. A number of respondents also were totally unaware that an AgriLink type service was available as they felt that access to Trading and Milk Account details would encourage them to use AgriLink. This may indicate that an awareness campaign by Glanbia, the service provider, might encourage increased use.

15. CONCLUSION TO AGRILINK STUDY

This study examined the Glanbia AgriLink website and the respondents to the survey could not be taken as representative of how Irish farmers are adopting ICT on their farms. However, the study did discover a number of issues which may be slowing the rate of adoption of ICT. The sample of farmers who did not use AgriLink have also embraced ICT but not to its full potential. Some 75 per cent of those farmers have a home PC and 60 per cent of them can operate it. Therefore there is a huge onus on IT support teams both in Glanbia and in training bodies such as Teagasc to provide bespoke training to help those farmers in the years ahead to embrace ICT more fully and so reap the benefits accruing.

16. CONCLUSIONS

This chapter has described the penetration of ICTs among the general population and among farmers in Ireland and has discussed the initiatives to develop ICTs for agricultural extension activities by the Irish government, the public extension service and private extension (agribusiness). It is apparent from this chapter that considerable strides have been made in ICT and e-Government initiatives in Ireland in recent years. Unfortunately, without the necessary infrastructural development which includes broadband, Irish farmers could continue to be marginalised when it comes to ICT adoption.

17. CONSTRAINTS THAT IMPEDE FARMERS FROM USING ICT

It emerged throughout this chapter that older farmers are less likely to use ICTs than farmers under 50 years of age. Older farmers have less understanding of the benefits of ICT adoption. This is partly due to their lower education levels. According to the Irish Society for Information Technology in Agriculture (ISITA) 2003, computer training is an issue for farmers in the adoption of ICT and research findings presented in this chapter also found that the lack of computer skills is an issue. Less than one third of non users of AgriLink, for instance, had completed any formal computer training while none had completed ECDL training compared with 57 per cent of users who had completed training. This lack of training will continue to be a barrier to increased uptake of ICT among non users.

While the skills necessary to engage with ICT continue to limit uptake, the lack of infrastructure necessary to use ICT to its full potential continues to be a barrier. The lack of availability of broadband to rural areas is limiting the impact of ICT among Irish farmers. While rates of broadband use among AgriLink users, for example, were high at 51 per cent, it was much lower among non users at 24 per cent indicating that lack of availability and lack of awareness of the potential of broadband were evident among non users. It may be recalled that recent research by Smith (2008) indicates that 75 per cent of rural areas in Ireland do not have an adequate broadband supply. This is significant since 20 per cent of non users of AgriLink did indicate that improved internet access would encourage them to use it.

18. RECOMMENDATIONS

The adoption of computer technology to enhance the flow of information to farmers is seen as a vital tool to allow farmers remain competitive on a global stage. What is clear from this chapter is that a core group of farmers have embraced ICT and are benefiting from its adoption. However, it is essential that barriers to the adoption of ICT by farmers who are not using it are overcome. Efforts must be made to focus all players within the agricultural industry to make ICTs more relevant to farmers.

The transformation which has taken place in agriculture due to e-Government has been unparalleled with the facilities to interact with the Ministry of Agriculture on a number of vital schemes for farmers. This process must continue. However, one area where Government has not been so successful is in broadband provision to rural areas, where the farming community are concentrated. Today, there is a major imbalance in the provision of broadband to urban and rural areas with many rural areas trailing in ICT penetration. The commitment given by Government to extend the benefits of the Information Society to all parts of the country including farming communities and the rural population must be delivered. Broadband is now accepted as vital to rural sustainable development and without it rural areas will not attract external development and farmers cannot make full use of ICT.

It is important to acknowledge that Teagasc has made significant steps in developing and promoting ICT applications for Irish farmers. However, it is evident that there is a desire among farmers to embrace ICT but without the necessary skills to do so, this will not happen. If farmers are to adopt ICT on a grand scale it will only happen if they are taught the skills which they can implement immediately in relation to their farming businesses. While agribusinesses themselves may not want to get involved directly in training, it is important that they are supportive of organisations such as Teagasc or other local organisations in delivering the type of training required by farmers.

References

Amarach Consulting, 2007. A Research De-Brief for Comreg. Prepared by Amárach Consulting. Accessed at http://www.comreg.ie/_fileupload/publications/ComReg0708.pdf (Accessed on 21st March, 2009).

Boyle, G., 2008. Report of the Director. *Teagasc Annual Report, 2007*, Teagasc, Carlow, Ireland.

Chambers Ireland, 2007. *E-Business Survey 2007.* Vol 1. Co Partnered Chambers Ireland and Comreg Dublin Accessed at: http://www.chambers.ie/ktmlpro/files/uploads/eBusinessSurvey per cent2007.pdf (Accessed on 18th February, 2009).

Comptroller and Auditor General, 2008. *Special Report: eGovernment (2007).* Government of Ireland. Accessed at http://www.audgen.gov.ie/documents/vfmreports/58_eGovernment.pdf (Accessed on 19th February, 2009).

Comreg, 2008. Irish Communications Market: Quarterly Key Data Report. Data as of third quarter 2008. *Document No. 08/101.* Commission for Communication Regulation, Dublin.

Connolly, K., 2009. Farm Management: Want to manage your costs? *Today's Farm*, March/April 2009, Teagasc, Carlow, Ireland.

CSO, 2008, CSO Output, Input and Income in Agriculture, 2007, Final Estimate, Central Statistics Office, Dublin.

Department of Agriculture and Food, 1999. *Ensuring the Future- A Strategy for Rural Development: A White Paper on Rural Development in Ireland*, Department of Agriculture and Food, Dublin.

Department of Agriculture, Fisheries and Food, 2008a. *Annual Review and Outlook for Agriculture, Fisheries and Food 2007/2008*, Department of Agriculture, Fisheries and Food, Dublin.

Department of Agriculture, Fisheries and Food, 2008b. *Fact sheet on Irish Agriculture - October 2008.* Economics and Planning Division, Department of Agriculture, Fisheries and Food, Dublin.

Department of Agriculture, Fisheries and Food, 2008c. Committee on the Uptake of Information Technology in Agriculture and in Rural Communities. *Interim Report*, December 2008, Department of Agriculture, Fisheries and Food, Dublin.

European Commission, 2007. *Europa: Activities of the European Union, Summary Legislation. Accessed at:* http://europa.eu/scadplus/leg/en/s21012.htm#eEurope (Accessed on 18th February, 2009).

Government of Ireland, 1999. *Implementing the Information Society in Ireland: An Action Plan (PN6727).* Government of Ireland. *Accessed at:* http://www.taoiseach.gov.ie/attached_files/Pdf per cent20files/InfoSociety Action PlanJan99.pdf (Accessed on 18th February, 2009).

Hanly, J., 2008. eGovernment and Online Services by Department of Agriculture, Fisheries and Food. *Paper presented at annual technical meeting of ISITA (Irish Society for Information Technology in Agriculture),* 30th October, 2008.

Jones G.E., 1979. *The Original Agricultural Advisory Service: Lord Clarendons's Practical Instructors in Mid-Nineteenth Century Ireland.* Proceedings of the 4th European Seminar on Extension Education, Department of Agricultural Extension, University College Dublin.

McEvoy, M., 2008. A Study of the Glanbia AgriLink website. *Unpublished Thesis,* Submitted in partial completion of the requirements for the B.Sc. (Rural Development), School of Biology and Environmental Science, University College Dublin.

Millward Brown IMS, 2008a. *Consumer ICT Survey: Quarter 4, 2007, A Review of findings.* Millward Brown IMS. Accessed at: http://www.comreg.ie/_fileupload/publications/ComReg0816.pdf (Accessed on 19th March, 2009).

Millward Brown IMS, 2008b. *SME and Corporate ICT Research; First Half, 2008.* Millward Brown IMS. Accessed at: http://www.odtr.ie/_fileupload/publications/ComReg0826.pdf (Accessed on 20th March, 2009).

Millward Brown IMS, 2008c. *Consumer ICT Survey: June 2008, A Review of findings.* Millward Brown IMS. Accessed at: http://www.odtr.ie/_fileupload/publications/ComReg0849.pdf (Accessed on 19th March, 2009).

Millward Brown IMS, 2009. *Consumer ICT Survey: Quarter 4, 2008, A Review of findings.* Millward Brown IMS. Accessed at: http://www.odtr.ie/_fileupload/publications/ComReg0907.pdf (Accessed on 24th March, 2009)

Moore, M., 2008. The Joy of Texts – All you need is a mobile phone. *Today's Farm,* July/August 2008, Teagasc, Carlow, Ireland.

Morrow, L., T. Kelly, and T. Kirley, 2003. ICT – Its Potential as a Channel for Enhanced Extension Services. *Paper presented at the AIAEE conference,* University College Dublin, June, 2003.

New Connections, 2002. A Strategy to realise the potential of the Information Society. Second Government Action Plan, Government of Ireland, Accessed at *http://www.taoiseach.ie/attached_files/Pdf per cent20files/NewConnectionsMarch2002*.pdf (Accessed on 19th February, 2009)

O'Donnell, S., H. McQuillan and A. Malina, 2003. E-inclusion: Expanding the Information Society in Ireland. *Report to the Information Society Commission, Dublin.* Accessed at: http://www.isc.ie/downloads/einclusion.pdf (Accessed on 18th February, 2009).

Phelan, J., 2009. Agricultural Extension: Is the Bubble Bursting?. *Inaugural Professorial Lecture*, School of Biology and Environmental Science, University College Dublin, 27th January, 2009.

Ross, B. and G. Waksman, 2004. Internet use in Agriculture: Progress Driven by Demand, not by Supply. In: *New Ag International*, Web address: http://www.actainformatique.fr/telechar/NewAg/progress_driven_by_demand.pdf, pp 22-28, (Accessed on 25th March, 2009).

Smith, P., 2008. No Broadband in 75 per cent of rural homes in Ireland. A National Survey of Irish Farmers' Association 945 branches conducted in 2007, *Irish Farmers Journal*, (Saturday 23rd February, 2008; Pg.8).

Teagasc, 2002. *National Farm Survey*, Teagasc, Dublin.

Teagasc, 2008a. Advisory Service Charges, http://www.teagasc.ie/quicksearch.asp. (Accessed on 19th February, 2009).

Teagasc, 2008b, Report of the Director. *Teagasc Annual Report, 2007*, Teagasc, Carlow, Ireland.

Wims, 2007. Analysis of adoption and use of ICTs among Irish farm families. *Journal of Extension Systems*, 23, (1), pp. 14-28.

Author Information

Dr. Pádraig Wims

Lecturer in Rural Development
School of Biology and Environmental Science
Agriculture and Food Science Centre
University College Dublin
Belfield, Dublin 4
IRELAND
Phone No.:+ 353 1 716 7784
e-mail: p.wims@ucd.ie

Dr. Pádraig Wims has been Lecturer in Rural Development at University College Dublin, Ireland since, 1987. His main academic interest is agricultural extension and rural development and his main research interest is in the contribution of ICT to agricultural extension and rural development and the application of ICT to rural business and rural communities and e-business in agribusinesses. Prior to joining University College Dublin, Dr. Wims worked in the Information Section of the Irish public agricultural extension service, on the development of a videotex service for Irish farmers.

❑❑❑

Chapter 7

ISRAEL

E. Gelb, B. Gal and D. Wolfson

1. INTRODUCTION

The contribution of Extension and research to Israel's agricultural productivity and profitability is a collaborative effort with research hailing back to the late 19th century. It facilitated technological innovation in agriculture, transfer of information, training of farmers and connectivity. These were in effect a dictate of continuously deteriorating agricultural terms of trade, crippling water shortages and severe climate and soil constraints.

The Israeli Extension Service's (IES) adoption of Information and Communication Technologies (ICTs) significantly enhanced extension's capabilities and scope of activities. ICT empowered extension's ability to develop, introduce and adopt innovative agricultural technologies and collaborate with research and other partners. Adoption of ICT as one of the tools of technological innovation dramatically improved the transfer and management of information, production chain efficiencies and integration within and with the agricultural sector. These were and are critical success factors contributing to Israeli Extension, Research, Services and Farmers ability to sustain a profitable and thriving agriculture sector. As such the overall results are a major contribution to rural viability and an illuminating case study of technological innovation adoption.

2. BACKGROUND – ISRAEL

Israel's population is as on 2008, 7.5 million with a total arable area of 440000 Ha. With scarcity of water, only 182,000 Ha. are irrigated - to a large extent with recycled water. Israel's topography spans areas from 1200 m. above to 400 meters below sea level within a small, compact geographical confine. This results in an extremely wide range of variable climatic and soil conditions in close proximity.

Israel's geographical location is supported by a generally moderate climate which can be translated to a comparative advantage in supplying out of season agricultural products to Europe, the America and Asia. Sustaining Israel's advantages is dependant mainly on product quality, continuing innovativeness, technological production excellence and optimizing use of scarce water, labor and land resources within the multi varied set of environments.

These combined with the major contribution of extension, to an ongoing development of a broad multitude of "agricultural products", translate into traditional and niche markets. In many cases the products are export-oriented, with high-value quality items for specialized, sophisticated and increasingly demanding niche markets. They in turn are now supporting Israel's agricultural sector viability. The product categories include food, raw materials, agricultural inputs, leisure products, primary and secondary services, resource management and derived products of education, science and culture. All these are served and in many cases dependant on ICT supported extension activities. Attaining these results dictates an ongoing specialization within the Extension service. The specialization requires an ever growing emphasis on innovation, professional competence and intimate collaboration with R&D all ICT supported.

Israel suffers from chronic water scarcity, with a short rainy season and frequent droughts. Examples of alleviating measures include a wide range of water conservation procedures, water recycling installations, sea water desalinization, innovative agricultural water conservation technologies, implementation of hi-tech irrigation methods and equipment, genetic engineering of plants for salt tolerance to name a few. All these are at the forefront of innovative ICT developments and ICT supported systems. Israel's ICT Supported Water and Irrigation Technology alternatives provide a useful model of the impact of Extension's ICT initiatives and their implementation. These, in turn, dictate a relentless pace of technological innovation and its implementation. Failure to keep pace with innovation has dire consequences for farmers individually and for the agricultural sector in general. A major measure of extension excellence in this context is minimizing the time lag between innovation availability and its profitable adoption. Adoption of Internet by Extension, detailed below, is another example pinpointing the importance of Extensions' involvement in minimizing such a technology adoption initial time lag.

Lehman and Regev (2008) review in detail the particulars of Israeli Agriculture and list areas of technological innovation priorities, efforts and excellence. They include water technology, genetics, environment management, and ICT. Success of these efforts is described in Table 1 which quantifies Israel's agriculture relative productivity since 1986. This accomplishment follows the results of the increase in food production dictated by the establishment of the State of Israel in 1948. This achievement was facilitated at the time by a dramatic increase in research and extension facilities. Eventually the outcome by 1962 was a surplus of agricultural production. The export of the high value agricultural produce component was then expected to balance the cost of importing agricultural products still needed: food-grains, oilseeds and sugar. Export of agricultural products (2008) exceeds \$2.13 Billion which is about 25 per cent of the total agricultural production which in turn contributes 1.7 per cent of the GDP and 3.5 per cent of total Israeli exports. About 65 thousand workers are employed in the agricultural sector representing less than 2 per cent of the Israeli workforce.

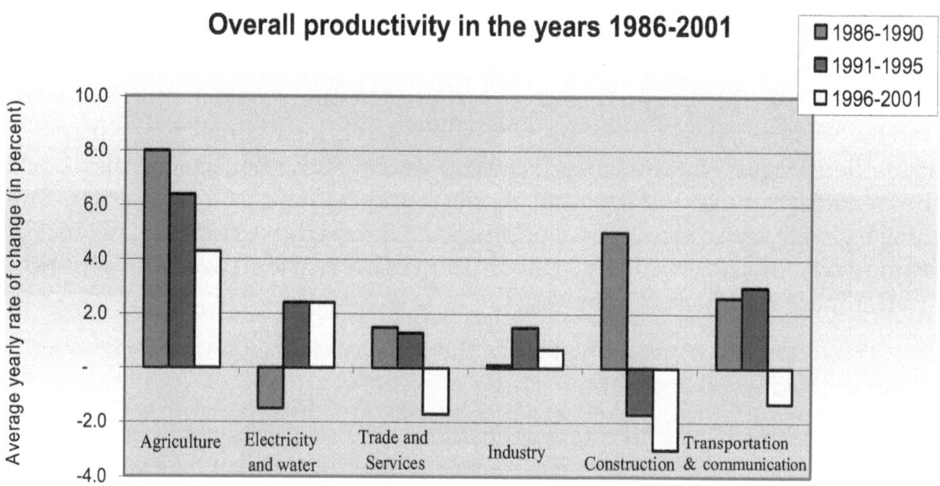

Figure 1 : Relative Productivity of Israel's Agriculture (Gelb and Levanon, 2008).

The Israeli agricultural productivity is to a large extent the product of collaboration between the Israeli scientific, educational, agricultural and rural sector's innovative environment. It includes

- Universities, colleges, regional research centers and agricultural vocational schools;
- Sophisticated technical, communication, regulatory and commercial ICT infrastructures and services;
- Intimate connectivity with all stakeholders and their organizations in Israel and abroad;
- Efficient and varied transportation facilities and other services;
- International collaborations.

3. ISRAEL'S EXTENSION SERVICE (IES)

The IES is a semi independent unit within the Israeli Ministry of Agriculture and Rural Development. The short and long term IES goals are derived from national agricultural and rural policies. They include, for example, replacement of labor by capital, water use efficiency, the prevention of groundwater pollution, increased exports, environmental concerns, sustainable rural viability and more. The 2009-2010 IES goals are focused on continued replacement of labor with capital, more efficient use of water, adoption of innovative best practices to peripheral areas, an emphasis on promoting the export of the agricultural sector's products, environmental conservation, ever improving product quality and product diversification. All this is within improvement of internal IES efficiency

The basic IES organizational structure concurs with the Ministry framework of regional offices supported by the various Ministry departments at the Beit Dagan headquarters located near Tel Aviv. Beit Dagan also hosts the headquarters of the national Agricultural Research Organization. The ministry's site www.moag.gov.il outlines and details the current organizational framework of the Ministry, the IES and the Agricultural Information environment in which it operates.

The following is a short history of the IES - condensed with special thanks to our colleague Elkana, 2001.

"...Israel has a long history of agricultural extension. It began in fact with the establishment in the 1880s of the first modern agricultural villages, and was provided by agronomists of the first agricultural school (established in 1870) and of the settlement bodies. At the start of the 20th Century, researchers of newly established research farms (later developed into stations) also provided extension. At the termination of the Ottoman Empire control and the taking over by Britain, the Jewish Agency Settlement Department established an extension unit that operated from the research stations. In parallel, the British Mandate Government set up a Ministry with professional departments that among their responsibilities, was the provision of extension.

Following the establishment of the State of Israel in 1948, the various professional departments of the newly established Ministry of Agriculture began providing, systematically organized extension to all agricultural producers.

In parallel, the Settlement Department of the Jewish Agency greatly strengthened its pre-statehood extension department into a full-blown extension service that provided extension to the tens of thousands of new immigrants who were settled in new agricultural villages. They needed very basic type of extension services: how to plow, seed, cultivate, and how to feed and milk a cow. This parallel system recruited from established villages hundreds of young farmers who would serve as village level instructors, living in the village among their farmers. Each village of 60-100 farm families would receive three of these para-professionals: a) Social – to teach how to organize and manage a democratic village cooperative and run it both socially and economically; b) Home economics and management, working with the women of the household; and c) Agricultural - a generalist, to teach and follow very closely all the agricultural operations of each farmer. The latter received professional backing from regionally based Subject Matter Specialists of the Agency, mostly ex-Kibbutz branch managers and agricultural high school graduates.

In 1960 the specialists of these two systems were united under one overall umbrella to form the "Unified Extension Authority" (UEA), with 12 regional centers. In 1965 the Ministry of Agriculture adopted the Agency Subject Matter Specialists and formed one national service – "The Extension and Professional Service" the IES of today. This organizational effort included the Soil and Irrigation Field Service with its 14 laboratories. It was a large unit established to enhance water use efficiency.

The model for IES was based on the USDA Extension service with an input of experience from the Netherlands providing a professional, specialized service to reach all the farmers through regional centers.

A variety of extension methods and communication media were developed with a strong emphasis on personal on-farm visits. Strong ties with research were established from the start including involvement of extension specialists in R&D. Over time farmers, through their local, regional and national organizations were associated as part time and eventually full time partners in planning of Extension steered by a National Extension Assembly, chaired by the Minister of Agriculture. Farmer representatives formed the majority in the council.

During this period Israel's agriculture underwent extreme changes - from an almost traditional sector to a highly sophisticated one, at the forefront of technology. IES underwent a process of increased specialization, devoting more of its time to generation of technology through applied R&D activities at farm and research levels. They included new concepts such as fertilizer use incorporated in irrigation, biological and integrated pest control and multi disciplinary orientation. In the duration many of the IES became involved in international programs with their derived expertise. Demand for more specialized extension went hand in hand with increased farmer proficiency. The extension competence included individual specialization, innovative and unique technologies, niche market considerations including the risk taking involved.

IES led collaboration between research bodies, growers, their organizations, service providers and client consideration were major contributors sustaining the viability of the Israeli agricultural sector. Regardless governmental funding constraints and political dictates resulted in drastic staff reductions from 1980 the Agriculture Research Organisation (ARO) was reduced from 1800 to 800 individuals and IES from 640 to about 150 today (2008). The inevitable need to update equipment, facilities and capabilities augmented this situation forcing both the ARO and the IES to seek additional sources of funding, partnership and commercialization. Partners included growers' associations, production and marketing boards, agricultural committees, regional councils, regional enterprises, packing houses, food processing industries, seed companies, fertilizer and chemical producers and agencies, irrigation equipment industries, machinery firms, and more in Israel and abroad......"

The IES has a long and successful record of promoting technological innovation in agriculture, in many cases being the facilitator of technology transfer from research to the field and providing feed back to research (Katz and Ben David, 1975). It is almost impossible to exaggerate the importance of ICT in maintaining these results and enabling smooth interactions between innovation developers, end-users, the various beneficiaries, the public at large and international collaborators. In fact in times collaboration between Extension and research is so close as to make differentiation most difficult. Gelb and Kislev (1982) detail another example where farmer participation in deciding and funding the research priorities enabled a "bottom up" contribution to the traditional "top down" public funding allocation framework. Participants are more varied and include the research personnel (which in some cases can be the farmers or other beneficiaries), the institutes funding or contracting the studies, those involved in the funding allocation process, end-users and the public at large. ICT is at the forefront of these interactions facilitating real time access to information in addition to accelerating the rate of knowledge and innovation transmission all these with Extension leading the way. Furthermore, ICT currently enables ongoing, real time, public accessibility to available information. This concept identifies demands and expectations from publicly funded research benefiting them with real time response ability to general feedback, comments, questions and suggestions from the public.

The Israeli extension officer is invariably a specialist with a large degree of individual "academic freedom". The officer's professional environment includes subject matter committees, other subject matter specialists *e.g.* at Extension headquarters and farmers - some with equal and even superior training. Ever since the IES was established there have been additional providers of extension including several with pioneering efforts in ICT Adoption. These extension entities have in the past, and are currently closely associated with IES. They include:

- Regional councils and cooperatives organized as local government and municipal entities. These farmer-led entities have also developed a range of economic services for their members such as packing houses, cotton gins and feed mills, increasingly maintain professional field staff to organize, guide, and coordinate agricultural activities in their jurisdiction. All these in turn provide regular extension.
- Regional R&D entities that maintain extension directly with "model farms" and pilot programs - interacting directly with end users.

- Agricultural-input company specialists dealing with *e.g.* seeds, animal feed, pesticides and herbicides, pharmaceuticals, fertilizers, irrigation equipment, greenhouses, nurseries, *etc.*
- Agricultural produce processing agro-industries.
- Private consultants.

Financing of IES in the past had evolved and adapted over time from a straightforward government funding model based on the recognition that extension is a "public good". The following lists some of the past supplementary sources of funds for extension in general and specifically for ICT.

- Farmer "taxation": All agricultural production is regulated *via* Agricultural Production Boards that levy annually 0.625 per cent of the value added of their produce. Of this sum 90 per cent goes to Agricultural R&D (*via* joint farmer and research committees) and 10 per cent to extension activities. This participation is a long standing agreement;
- Funding by national and regional growers' associations for standard extension services and or specific equipment - *e.g.* laptops and or mobile phones for Extension field staff;
- Funding by national and regional growers' associations for enhanced extension services in a "contractual 24/7 extension commitment";
- Funding by farmer "commodity services" *e.g.* packing houses, feed mills, *etc.*
- Partnerships with farmers such as the Irrigation and Soils Field Service which is funded by users through municipal taxes paid to regional councils and by direct payment for laboratory tests;
- Direct Farmer payment per extension visit over and above "standard services";
- Direct Farmer payment to extension personnel on a private basis after their official working schedules;
- Direct payment to IES and or individuals for participation in related activities by rural regional and or agricultural schools, participation in various campaigns *e.g.* environmental conservation activities, programs by other Ministries, entities, *etc.*
- Funds generated by collaboration with and participation in International ICT Adoption programs - *e.g. via* EFITA/AFITA (European/Asian Federation for Information Technology in Agriculture).

This multi sourcing of funding provided IES with ongoing flexibility which in turn enabled IES to access the latest technological innovations. In the case of ICT, IES was always able to access the latest ICT - hardware, software and methodology - even if only for demonstration and evaluation. This head start was and is in the common interest of farmers, service and equipment providers, research and IES. A unique advantage of these partnerships and options was IES ability to "leap frog" convention to cutting edge ICT innovations and avoid a counterproductive and cumbersome linear progress. ICT supported trailblazers, for example, initiated irrigation methodology to integrate fertilization and plant protection measures in water applications, innovative genetic adaptations, innovative automation, modeling which enabled subject matter integration, connectivity and more.

4. THE ISRAELI EXTENSION SERVICE AND ICT

By the early 1980s a need for a more formal evaluation of the nature and impact of ICT in and on extension was recognized, in Israel as well as in other countries. The Florida Cooperative Extension Service (1983), for example, initiated a series of conferences evaluating Technological Change - specifically focused on Computers in Agricultural Extension Programs. Others were to follow worldwide. Although no longer a novelty ICT Adoption for Agriculture and Rural Development still remains a contiunuously studied critical issue at regional, national and international levels. Ongoing evaluation efforts are supported *via* a multitude of ongoing specialized and or combined studies, fairs, workshops and conferences. A recent example is the 2008 IAALD/AFITA/WCCA conference - see http://iaald-afita-wcca2008.org. Appendix-A summarizes its recommendations relevant to ICT Adoption and extension (Gelb *et al.*, 2008).

The 1983 Florida effort was followed by a bi-national US – Israel workshop (see Gelb *et al.*, 1985). The workshop focused on the role of ICT for extension in general *via* each country's specific experience. The Israeli emphasis was on the 1979 Israeli Extension Service (IES) "Computer Supported Extension" (CSE) effort. It was defined as "*... a major effort to integrate advanced quantitative methodology into the extension supported decision-making process – from policy making to field recommendation levels...*". The extension organizational outline, physical and information environments at the time (1985) are presented in Appendix-B replicating the original presentations. The following details several aspects of CSE:

The CSE supported categories were:

- Data and information - input and management;

- Extension service programs - statistics, Decision Support Systems, graphics, communications;
- Agricultural production automation and process control;
- Production planning, monitoring and evaluation;
- Crop and production models and simulation;
- Optimization;
- Expert systems.

The basic guidelines for CSE implementation were:

- Adopt CSE gradually at all IES levels without structural changes;
- Computer facilities will be accessed, shared and not duplicated;
- In-service ICT training will be provided;
- Incorporate ICT in every possible extension activity;
- Ensure maximum CSE flexibility.
- CSE will be incorporated into existing collaboration with farmers, research and services.

Some major issues, constraints and insights emerged early on during CSE implementation and were reported and discussed in the workshop. They included:

- The ability of management to guide regional activities and support them;
- Funding procedures and funds for equipment procurement;
- Lack of tools for evaluation of improved extension staff effectiveness;
- Guidelines for allocation of extension priorities;
- Over-enthusiasm and unrealistic expectations;
- Benefits inherent in openness and collaboration;
- A need for retraining, education and reassessment of programs and their goals;
- How not to stifle initiatives, unplanned but successful scenarios and insights;
- How to cope with restructuring within the organization's balanced hierarchies, management and individual conservatism and in-service inadequacies;
- How to define and meet personal, political and organizational expectations.

The interim CSE results were reviewed during the workshop discussions. The criteria for the review included:

- Improved service to the farmer including information delivery efficiency;
- A better farmer understanding of production and biological systems;
- Improved professional communications - between extension and farmers, both with themselves, with each other and with research;
- Additional dimensions to extension activities which included remote sensing and control, real time imaging, simulation, signaling of priorities to the private sector and integration with the education systems;
- A basis for professional development of programs and IES personnel

The lessons and notes of caution from CSE implementation that were reported emphasized the following:

- Employ ICT to foster the closest collaboration with research, producers and their organizations;
- Ensure maximum extension program flexibility;
- Focus ICT efforts to optimize scarce human capital and funds;
- Verify the primacy of the human element in all programs.
- Be sure to consider farmer's interests;
- There are conflicting interests in CSE implementation which include non convergent policies;
- Knowledge is power;
- Market forces and interests can conflict with farmer's concerns and extension recommendations.

A question which was discussed in the workshop but left open for further joint and individual deliberation was who should pay for this ICT derived public-good benefit - as a case study of ICT Adoption for extension. A specifically useful and focused evaluation regarding the various aspects and constraints of ICT Adoption for extension was derived from an EU sponsored workshop in 1997 (EUNITA, 1997). This workshop analyzed in close detail the initial adoption and use of Internet for Extension (Gelb and Bonati, 1997).

ICT supported and embedded systems in Israeli agriculture illustrate the role of extension and research in ICT development and its' Adoption. Characteristically they also open a promising venue for research to measure the rate of ICT Adoption and its benefit. In detail Gelb and Kislev, (1982) quantify farmers' (adopters) preferences for technological change *via* their financing of agricultural research and extension. Gelb and Levanon, (2008) elaborate these interactions in Israel in detail over time. To go into further ICT Adoption detail Gelb *et al.*, (1976) outline farmer recognition of the benefit of a dedicated computerized predictive plant protection model and its economic benefit - the Benefit Cost Ratio (BCR) was 4.5. Gelb *et al.*, (1996) elaborate by measuring the specific economic value of information as manifested in ICT. The unique information measured was milk yields and estrous collated in a dairy milk parlor Decision Support System (DSS). The initial benefit was found to be close to $70 cow/lactation with a system repayment period of three years. This measured ICT benefit was over and above the contribution to improved dairy management derived from adopting a computerized Dairy Management Information System (MIS).

Table 1 presents three examples of the impact and results of the IES involvement in innovative ICT development and adoption. The programs were shared by research, extension and end users. Appendix C and D elaborate. Table 1 details the three. In each one IES was involved in the four stages of initiation, system development, pilot evaluation and widespread adoption of the result. The three examples reflect a regional collaborative effort, a collaborative effort to develop a tailored Management Information System (MIS) and an innovative plant protection model

ICT adoption by the Israeli Extension service combines substantial home work, local and international collaboration, trial and error. All these dictate the need to maintain IES comparative advantage as an agent of change, an effective "middle man, honest broker" positioned between new, old, useful or irrelevant innovations and their uptake. This responsibility is constantly being tested under the confines of ever decreasing public funding and resulting staff reductions (see Fig.2). In addition, there was and still prevails the ever growing influence of the powerful sentiment that ICT development and adoption should be left to the invisible hand of the market (Prahalad, 2004).

A unique IES feature of ICT Adoption since 1983 was the establishment of an informal, internal interest group to evaluate and promote ICT issues. The issues included familiarization with ICT innovations, their relevance for agricultural research, management and production, adoption of current and innovative ICT, identification of knowledge gaps and the nitty-gritty of equipment and ICT training specifics. Understanding the contribution of ICT to extension methodology and effectiveness, including the impact of innovative communication options, were of decisive significance in the forum discussions. The insights are discussed in detail below. The composition of the group involved a Subject Matter Specialist from each professional department. The assumption at the time was that it would be more effective to ICT train an agricultural Subject Matter Specialist than attempt attaining agricultural subject matter competence by an ICT specialist.

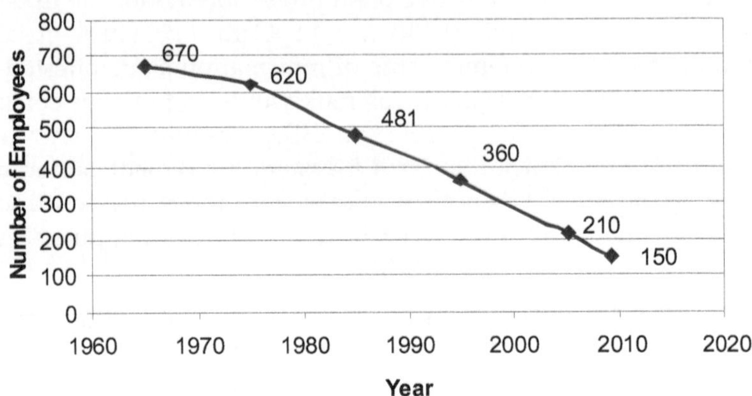

Figure 2: Israeli Extension Staff Members (Wolfson, personal correspondence)

Table 1: Examples of ICT products and their funding (Gelb and Levanon, 2008).

Product Name	Type	Management Methodology	Funding source	Net Benefit
Afimilk	Management Information System (MIS)	Exception and deviation	Farmers	US$200.00 cow/lactation
Phytek	MIS	Rates of deviation change	Farmers and Private	>10 per cent irrigation efficiency
Wheat Disease Control Advisor	Expert System	Disease control decision rules and recommendations	Farmers and Public	$92.70/Ha

This interest group proved to be an invaluable forum. It provided an ongoing source of experience, personal support, utilization of experience, insights and invaluable contacts when and as needed. Its composition ensured the sharing of ICT Adoption commonalities between the various agricultural crops, extension methodology and activities, ICT development, innovation uptake and evaluation of the whole range of ICT problem-solution results including trials and errors. These critical forum contributions minimized the occurrence of repeated mistakes. Consequently the benefit from Extensions' "scarce" human capital was optimized. The forum identified and focused on the major ICT development and adoption problems while benefiting from organizational memory and hands on experience. In practical terms, by 1995, the annual interest group summary agenda included: discussion and evaluation of diversification to laptops; collaboration with software programmers as part of collaboration with research and farmer organizations; defining collaboration with local knowledge bases initiatives; ICT supported extension delivery routine - including Farmer ICT basic training; initial evaluation of site specific Crop Management, GIS, GPS, e-Marketing, two crop models, two ICT managed irrigation and fertigation systems, greenhouse management systems, process control pilots, distance learning; international collaboration priorities and internal in service ICT training (Source: The 1995 Interest Group summary agenda). The 1995 cutting edge technology agenda at the time is a far cry from the current ICT reality and ICT needs in Israel today. Essentially ICT proficiency, connectivity and digital inclusion are non issues in the sense that ICT is a commodity. In practical terms, Extension does not have to teach farmers how to use ICT supported applications and systems. IES can now focus on ICT initiatives and activities which are a reply to the question: "Now that you have ICT proficiency, systems and applications how do you ensure your benefit from their uptake?"

The above CSE overview provides a useful baseline yardstick for understanding past and current IES ICT Adoption. It should be used in evaluating the initiation, planning, goals and implementation issues discussed below *with a strict proviso and note of caution – these background notes are neither recommendations nor recipes for successful ICT Adoption for extension. They are not intended to, and do not, provide recommendations or suggested preferences for a comparable course of action or policy priorities. These are left to the reader's discretion and have to be decided and timed based on local specifics, experience and expectations.*

5. CASE STUDIES AND ISSUES

With the above background in mind the following will attempt to highlight the issues that confront the IES. This will be done by the following:

1. A review of IES: a. e-mail and b. Internet adoption;
2. Detailed comments from discussions with Extension Subject Matter Specialists;
3. An outline and comments on current issues confronting the ongoing adoption of innovative ICT.

5.1. Adoption of e-Mail

IES officers are spread throughout Israel - see Appendix B 1. The basic framework is of regional Extension representative offices with local agents and specialized subject matter departments in the national headquarters in Beit Dagan, near Tel Aviv. These serve the regional officers and provide professional backup when needed. This routine, in turn, depends heavily on efficient, real time if possible, communications.

The concept of a unified and comprehensive communication environment was conceived many years before the various ICT became available. This concept was compatible with IES's comparative advantage as a technological innovator and an agent of change. These characteristics were unique for a government office and indeed IES was among the Israeli government pioneers.

An attempt to harness the IES to local networks (*e.g.* regional telephone facilities) proved unsatisfactory. The availability of the academic BITNET model at the time and the advent of Internet for public use were promptly utilized and adopted by IES individuals. This initiative preceded the decision to provide personal computers for extension officers. The goal was to provide maximum real time computerized connectivity to at first all extension officers and eventually to extension clients. Initially IES management decided to utilize these communication facilities within the service based on the early activities of the pioneering extension officers (a classical "bottom-up" initiative). The purpose was to facilitate communication between the extension officers, the regional offices and IES headquarters, regional offices and the extension field officers and the regional offices between themselves. Shortly afterwards the network was opened to others, clients, service providers, collaborators and associates in Israel and abroad. In the 1990s the service communication equipment was based on land lines, dial-up modems and stationary computers in the various IES offices. The officers did not have communication facilities, mainly modems, at home. At this point the main investment in equipment was in dial-up modems and establishing a pilot program. As mentioned IES was among the first governmental units to revert to electronic communications prior to development and adoption of the current universal government e-mail network. A typical resulting bureaucratic dilemma was "should the government buy dial up modems for Extension staff so they could work from *e.g.* home?"

The Floriculture and Production Economics departments were selected to evaluate e-mail communications, language constraints and program effectiveness. Till then the departments relied on surface mail which comprised flyers, seasonal recommendations, invitations to field demonstrations and interest group meetings. These were held routinely 2-3 times each month for the Extension officers and several times annually for farmers. It was realized early on that e-mail, in order to succeed, would have to replace "snail mail". A prerequisite for this cut off was absolute e-mail reliability. Accordingly, Extension officers were trained to attain that goal. Preparations included meticulous training of personnel, installing personal modems, purchasing personal computers and standardizing a communication program with e-mail facilities. With all that accomplished the system was tested to see that it worked. It should be noted that all this was done based on the Israeli land line services using Windows 3.0. The Internet vendor at the time used a bi-lingual - English-Hebrew version of an e-mail program, Cameleon, which was compatible with Windows 3.1 as well.

The human element adjustment of the ICT transformation proved to be challenging. The main problem was internalizing the superior connectivity, derived routines and information exchange methodology. One obstacle derived from the fact that the Floriculture department target group consisted mainly of professionals uninitiated in ICTs and their utilization. Training had to include basic ICT familiarization and technical proficiency to facilitate the initiation and operation of the pilot program. Most of the participants internalized the new work patterns while realizing that without participation they will be "out of the know". This proved to be a powerful incentive - sufficient for overcoming the introductory problems and enabling the success of the pilot. Most of the older Extension officers mastered e-mail reading although e-mail writing remained an impediment for years to come. Typing was not a common skill at the time.

The format of both e-mail routine and methodology adoption was initially top down dictated. The sequence was to ensure reading electronically transmitted material to be followed by between-officer communication. It was expected that most of the content would be focused on problems encountered during visits to the fields. The communication format was assumed to be a description of a problem and a request for an opinion and or a recommendation. This concept was to include eventually pictures in addition to written communications. Over time this differentiation was trivialized with the wide spread availability and use of the cameras embedded in mobile phones, dedicated cameras and the availability of advanced Internet capabilities. This e-mail picture relay option became standard and a common communication tool between the Extension officers and their clients. The phones, however, did not replace the more sophisticated visual aids necessary to communicate and share more accurate details of specific problems such as disease details, nutrient deficiencies, *etc.*

By 1995, personal laptops were added to the Extension officer's toolbox - again a pioneering effort for government officials - mainly due to their cost. The early adopters within the service were the first to utilize this option - sending field data - numerical and visual within the IES.

Universal acceptance of computerized connectivity and its ubiquitous availability made e-mail a tool of first-choice convenience. Cooperation and collaboration were facilitated within the IES, with clients, service providers and between themselves. The issues involved remained focused on ICT proficiency, electracy and subject matter content. These are reviewed below.

5.2. Development and Adoption of the IES Internet site

In a sense the development of an IES Internet site complemented the e-mail connectivity evolution. Establishing it augmented IES activities. The site provided all Extension written materials and a communication service in a simple, cost effective format. A review of the early stages is detailed in Appendix E. A crucial point was to maintain a dynamic site format enabling end users to conveniently find what they needed rather than what Extension thought they should have in dictated information formats. Site categories were regional information, subject matter information, organizational details (*e.g.* a list of available courses and publications) and recommended contacts. Since budget constraints ruled out a permanent core group of Internet specialists to update the site and its' content the site development was outsourced. It was soon realized that the key site issues required assigning content priorities, indication and measures of information relevance, feed back facilities, a definition of needs, demands and their relevance. Technical issues were surmountable including an option to initiate and support common interest forums. The state-of-the art at the time dictated site rigidity, which over time (2009) became a non issue.

Being a government agency imposed an unexpected forum impediment. Due to security constraints, accessing a forum on the IES site dictated circumventing the standard site formats. Regardless it was soon realized that these technicalities were easily overcome. The critical success factors in establishing and maintaining an Internet forum were dependant on the forum leader, the forum participants' interest in it and with an emphasis on the providers of information to the forum. In fact only one IES site forum remains active till today.

Looking back on Internet's contribution to IES, the following main points are maintained to continuously evaluate the IES Internet site's efficiency and contribution to IES goals:

- Technical uptake problems;
- User benefits from using the site;
- Drawbacks from using the site;
- Critical success factors;
- Organizational aspects and responsibilities;
- Obstacles to use Internet for extension;
- Knowledge gaps and new trends if any;
- Proposed and or potential development options;
- Additional issues of impact and or influence.

5.3. From a Land-line phone in the office to an Office in the Hand-held Mobile

The IES has a highly specialized staff, at both local and headquarter levels. This currently reflects the professional level of many farmers as well. Adoption of ICT by research, extension and farmers meshes with the interaction of all stakeholders. Consequently transferring, diffusing, teaching and introducing information and innovations to the farms are now a two, three or more way activity and undertaking. The result is that the traditional two-way model that restricts researchers to producing research results with Extension disseminating them is now only one possibility among many. Others include regional Farmer-research activities, collaboration with non public extension by agri-business services, *etc.*

A recent random series of interviews with field extension agents elicited the following list of ICT contributions to their effectiveness:

- Real time, 24/7 direct and conference contact with farmers, clients, colleagues, research, markets, in Israel and abroad;
- A higher degree of independence in written communications and addressing recipients and dependency on secretarial services;
- Ongoing input, access, formal organizing and recording of data in the various digital and keyed in manner - including direct and remote numerical, visual, audio and other forms;
- Documentation of input including recommendations to farmers, reports of all sorts and their authentication;
- Real time access to data, records, information, colleagues and clients;
- Automatic extension functions - for example receiving and forwarding alerts and recommendations; data input, follow-up of instruction results, *etc.*

- Ongoing follow-up of trends in the field, in the market and results of *e.g.* plant protection or irrigation recommendations;
- Efficient on-line statistical analysis, calculations, planning, photo analysis, *etc.*
- Efficient data presentation and accesses to remote presentation sources.

An individual mobile phone specifically facilitated in addition:

- Extension staff availability;
- A short response time to farmers and other clients;
- Effective field trip and time utilization;
- Diminished dependency on "on the spot" presence;
- Facilitation of impromptu consultations;
- Ability to real time correct- incorrect information & or recommendations.

All these resulted in the opinion that the Extension officer is a "better" professional backed up by more and better information.

ICT availability has drawbacks which were attributed to an overload of information and request for advice which impeded extension effectiveness, farmer receptiveness and imposition of an additional work load. The following list may be mentioned.

- An overload of information;
- A possibility of widespread dissemination of incorrect information, non optimal recommendations and in the extreme spread of malicious disinformation;
- An unproductive demand on (officer, farmer, research, client) scarce time;
- A dependency on ICT involving loss of physical contact, intuition and precision of decisions.

5.4. Discussion and Opinions: Ongoing Challenges Facing the Adoption of ICT for Extension

Since the initial IES management decision in 1979 to engage in CSE has changed, some crucial innovations and related issues became none issues, basic proficiency constraints became irrelevant, the mix of agricultural products and their clients have changed; the profile of the "average" farmer and farmer extension needs have changed many times over; the international agricultural production terms of trade have changed; ICT capabilities have in many cases surpassed imagination, extension goals, specialists and specialization have changed - the list is almost endless - doing Rip Van Winkle proud.

Reviewing ICT Adoption for extension in Israel by the authors identified several fundamental issues - with an ongoing need for consideration, discussion and inclusion. The following lists show several such issues as reflected in the current state of the IES ICT state of the art and issues yet to be resolved. These are offered to the reader for reflection and comments, all to be welcomed by the authors at their coordinates below.

- Innovative ICT supported systems are constantly enabling Extension and farmers to continue doing their respective activities more efficiently, with more sophisticated interactions and superior connectivity - at a cost, for example, use of non optimal ICT. Correct selection of ICT systems and priorities and their use are imperative;
- Expectation from new ICT *vs.* adoption results can render initiative priorities irrelevant over time. An example is the planned training for adoption of e-mail, justification for the use of laptops, relevance of spreadsheets for farm management, permission to work from wireless served locations, *etc.*
- ICT can produce an overload of information, an overload of data entry dictates, non essential connectivity, distraction from essentials, loss of intuition in decision making, reliance on "popular" vs. verified information, shared but non specific field trial summaries, *etc.*, Verified routines to handle these pitfalls must be part and parcel of Extension responsibility;
- ICT supported Extension as the go between Research, agents of change and early majority farmers is intensified and can be multi-directional. Specifically it may be farmers' feedback that is the source of innovation; research may be guided by farmer real time needs; the source of innovation may be from abroad jointly evaluated by farmers, extension and research, *etc.* This changing multi-directional flow, its routine and evaluation needs constant verification - especially considering IES's comparative advantage as an agent of change itself;
- IES, due to Israel's small size, is characterized by familiarity - enhanced by ICT supported connectivity. For example: do ICT dictate to Extension agents their work routines *e.g.* 24/7 availability; to what extent does such intimate connectivity ensure individual ICT proficiency back up is this interdependence an advantage, an uncalled for burden to the initiated; is this to be encouraged?

- Convention suggests that "farmers are conservative" and "slow in innovation uptake". Yet to be resolved is the compatibility of this convention with unique farmers' characteristics as ICT adopters. Variations abound;
- Organizational inertia and management resistance to change in various degrees are inevitable in any bureaucracy. IES basically avoided them by balancing bottom up ICT initiatives (*e.g.* IES's e-mail adoption) with top down priorities through the organization's internal interest group. Unfortunately, this model has been abandoned - yet to be resurrected while considering management's three ICT Adoption options: to initiate, support or discourage future ICT adoption activities;
- ICT utilization tends to create communities - some of which may be within IES responsibilities. These include special interest groups, collaborative production efforts, rural communities, agricultural education groups, agricultural services including with or without international affiliation, complementing services and more. Although publicly funded, IES may not be in a position to serve these communities and or be in a position to initiate new ones as needed. This dilemma essentially entails various collaboration efforts with unexpected partners - in Israel and abroad;
- How to measure the ICT benefits as attributed to extension results and Extension agent efficiency?
- How to evaluate and prioritize information systems for extension and their inherent information component? Namely - how to prioritize critical extension provided information with needs - from the farmer and or public policy points of view?
- How to sustain familiarity and competence with newly developed ICT within IES derived extension methodology, Extension personnel ICT competence and "keeping up" compatibility with farmer (ICT) agents of change? The problem is aggravated by external ICT options such as the free use of a g-mail rather than a Ministry provided e-mail address or large scale agri-business ICT facilities not necessarily made available to IES.
- How to ensure compatibility of farmer needs with ICT development priorities, funding and allocation of extension subject matter expertise?
- To what extent should Extension be involved in technical ICT development - *e.g.* should Extension have computer programming capacity?

- To what extent is IES effectiveness dependent on ICT proficiency and compatibility with innovative farmer operated ICT. One example could be related to farmers practicing precision agriculture? In this sense does ICT proficiency today reflect farmers' competence, profitability and connectivity with Extension?

6. SUMMARY

The Israeli Extensions Service's (IES) experience in adoption of Information and Communication Technologies (ICT) provides a comprehensive example of technological innovative adoption by a public service. Involvement of an internal Extension service interest group proved to be significant and demonstrated the potential benefit of "bottom up" participation in the innovation adoption process. The IES unique pattern of collaboration between research, extension and end users suggests areas of imitation and study of the efficiency of Extension as an intermediate agent of change. Identification of collaborating agents of change and their contribution to continuing successful ICT adoption by farmers, Extension and research cannot be overemphasized. With all that said ICT adoption remains an ongoing challenge for Extension and a prerequisite for attaining Extension's goals.

References

Elkana, Y., 2001. Extension services in Israel: A comprehensive report. Technical Training for Rural Development: Looking to the 21st Century, Baoding, China,

EUNITA, 1997. European Network for Information Technology in Agriculture, 1997. *Final Report.*

Gelb, E., Pollack, D., Spar, G., Reisman, A., and Dinoor. A., 1976. Economics of Information: Leaf Rust case study. *Case Western Tech. Report No. 375.*

Gelb, E., and Kislev, Y., 1982. Farmer's financing of Agricultural Research in Israel. *Research Policy 11*, 321-327,

Gelb, E., Kislev, Y., and Voet, H., 1996. A computer in the milking parlor. *Working paper 9805*, The Center for Agricultural Economic Research, Rehovot, Israel.

Gelb E., and Bonati, G., 1997. Evaluating Internet for Extension in Agriculture.

http://departments.agri.huji.ac.il/economics/gelb-main.html

Gelb, E., and Levanon, D., 2008. *Public Funding of Agricultural Research, Agricultural Research.* Lowenstein, Thottappilly., (Ed.) Springer ISBN-978-1-4020-6057 (e-book)

Gelb, E., Maru,A., Brodgen,J., Dodsworth E., Samii R., and Psece,V., 2008. Agricultural Development and Rural Viability GFAR. *Pre-Conference workshop summary*, IAALD/AFITA/WCCA (2008) at http://iaald-afita-wcca2008.org.

Gelb, E., Schmidt, J., and Rauschkolb, R., 1985. Computer Supported Extension - US, Israel. *Proceedings - US - Israel Workshop on Development of Computer supported Agricultural Extension*, Pg 30 – 64.

IAALD/AFITA/WCCA, 2008. 12th World Congress of the International Association of Agricultural Information Specialists (IAALD). *6th Conference of the Asian Federation of Information Technology in Agriculture (AFITA), 6th World Congress on Computers in Agriculture (WCCA)*, http://iaald-afita-wcca2008.org.

Florida Cooperative Extension Service, 1983. International Conference on Computers in Agricultural Extension Programs-1983. The Florida Cooperative Extension Service, the Institute of Food and Agricultural Sciences, the University of Florida, Gainesville.

Katz, S., and Ben David, J., 1975. Scientific Research and Agricultural Innovation in Israel. *Minerva*, 13: (2) 152-182.

Lehman,N., and Regev, A., 2008. Agriculture in Israel: Fact and Figures. Ministry of Agriculture and Rural Development, Israel.

Loebenstein, G., and Putievsky, E., 2007. Agricultural Research in Israel.

Prahalad, C.K, 2004. The fortune at the bottom of the pyramid - Eradicating poverty through profits. Wharton School Publishing.

APPENDICES

Appendix A: IAALD/AFITA/WCCA (2008) ICT Adoption Recommendations Compatible with Israeli IES Experience

1. ICT Infrastructure for rural areas must be part and parcel of all national infrastructure planning and programs;
2. ICT Adoption activities as a priority must encompass policy makers, ICT research and developers, end users/stakeholders and their communities;
3. Digital inclusion must be addressed as a policy priority and an infrastructure issue with the understanding that ICT penetration based solely on market forces is unlikely to eliminate digital exclusion;
4. Successful ICT Adoption is conditional on human capital development. This includes prioritizing extension and end user involvement in policy formulation and resource allocation;
5. Ensuring ICT compatibility with end user needs and national policies are to be included in all agricultural and rural development programs;
6. Ensure Connectivity (Individuals, communities) as a priority;
7. Create ICT Adoption Partnerships;
8. ICT Adoption is to be a priority in Rural viability development programs;

9. Management of information and its quality are as important as are the measures to assure the uptake of the ICT providing it;
10. Facilitate End-User Involvement in all ICT development and uptake programs.

Appendix B: 1. The Organizational framework of the Israeli Extension Service.

The framework illustrates the Central Office specialized departments with experts supporting the subject matter officers in the regional offices *e.g.* Galilee. This framework outline has basically remained unchanged since 1965.

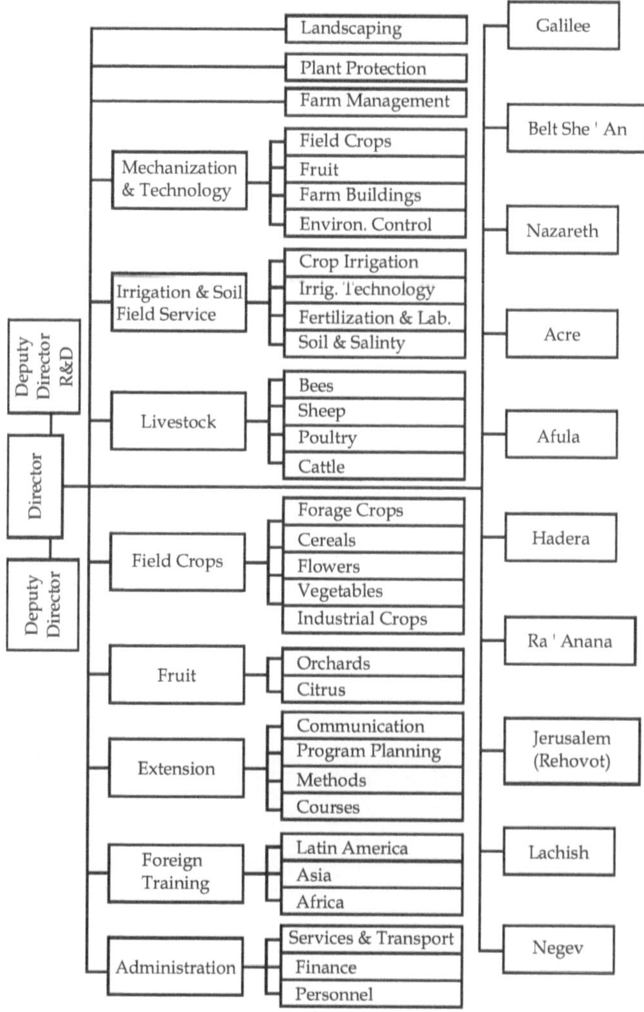

(Source: Gelb, E., Schmidt, J., and Rauschkolb, R., 1985).

Appendix B 2: Israel'sAgricultural information Environment

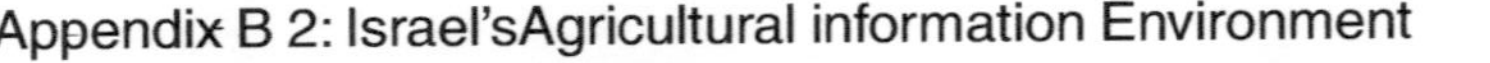

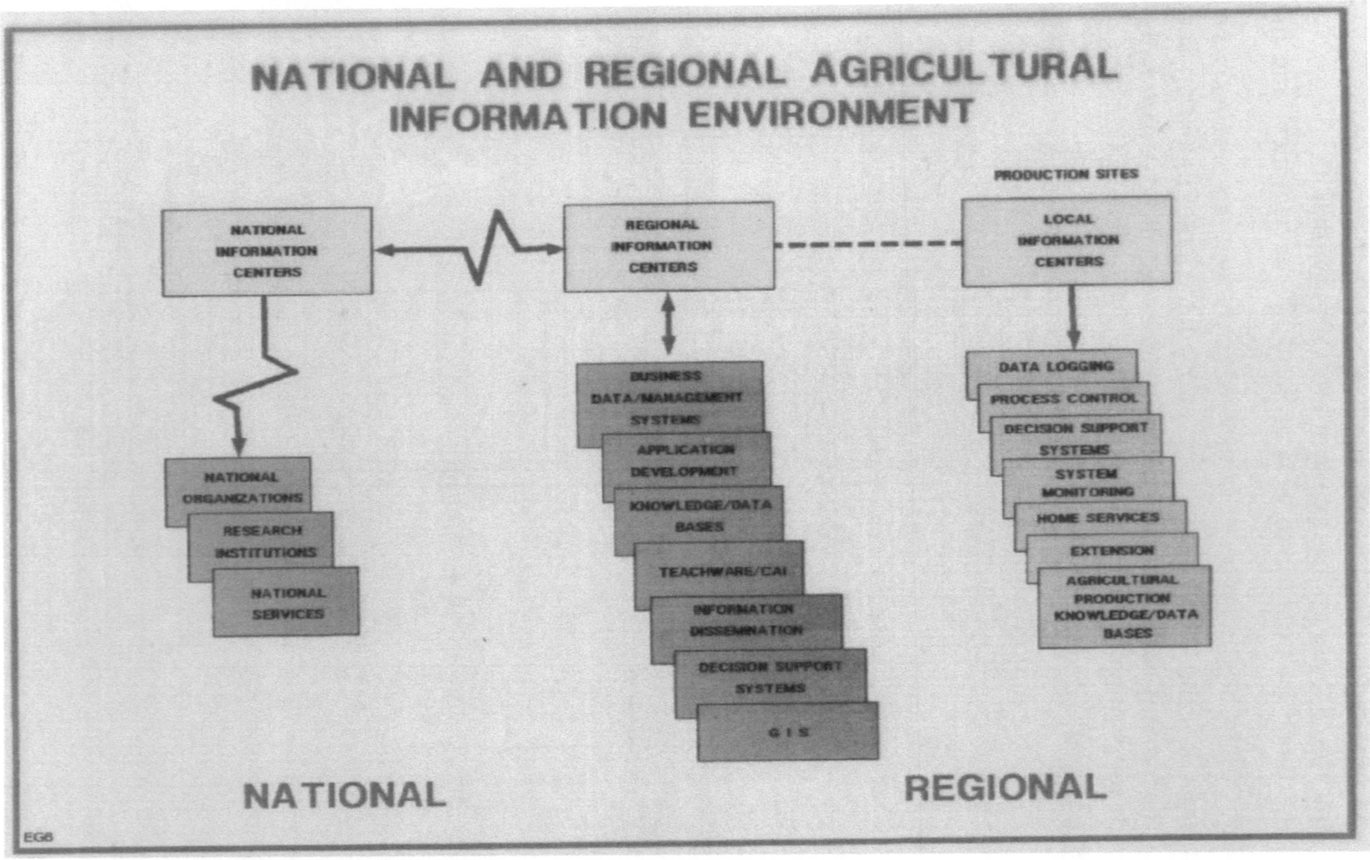

(Source: Gelb, E., Schmidt, J.,and Rauschkolb,R.,1985).

Appendix B:3. A Regional Agricultural Production Information Infrastructure

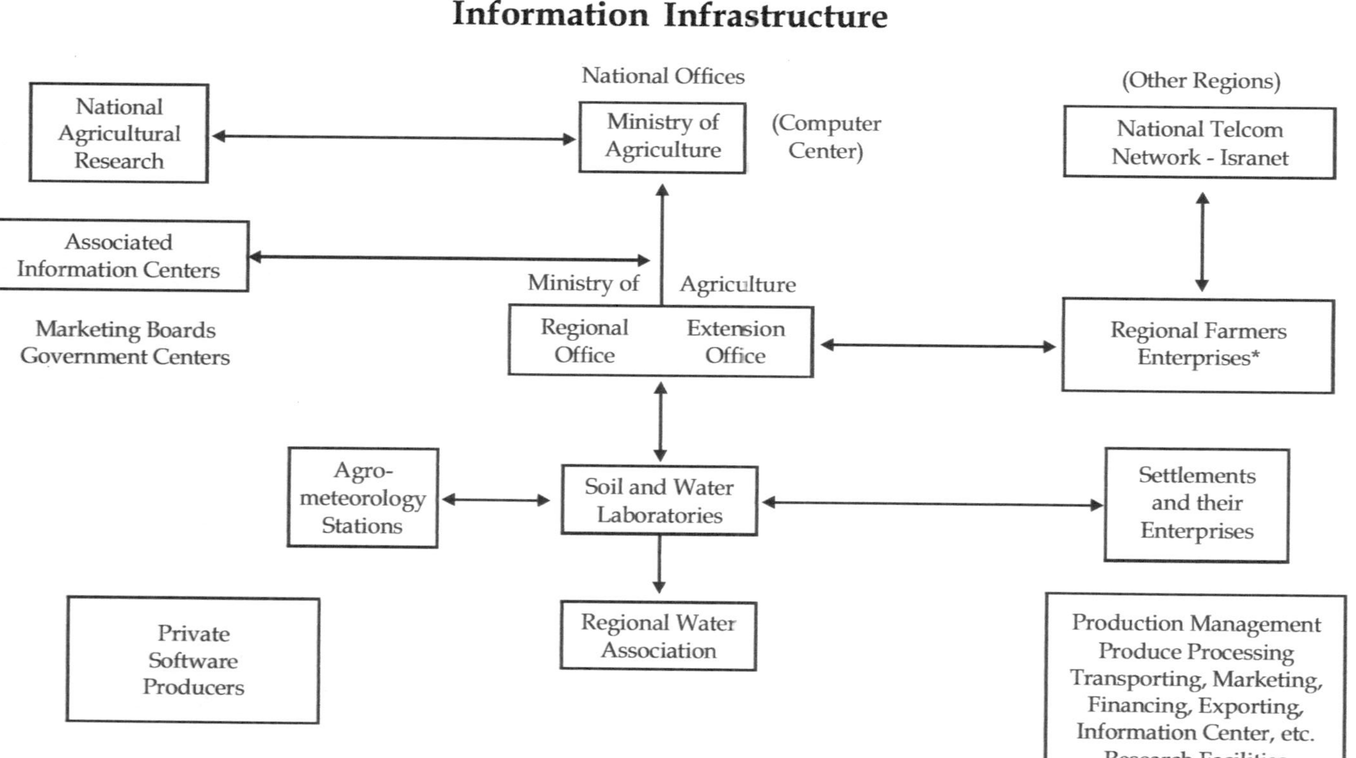

(Source: Gelb, E., Schmidt, J.,and Rauschkolb, R., 1985).

Appendix C: ICT Supported R&D in the Arava Regional Research Program

The Arava is an area in the south of Israel, characterized by desert conditions, hot climate with a limited supply of water suitable for agriculture. The Arava Regional R&D ICT Center was initiated in the early 1980s and formally integrated in the regional R&D center in 1992. It has initiated over the years, and operates a large number of various ICT supported services. The center maintains the regional ICT infrastructure, provides technical services, supports tens of research projects held in trial plots, dedicated structures, greenhouses and other environments and manages the input of data regarding irrigation, crops, pests, product quality and more. The center employs three professionals - two technicians and a director who provide the ICT services to the Arava regional council. The center's main activities involve technical support, data management and information dissemination. A major effort is spent on early detection of ICT development flaws and system constraint in view costly remedial measures and scarce human capital.

The research management methodology and research decisions follow the national framework of decision making (Loebenstein and Putievsky, 2007) with the regional elements considered and added on by the regional farmer's research committees (Gelb and Kislev, 1982). The regional research management is totally dependant on ICT supported tools as is the region's interaction with the national research programs. The bi - directional flow of data, information, research results and interim feed back to farmers, extension, scientists and service providers have made the region an agricultural success story.

The regional agricultural ICT infrastructure, 24/7, includes an extensive communication network and services, data, information and knowledge bases, computer supported control systems - *e.g.* greenhouse environmental controllers, irrigation, water recycling, fertilization management and fish pond management, data loggers, agro meteorology facilities and various computers and computer embedded devices.

Information management is the center's major activity. It involves collection of a wide variety of data types from a wide range of sources, verifying the data and processing it into reports. The information is then further processed for support of regional and individual operative decisions, input to knowledge bases and backed up for future use. Regional and national users and beneficiaries of the data include farmers, extension, scientists and service providers. The Information and Communication Technologies operated by the center and farmers involved are the most developed and innovative systems and equipment available on the market. Communication interactivity and information dissemination are both passive and active. This includes routine information *e.g.* weather and pest reports, current market data, extension recommendations and in turn research results and regional organizational information.

The extensive agricultural research activities in the Arava are managed by the Arava regional R&D center in collaboration with the regional ICT center. Both are publicly funded - mainly by the national Agricultural Research infrastructure and the region. Research results are evaluated by both entities mainly with the help of and dependant on ICT

ICT supported research programs and production feedback activities include a wide variety of subjects:

- A fruit fly eradication program supported by a computerized Geographical Information System (GIS);
- Regional greenhouse knowledge base management based on real time data input;
- On Line Research and production cost accounting;
- Online agro-meteorological stations with real time accessing;
- Real time product quality assurance with inventory control;
- Trial plot and research data monitoring;
- Plant protection policy and implementation follow-up;
- Regional agricultural parameter monitoring and follow-up;
- Ornamental fish growing research management;
- Growth, production, marketing and climate models - development and verification;
- Water quality monitoring and ongoing evaluation;
- Regional water balance monitoring.

The details in this appendix were gleaned from the 2006 report by I. Tzabari, the Director of the Arava R&D ICT activities.

Appendix D: The Role of Extension in Developing a Management Information System (MIS) - The Cohen Nursery Case Study

Expansion of the Cohen Nursery to 32,000 m^2 while exporting 34 million cuttings of 200 different flower varieties dictated the need for a MIS. Extension identified at the time the benefit to be gained by a Management System with a quicker response to customer orders, reduction of the necessary buffer stock and lower inventory margins. Specifically Extension pinpointed the main marketing and production problems - the follow up of orders; production to fit orders; shipment monitoring; invoicing control and accounting. Extension's contribution was to help with the system analysis, the definition of specifications, provision of horticultural knowledge and a sophisticated way of dealing with order confirmations. Extension did not finance the MIS development and was not paid - representing a public sector investment.

The software company delivered the initial MIS in 1993. Due to a change of programmers in 1995 they had to reprogram the MIS with a different software company. Being experienced they defined their specific needs clearly thereby considerably reducing development time and cost. The new program developers, in turn, intended to use their experience with the Cohen nursery to provide the new MIS to other nurseries - with their development costs covered - at a lower cost. Both these goals justified the continued involvement of Extension personnel. The reasons for the decision to expand Extension's involvement were:

- To help to avoid the repetition of conceptual and horticultural mistakes;
- To encourage the development of a variety of MIS products;
- To assist software developers in developing a product compatible with needs;
- To encourage more software developers to enter the market;
- To reduce the cost of the end product to the farmer.

The revised MIS was demonstrated to other leading nursery managers who were in the process of either adopting this specific MIS or developing their own. Management standards in nurseries were being refined accordingly. Evaluation of the results indicated that "tailoring" was feasible. The C/B analysis will be possible after a longer period, which will in turn enable a measure of Extension's contribution and its benefit. Extension's role was proved to be vital by ensuring a superior customized IT product with a proven benefit but the main question remains: Should the public pay, and till when, for developing farmer's management tools and training them to use them?

The Cohen nursery MIS experience indicates that the result justifies Extension's early involvement, initiation and evaluation of these systems. From this point onward Extension's experience and comparative advantage in matching MIS requirements with available program characteristics were evident. The Cohen nursery MIS development gave Extension a chance to intimately acquaint itself with issues involved in developing customized, small scale MIS without the need to be technically involved in programming. This methodology can be up-scaled, modified and shared with other types of programs.

To be specific the development steps taken were to:

a. Identify a need;

b. Identify available professionals that can carry through ICT development and adoption;

c. Identify and engage them for implementation and necessary follow up;

d. Transfer the results to the largest group of potential beneficiaries - within an existing local framework;

e. Involve beneficiary's participatory financing.

A critical success factor is identification of agents of change – in this case the Cohen nursery. If however this type of involvement is expected of the Extension Service

a. What proficiencies should Extension acquire or have?

b. Should that specialized service be a source of income for Extension?

c. What is the limit of Extension's responsibility for the quality of the MIS?

d. What will be the demands for future involvement in development and implementation?

The answers to these and other questions are crucial at this point in time (1995) for Extension Services - in Israel and other countries - undergoing structural reviews and changes, dwindling public funding for rural sector development and donors under growing pressure to see through their projects till they bear fruit.

(Source: Wolfson and Gelb – 1995 internal IES report)

Appendix:E. Internet Services for Agricultural Extension in Israel (An Israeli Situation Review from 1996)

Since the beginning of 1997, the IES has been running an intensive "Adopting Internet" program. The purpose of this program is to turn Internet applications into a service routine which means including all 230 service Subject Matter Specialists. In 1996 only about 40 per cent of the specialists had access to Internet but only half of them used it as a routine.

There are hundreds of agricultural Internet sites available Worldwide, but virtually none of them benefit the Israeli farmer in his daily production - mainly because very few are written in Hebrew and relate to Israeli specifics. Most of the available sites provide general information while others developed by agricultural commodity suppliers are designed to serve their clients.

Additional local sites are under construction, designed to provide necessary information and knowledge to the Israeli farmers. All of them are funded by farmer organizations with the support of IES Subject Matter Specialists ensuring an Extension component on site. An issue yet to be reconciled is the IES responsibility, if at all, for updating the information on the site and its impartiality. Sites providing Forums, Chat rooms, Advertising boards and Hebrew e-mail are being developed and will eventually suit IES's responsibilities and goals.

The current list of suitable Internet sites includes 4 background sites; a regional information site, two research institute sites and 3 agri-business related sites. None, however, provide specific agronomic recommendations or local product price details. Weather forecasts and news in general are available along with data from the Central Bureau of Statistics.

Internet use can compensate for the fact that an Extension specialist cannot provide all the professional information required by modern agriculture. Especially since Extension is expected to provide current and comprehensive answers continuously on demand. Using the Internet as a communication infrastructure and information system is most cost effective, especially compared to establishing an in- house alternative. With this in mind, and the needs and the opportunities so obvious, adoption would seem to be a simple task. In fact, 8 months since providing the specialists access to the Internet, uptake is still slow, difficult and inefficient.

Our short experience of using the Internet indicates that it is a perfect tool for Extension Service needs. It enables the following:

- Exchange of information between specialist/specialized groups;
- Exchange of information between specialist and the farmers;
- A short response time;
- Integration of disciplines;
- Accessing information from outside organizational sources, using the "net" as a source of information.

We found the critical success factors for Internet adoption to be:

- Technological infrastructure;
- Computer and Internet literacy;
- Relevant and timely information availability;
- Change in work habits;
- Focused organizational effort.

(Source: Y. Levi, 1997, condensed from a presentation at the Alberese workshop http://departments.agri.huji.ac.il/economics/gelb-sum-12.pdf)

Authors Information

Dr. E. Gelb

Center for Agricultural Economic Research,

Samuel Neaman Institute for Advanced studies in Science and Technology

9 Hagalil Street, Rehovot, 76601, Israel

Phone No: +972 8 9411677

e-mail: gelb@agri.huji.ac.il

Dr. E. Gelb, Former CIO Ministry of Agriculture and Rural Development, Israel. Responsible since 1980 for Integration of ICT, research, information systems and applications into agricultural production, rural services, rural development, extension and education; ICT policy coordinator; International cooperation and coordination. Currently an ICT consultant on agriculture and rural development.

Dr. B. Gal

Head of the Extension Service Farm Management and Economics Department

Ministry of Agriculture and Rural Development, Israel

P.O. Box. 28 Beit Dagan, 50250, Israel

Phone No: +972 3 9485349

e-mail: bragal@shaham.moag.gov.il

Dr. B. Gal, Head of the Extension Service Farm Management and Economics Department. Responsibilities include: Advising farmers, Farmers' Organizations, Extension, and Ministry of Agriculture departments; Promoting Information Technology (IT) adoption in the agricultural sector as well as within the organization;. Preparing cost accounting budgets and investment evaluations Conducting courses in Farm Management and Economics.

Mr. D. Wolfson

Senior Extension Specialist, Floriculture Division

Ministry of Agriculture and Rural Development, Israel

P.O. Box. 28 Beit Dagan, 50250, Israel

Phone No: +972 3 948 5649

e-mail: dwolfson@shaham.moag.gov.il

Mr. D. Wolfson is a Senior Extension Specialist, Floriculture Division. Focused on Extension. Training Coordinator and Product's Quality Coordinating. Formally an Extension Specialist, Regional Office of Extension Service of the Ministry of Agriculture, Rehovot Israel - specializing in cut flowers and potted plants production; Plant Protection Supervisor, Shemi nursery. Shedema Israel; Research Assistant, Department of Plant Protection, Agricultural Research Organization Volcani Center, Bet Dagan Israel.

❑❑❑

Chapter 8

JAPAN

Masami Yamada

1. OVERVIEW OF AGRICULTURAL SCENARIO AND EXTENSION SYSTEMS IN JAPAN

1.1. Japanese Agricultural Policy and its Short History

The modernization of Japan's agriculture was started through mechanization such as cultivators, rice planting machines, tractors and combines in the 1960s. Mechanization led to a large decrease in hours spent on rice cultivation, and the redundant labor force moved from agriculture to the manufacturing industry. This movement resulted in the modernization of rice cultivation, high productivity and high income for farm households.

On the other hand, a decreased demand for rice caused by westernization of the diet, *e.g.* greater consumption of animal products, became a serious problem. The national government decided to regulate rice production, and the acreage of paddy fields equivalent to the excess rice production has been used to cultivate other plants since 1978. Nowadays, the permitted area for planting rice is about 60 per cent of the paddy fields, with the other 40 per cent used to cultivate other plants such as barley, soybean, buckwheat and so on.

Through these agricultural changes, other problems arose as follows:

- A lower self-sufficiency ratio for food through an increase in imports of agricultural products
- Rapid aging of farmers resulting from the lack of young successors
- Decrease in agricultural land acreage through conversion to residential use
- Diminished vitality of agricultural communities resulting from migration to cities
- As before, small average acreage per farm household

To handle these anxieties, the national government established a new agricultural basic law called the Basic Law on Food, Agriculture and Rural Areas in 1999. The law has four basic principals as follows:

- To provide food to citizens stably
- To conserve diversified roles in agriculture and rural areas
- To support sustainable development of agriculture by guaranteeing agricultural resources and successors
- To revitalize rural areas by improving production conditions and living environment

The new basic law was not only made from the standpoint of farmers but also citizens, including consumers. The principals of the law show an ideal image of food, agriculture and rural areas.

Every five years, the government discloses as a basic plan the direction for the policies which promote the principals of this law, based on changes in the agricultural situation. The present basic plan is the second plan since the basic law was established. The third plan will be disclosed in the spring of 2010.

1.2. Agricultural Extension System as a Realization of Agricultural Policies

To realize agricultural policies, agricultural extension systems which approach farmers directly on site perform an important role, as well as government subsidies. There are two main agricultural extension services in Japan. One is a public cooperative work between the national and prefectural governments by extension advisors. Another is Japan Agricultural Cooperatives (JA: the biggest farmers' association in Japan) farm guidance service by JA farm advisors. In addition, private companies play a small part in private extension (Yamada, 2008).

The public extension service is a cooperative work between the national and prefectural government and is accomplished by 8,600 (in 2006) extension advisors. The advisors who work on site have a national government qualification, but they are employees of the prefectural government and they promote not only prefectural policies but also national policies.

On the other hand, the 14,000 JA farm advisors are employees of JA. They offer comprehensive advice for farmers from cultivation techniques with agricultural materials to shipping of farmers' agricultural products. They organize farmers' groups and are indispensable for small farmers who do not have their own sales channel.

These public and JA extension activities are very important to promote national policies to farmers and consumers. However, the basic problem of the extension services is a decrease in the number of extension advisors and JA farm advisors, which decreases 24 per cent and 26 per cent respectively in two decades. This means that they carry out their effective activities with fewer advisors.

2. THE IMPORTANCE OF INFORMATION AND COMMUNICATION TECHNOLOGY (ICT) IN AGRICULTURAL EXTENSION

2.1. The Importance of Information in Agricultural Extension

Farmers require a great deal of information such as plant nature, soil fertility, weather forecast, chemicals, labor and so on, for cultivation, information about the changing trends in the marketplace for selling products and information about the direction of agricultural policies for farm management. Agriculture is an industry which needs much up-to-date information.

The importance of such information has not changed through the years. However, to use the information effectively, an effective system for gathering and providing information is required. Fortunately, extension advisors and farmers can now get necessary information from ICT devices such as a personal computer, fax, mobile phone, *etc.*

2.2. The importance of ICTs in Agricultural Extension

There are 2,000,000 farm households and 8,600 public extension advisors in Japan, which means one advisor for every 230 farm households. It is very difficult to advise all of them using conventional extension methods such as workshops or conversations. In addition, about 80 per cent are part-time farmers who work other jobs on weekdays, making it difficult to provide information using conventional methods. Farmers, on the other hand, want to get information when they need without having to attend talks.

From these points of view, the utilization of ICT devices such as the Internet is important to satisfy both advisors and farmers. Furthermore, an information providing service using ICT devices makes it easy to provide information not only to farmers but also to consumers, who were not a target of the agricultural extension service. This is expected to promote mutual understanding between farmers and consumers.

2.3. Advantages and Disadvantages of New Extension Methods Using ICT

Providing information *via* a computer system is not enough to communicate concrete content as compared with conventional extension methods such as talks, demonstrations, demonstration farms, *etc.*

On the other hand, there are many advantages of a computer system, which includes web pages, electronic conferencing, e-mail, *etc.*

First, it is possible to store many messages and it is possible not to communicate simultaneously. A farmers' group mailing list is a good example: they do not need to have actual meetings and the members can attend the electric meeting over the computer at their convenience and express their opinions anytime.

Second, there is no distance between senders and recipients. In other words, it is possible to exchange information widely. Farmers do not need to gather in a conference room and they can communicate with any person in the world.

Third, a computer system is interactive. Mass media, one of the conventional extension methods, is a one-way method of providing information. However, a computer system can promote interactive exchange between the sender and the recipients.

Given these three advantages, it will be possible to reduce the extension activity cost per farmer using a computer system. A meeting is a good example: it requires a great deal of time and money for farmers to gather in a meeting room, yet attending an electric meeting over a computer system is effective and less costly (Table 1) (Fukuda, 2004).

Table 1 : Comparison of extension activity method between conventional and ICT methods

		Conventional Extension Methods			ICT Extension Methods		
		Workshops	Demonstration Farms	Dialogue	FAX	Web pages	E-mail
Extension activities	Introduction of new technologies	VS	VS	S	VS	VS	S
	Promotion of business solution	S	VS	VS	S	-	S
	Cooperation with relevant organizations	-	-	S	S	S	S
Characteristic of extension methods	Promptness	S	-	S	VS	S	VS
	Wideness	-	-	-	S	VS	S
	Asynchronism	-	-	-	S	VS	VS
	Concreteness	S	VS	S	S	S	S

-=unsuitable, S=suitable, VS=very suitable
(Source: Fukuda, 2004)

2.4. Popularization of the Internet Useful for Agriculture

The Internet and mobile phones have been popular since the later half of the 1990s. Now, the rate of the Internet users is almost saturated at about 70 per cent of the population. These data suggest that the Internet is popular even in rural area and among farmers.

However, the Internet connection speed is not sufficient in rural areas. This makes it difficult to get information which requires a great deal of data, such as video information and this should be improved.

While there are still some problems, it is very easy to construct an information system using the Internet and the system can become a strong tool not only for agricultural extension but also for farm management.

2.5. The Role of the Public Extension System for Providing Information

The growing popularity of personal computers and mobile phones is creating the conditions to get necessary information from the Internet even in rural areas. How to provide information necessary for farmers is very important. Generally speaking, informational materials on machinery, fertilizers, chemicals, *etc.*, is provided by private companies, because that information leads to the companies' benefit. On the other hand, providing information about disease and pest forecast, rural activities, research results, agricultural policy, *etc.*, is an important role of the public extension service. This kind of information providing system is also considered from a cost-effectiveness point of view.

3. INFORMATION TECHNOLOGY (IT) SCENARIO IN JAPAN

3.1. Short History of National IT Strategy

Before the Internet was popularized, personal computer communication system had been used in the academic sector since 1980s. Use of the Internet spread rapidly after the appearance of commercial internet service providers in the mid 1990s. While Internet users connected by dial-up, ADSL (Asymmetric Digital Subscriber Line) which do not need dial-up started in 1999.

ADSL made a contribution to the rapid increase of the Internet users because of being continually connected and free of telephone charges. Recently, super high-speed fiber-optic communication called FTTH (Fiber To The Home) has rapidly become popular in urban areas.

Based on the information infrastructure, a package of policies which were established from 2000 to 2001 strongly promoted an information society. The national government set up the Information Technology Strategy Headquarters and the Information Technology Strategy Council in July 2000.

The government also passed the Basic Law for Building an Advanced Info-Communications Network (Basic IT Law) in 2001. Shortly thereafter, the government launched the e-Japan Strategy as an initiative aimed at making Japan a front-runner in information technology.

In the five years of e-Japan Strategy (2001-2005), internet available mobile phones and the e-commerce environment spread rapidly. By 2005, the country had established the world's fastest and cheapest broadband environment (Figure 1).

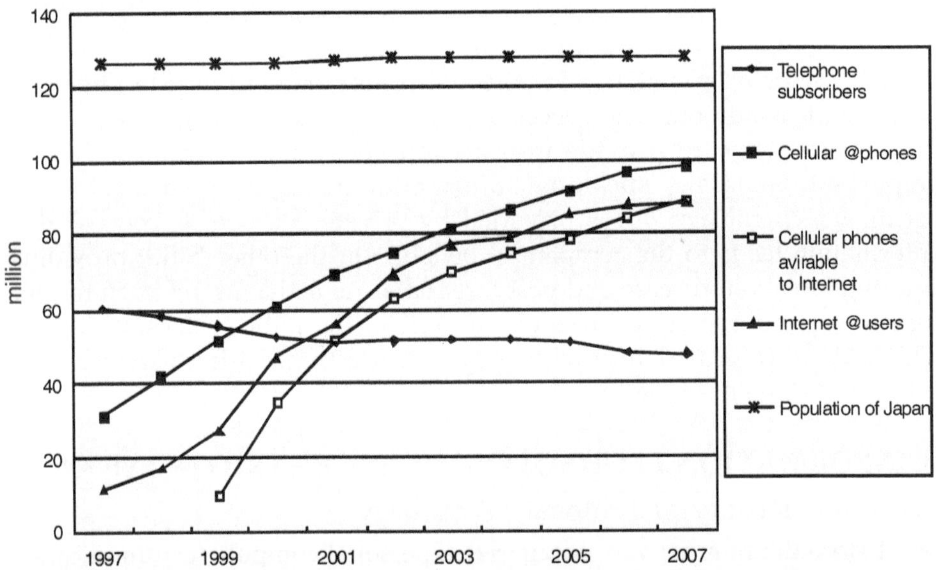

Figure 1 : Development of Information and Communication Service in Japan (Souce: Ministry of Internal Affairs and Communications)

3.2. IT Strategy in the Agricultural Sector

On the basis of the government's e-Japan Strategy, the Ministry of Agriculture, Forestry and Fisheries of Japan (MAFF) created an IT Strategy Study Group and corresponding work team and in 2000 discussed IT Strategy in Agriculture, Forestry and Fisheries in the 21st Century. The concrete image of the strategy is as follows (MAFF, 2000):

3.2.1. Utilization for Farm Management

- Farmers develop their information literacy for accumulation and use of data on fields and to gather information useful for production and farm management such as market status, weather, cultivation technology, consumer attitudes, *etc.*
- Farmers can make their management plans using managerial diagnosis software, use information devices in their fields to receive, transmit and store information and receive technical advice from extension advisors.
- Farmers can broaden their sales channels by selling directly over the Internet and can develop small quantity production of a variety of products, a characteristic of agriculture in Japan.

3.2.2. Provision of Information to Consumers

- Consumers can get information about agricultural products and food safety quickly and easily, increasing consumers' opportunities to choose products rationally.
- Urban residents can get a variety of information about rural areas such as special local products, green tourism, *etc.*, quickly and easily, which contributes to healthy and happy living and children's emotional education by coming in contact directly with nature and culture in rural areas.

3.2.3. Improvement of Government Services

- A network among local agricultural institutes can improve government service and allow sharing of data on fields and farmers' data.

3.2.4. Improvement of Food Distribution Systems

- Distribution cost is reduced by e-commerce and an efficient distribution system.
- A system integrating transactions, payment and distribution can be created for business-to-business transactions, particularly for fresh food.

3.2.5. Resource Management

- Utilization of satellite data about fields and forests promotes improvement of GIS (Geographic Information System) and government services can be improved through more sophisticated resource management.

3.2.6. Improvement of Convenience of Living in Rural Areas

- Rural residents can enjoy services for a high quality of life, even in remote areas.
- Rural residents can receive high quality health and welfare services at home or through a remote health management system.
- Rural residents can enjoy high quality education and cultural activities *via* a remote education system.
- Rural residents can discover urban residents' or consumers' needs regarding rural areas.
- Rural residents can get a good deal of information about government, community, disaster prevention, *etc.*, at home.
- Promotion of communication with urban residents allows for exchange with cities.

The present goals of these ICT plans are summarized in the following three points.

- Motivated farmers can make it possible to develop sustainable farm management using a great deal of information such as market trends, weather data, consumers' needs, *etc.*
- Along with a reduction of distribution costs and provision of products with reasonable prices, consumers too can have increased opportunities to choose products referring to information about their safety.
- Rural residents can get the same level of information from the Internet, and they can lead affluent lives in a rich natural environment.

4. NATIONAL POLICY ON ICTs IN AGRICULTURAL EXTENSION

4.1. Establishment of a National Agricultural Extension Information Centre

Since the extension service was started in 1948, many kinds of information such as on-site activities, technology, farm management and circulation have been gathered. Rising momentum for the effective utilization of this information in many extension centres led to the establishment of an extension information centre in the Japan Agricultural Development and Extension Association (JADEA) in 1975.

The original work of the extension information centre was to provide relevant information for extension activities and government policies. The centre published printed materials which included recent agricultural policies, information on rural living or title lists of important documents every month.

Several thousand examples of the achievement of extension activities in every year were collected from local extension centres with extension advisors' cooperation. The centre provides this information on request to local extension advisors and sometimes provides information *via* printed material about very important subjects for all advisors.

4.2. Nationwide Extension Network System by PC Communication

Technology using personal computers as a communication tool became popular around 1985. The JADEA started to research the establishment of a PC network when a commercial network called PC-VAN was started in Japan. After two years of experimentation, a PC network system called F-VAN (F is a capital letter of Japanese "*Fukyu,*" which means extension and VAN means Value-Added Network) intended for use only in the agricultural extension service was started in 1990 (Figure 2).

The F-VAN had four kinds of services: electric bulletin boards, e-mail, provision of information and databases. Basically, the system was characterized by a communication-based system such as an electric bulletin board. The bulletin boards were divided into national level boards and prefectural level boards. The former consisted of specialized boards for extension work and conversation boards where extension advisors could exchange their opinions and the later consisted of prefectures' original boards.

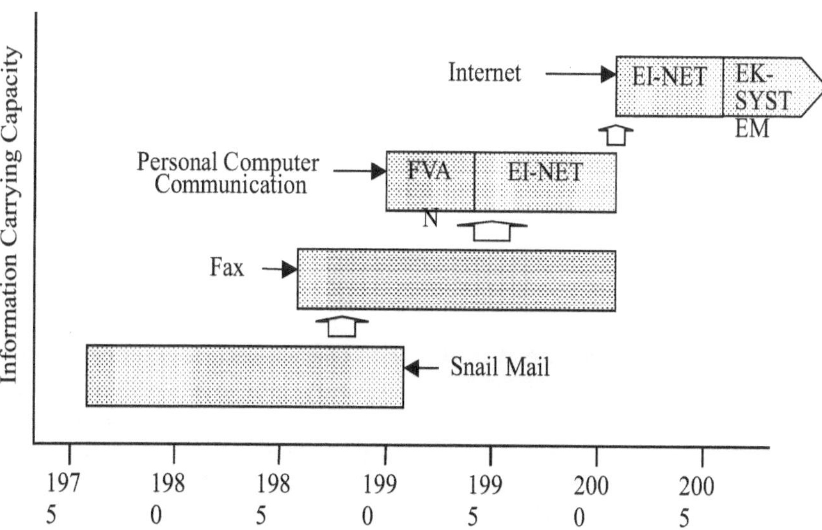

Figure 2 : Changes in the Methods of Providing "Case Information of Extension Activities" (Source: Fukuda, 2005)

The provision of information service included government policies, agricultural materials and weather data. The database service included information on extension activities, research results by research institutes, investigative research results by extension specialists and computer software.

Upon the official start in 1990, the number of participating terminals was 37 prefectural information centres in 47 centres and 247 extension centres in 604 centres. After three years, the number of terminals increased to 47 prefectural information centres, 487 extension centres, 15 MAFF organizations and 13 manufacturers (Yamada, 1999).

Amidst the rapid development of communication technology and information technology, the F-VAN was replaced by a sophisticated network system called EI-NET (Extension Information NETwork system) in 1994.

The characteristics of the EI-NET system are as follows:

- Many kinds of information were stored as image data or character data and users could receive image data by fax automatically or character data by PC communication or fax.
- External databases such as research results, weather information, market information, overseas information, *etc.*, were available.
- The e-mail system was enhanced to permit active communication among members.
- The user interface was enhanced *via* the development of special communication terminal software.

Another unique characteristic of EI-NET is that it was a local network system which promotes communication among extension advisors and farmers in an extension centre's area and farmers participate directly in the management of the system.

4.3. Nation Wide Extension Information Network Service *via* the Internet

As the Internet became popular, EI-NET system was connected to the Internet in 2001. Then communication became possible with Internet users, including research institutes, universities, companies, farmers and overseas users and users were able to exchange information with a variety of concerned parties.

As for program development, analysis and diagnosis software for farm management was developed over three years starting in 2000 and it was utilized as a useful means to analyze farm management.

In 2006, a new system called EK-SYSTEM (Extension Knowledge SYSTEM) was started. The system took over 60,000 data items accumulated over 30 years and it has a significantly improved online retrieval system as compared to the former EI-NET system.

A questionnaire survey on actual usage of the EK-SYSTEM was carried out through extension centres in 2008. The results showed that 16 per cent of extension advisors use the system more than once a week, whereas 25 per cent almost never use it. These results mean that the usage of the EK-SYSTEM depends on each advisor. Some advisors do not connect to the EK-SYSTEM, but they use it through another advisor.

Commonly-used services in the EK-SYSTEM are specialized services for extension advisors and their usage rates are high. For example, a database which includes extension activities and inspection reports and information exchange and consulting in specialized fields, show high usage rates (Table 2).

Table 2 : Services provided by EK-SYSTEM

Service items	Usage rate (%)
Database (Extension activities, inspection reports *etc.*)	45.8
Information exchange and consulting in specialized fields	40.3
Information from JADEA	34.7
List of names of extension advisors	31.9
Information on agricultural policies, *etc.*	27.8
Farmers' bulletin board	20.8
Information on computer software	6.9

(Source: JADEA, 2008)

Meanwhile, the main service users expect from the EK-SYSTEM are an enhancement of information exchange or consulting with specialists, agricultural policies concerned with extension service and technical information. In any case, the users want to get specialized information which is useful for their extension activities, information not available from other websites (Table 3).

Table 3 : Service users expectation from the new EK-SYSTEM

Expected services	Expectation rate
Information exchange or consulting with specialists	55.6 per cent
Agricultural policies about extension services	47.2 per cent
Enhancement of technical information	40.3 per cent
Information exchange with research institutes	38.9 per cent
Enhancement of technical information	31.9 per cent
Information of a variety of reports and conferences	26.4 per cent
Online training for improvement of skills	23.6 per cent
Posting training materials for producers	23.6 per cent
Enhancement of information for producers	19.4 per cent
Opinion exchange on specific subjects	18.1 per cent
Enhancement of web links especially for agriculture	15.3 per cent
Enhancement of information for consumers	13.9 per cent
Exchange with consumers and producers	11.1 per cent
External databases, such as weather information	9.7 per cent
Network system service for mobile phones	9.7 per cent
Other	2.8 per cent

Source: JADEA, 2008

4.4. e-Learning

e-Learning is an educational method which uses a personal computer or computer network. The advantages of an e-learning system as compared with classroom teaching include the ability to receive an education even from a remote place and to use special computerized teaching materials.

In 2006, an extra-governmental organization in MAFF started an e-learning training system that supports young people who want to begin farming. Students of the system can also get advice from agricultural specialists by e-mail and learn about opportunities for hands-on farming. In 2006, 47 students were enrolled in the fruit vegetable course, 29 in the leaf vegetable course and 11 in the poultry raising course, for a total of 87 students.

Another e-learning system for local extension advisors was started in 2005 to replace national level training seminars with four courses: management, community farming, agricultural pests and marketing. In 2007 and 2008, two new courses were added, a course on support for entrepreneurs in rural areas and a coaching program, for a total of six courses. These courses are highly popular among the extension staff, with about 240 students in the first year, 2005 and about 450 students in the second year, 2006. This kind of system will also be provided for farmers in the near future.

4.5. Information Dissemination Service to Regional Farmers by Prefectural Extension Centres

Information provision by prefectural governments to farmers and consumers *via* the Internet is very common. Even in the survey on August 2003, 457 of the 478 extension centres or about 95 per cent, had a website for their centre (Fukuda, 2004).

Common contents of the extension centres' websites are as follows:

- Public relations: Outline of agriculture and activities in the extension centre area
- Technology: Cultivation or feeding technologies, examples within the managed area, growth conditions, information on diseases and pests, weather, agricultural chemicals, *etc.*
- Management and living: Agricultural policies and management, leading farmers, information on women and daily life
- Other: Electric bulletin boards, overseas information, web links, *etc.*

The general purposes of the extension centres' website are to inform citizens about the centre's activities and to support farmers in extension activities. Technical support includes cultivation or feeding technologies, growth conditions, information on diseases and pests, weather, agricultural chemicals, *etc.*

For example, a site map of the "Fukui Agricultural Network" by the Department of Agriculture, Forestry and Fisheries in the Fukui Prefectural Government is shown in Figure 3 (http://www.agri-net.pref.fukui.jp/). There is a great deal of information about rice, because rice is an important agricultural product accounting for 70 per cent of production in Fukui Prefecture. This information is updated at a frequency of between several dozen times to a hundred times per month.

Top page

Technical information
This month's agricultural technologies
Forecast of disease and pest
Emergency technical countermeasure
Rice cultivation information system
Rice information
Agricultural weather information system
Information on barley, soybean, buckwheat, vegetables,
Flowers, orchards and livestock
Regional technical information

Prefecture's agricultural policies
Department of agriculture, forestry and fisheries
Agriculture, forestry and fisheries planning division
Food safety division
Market development division
Agricultural technology management division
Horticulture and livestock division
Farm area development division
Other information or policies

Research information
Research bulletin (Field Report)
Research results transferred to extension
Agricultural institutes' research report
Fertilizer dose diagnosis for *Koshihikari* rice
Materials about research results

Regional information
Agricultural information in the prefecture
Specialty products, traditional food
Report on local cuisine / food products
Agri-tourism, eco-tourism (green tourism)
Food safety information
For people who are interested in farming
Regional information

Web links
Extension centers?Fukui, Sak ai, Okuetsu, Tannan, Nishu, Wakasa? / National institutes / Weather bureau / JA / Markets / Municipalities / Academic institutes / Mass media / *etc.*

Figure 3 : Site map of "Fukui Agricultural Network" by Department of Agriculture, Forestry and Fisheries in the Fukui Prefectural Government

5. PRESENT SITUATION OF INFORMATIZATION IN AGRICULTURAL PRODUCTION SITES

5.1. ICT Devices in Farm Households and Internet Use

MAFF carried out a questionnaire survey on 6,000 farm households in 2005 and the 2,037 answers show the present situation of informatization in farm management (MAFF, 2006).

About two thirds of farm households have a personal computer (61.2 per cent) and mobile phone (70.9 per cent) as ICT devices. Internet utilization *via* personal computer is 69 per cent, and *via* mobile phone, 45 per cent. Usually, personal computers have large data storage capacity and a wide screen. Therefore, they are convenient for downloading image data or large data. On the other hand, a mobile phone is small enough to be portable, so it is useful for checking ever-changing information such as weather or market conditions.

As for communication lines, 39 per cent of farmers' personal computers are connected *via* an ordinary telephone line and 58 per cent are connected *via* high speed communication lines such as ADSL, CATV or optical fiber. An increase of connections to high speed lines is expected and communication *via* image data, sound data and video data will become popular.

5.2. ICT Devices for Farm Management

The utilization ratio of ICT devices such as personal computers or mobile phones for farm management is 24 per cent of farmers. Concretely, 21 per cent of farmers use personal computers and 8 per cent of farmers access the Internet *via* mobile phones.

The unexpected low utilization rate of ICT devices for farm management is considered to depend on the high rate of part-time farm households (77 per cent) that do not need advanced farm management. This fact suggests that most full-time households use ICT devices for their farming.

The purposes for personal computer use for farm management are business management (59.1 per cent), production management (50.6 per cent) and retrieving market or weather information from the Internet (44.9 per cent).

Internet capable mobile phones are small enough to be portable and can connect to the Internet at any time. The rate for retrieving market or weather information from the Internet by mobile phone, 62 per cent, is actually higher than the rate for PC's.

The biggest advantage of farmers' ICT use for their farm management is the ability to collect many kinds of information (51.6 per cent). This result shows that ICT devices are a good tool for information collection. The next highest advantage (32.7 per cent) was the ability to manage farms effectively. It is thought that farmers utilize personal computers to improve their farming *via* bookkeeping, activity diaries, *etc.*

On the other hand, the highest reason why farmers who have access to ICT but do not use it for farm management is "No need or interest for farm management" (43.1 per cent) and the second is "Looks difficult to operate" (22.6 per cent). These results suggest that most small farmers do not have high motivation for their farm management, because the farmers do not sell their products by themselves; instead, JA sells their products.

At any rate, ICT devices are absolutely necessary for full-time farmers to use in business management, production management and information collection.

The usefulness of mobile phone mail service for exchanging information with local extension centres has been reported (Fukuda, 2009). The report said that mobile phone mail is widely used not only by young farmers but also by older farmers, suggesting that the mobile phone mail service from extension centres has a role to play as a springboard to facilitate close communication with extension advisors.

5.3. Attitude of Advanced Farmers towards Various Information Sources

Farmers get a lot of information from conversations, newspapers, magazines, TV, radio, the Internet, and so on every day. It is very important to know the roles of information sources for advanced farmers. Kojiya and Ueda (2001) studied the ranking of information sources by advanced farmers who have their own website (Kojiya and Ueda, 2001). The study was carried out *via* an e-mail questionnaire and had 199 replies.

The result of principal component analysis shows that the many information sources used by ICT advanced farmers can be divided into three groups (Figure 4).

Group I information sources are bidirectional and reliable. The sources are highly credible and useful for farm management.

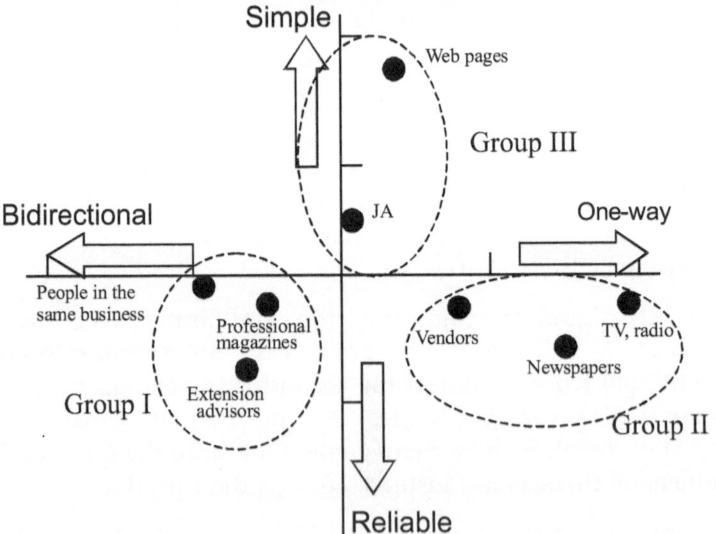

Figure 4 : Characteristics of information sources of advanced farmers (Analysis of principal component) (Source: Kojiya and Ueda, 2001)

Group II sources are reliable, but as they are one-way sources of information, it is difficult to consult regarding farm management.

Group III are simple sources of information. Web pages are one such source in this group, which is fast and simple rather than reliable.

This classification result suggests that ICT advanced farmers rely on conversation based communication with farm advisors or people in the same business when they make a decision. This implies that in such cases, information from web pages, where it is easy to gather a great deal of information, is used as a supplementary measure when making management decisions.

6. A CASE OF REVITALIZATION OF A MOUNTAINOUS AREA BY ICT

6.1. "*Tsumamono*" Ordering System by ICT

There are many cases which use ICT to vitalize regional agriculture in Japan. This section introduces a depopulated mountainous area vitalized by ICT.

Kamikatsu Town in Tokushima Prefecture is a very small mountainous town with a population of 2,000, a population which, like other rural areas, is aging. The ratio of elderly people at 65 or older is very high, standing at 49 per cent in Kamikatsu Town, whereas the nationwide average is 22 per cent.

Agriculture in this area is known for production and shipping of *Tsumamono* (seasonal leaves and flowers used as a garnish for Japanese dishes). *Tsumamono* is usually used in fine dining, and the amount and the type depends on chefs' ideas. For this reason, farmers always need to prepare a wide variety but small quantity of *Tsumamono*. Therefore, a new system was required to respond to rush orders and specific orders. An information system which transmits rush orders and specific orders quickly to the farm producers in the area and collects the required amounts was established.

The flow of the ordering information system is shown in Figure 5. At first, orders from the market to JA are transmitted to all farm producers by broadcast fax from the network system. The information centre takes calls from producers willing to accept the order on a first-come-first-served basis. The producers pack the order for *Tsumamono* with a bar-code and bring it to a shipment station. In the station, the bar-code is scanned and shipment information is managed on the network system. The *Tsumamono* is shipped to a market in the consuming region such as Kyoto, Osaka and Tokyo by airplane, and price is determined at the market.

Although most of the farm producers in this region are elderly, they effectively use this network system (Tsubota and Kuniho, 2005).

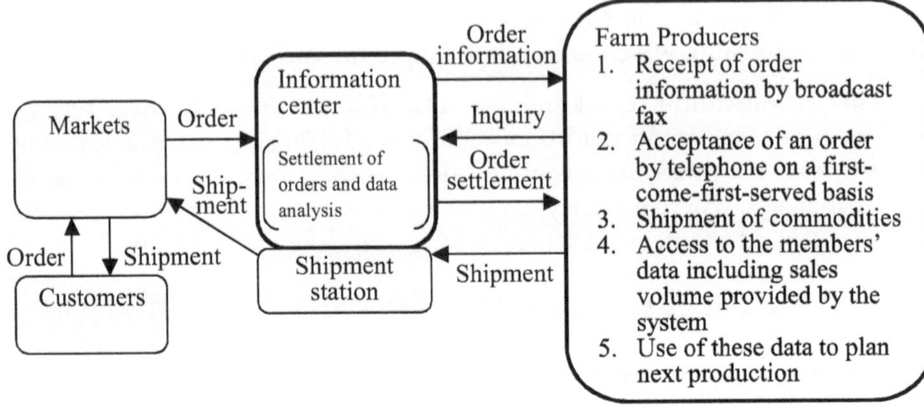

Figure 5 : The *Tsumamono* ordering information system in mountainous area (Kamikatsu Town in Tokushima Prefecture)

6.2. The Effect of the Introduction of the Information System on the Region

This system was developed for elderly members and it can be operated only with a trackball and numerical keypad. Workshops are organized frequently, making this a "System usable by everyone." Furthermore, by arousing a spirit of competition, this is made into a "System everyone wants to use."

In the network system, members are able to check not only on sales trends, shipping amounts and changes in unit prices, but also on each member's shipping performance, market price and rankings of monthly total amount of proceeds. Even if the elderly dislike personal computers, they have ended up using the system a few times a day because they can understand that the data is directly connected to their benefit. The system also heightens each member's motivation.

The establishment of the information network system promotes members' activities and makes it possible to respond to unexpected orders promptly. The system has affected the total sales of *Tsumamono,* which have increased from 142 million yen (about USD 1.4 million) in 1999 to 226 million yen (about USD 2.3 million) in 2005. The system provides the elderly with motivation and revitalizes the mountainous area.

Many kinds of network systems like this are functioning in Japan.

7. A CASE OF ICT UTILIZATION FOR DISTRIBUTION OF AGRICULTURAL PRODUCTS

7.1. Nationwide Network SEICA Bridge between Producers and Consumers

In the field of agricultural product distribution, there is much direct selling to consumers by individual farmers or private companies *via* the Internet. These businesses are increasing year by year.

The network system introduced in this chapter is not simply a system for selling products, but is an original, public online system for consumers to search for agricultural produce information (Sugiyama, 2008). The network system is called *Seika* ("green goods" in Japanese) Net Catalog SEICA, and is a web-based public database that started on August 2002. The system provides agricultural produce information for each product at http://seica.info/ by input of an eight digit SEICA number *via* a personal computer or mobile phone (Figure 6).

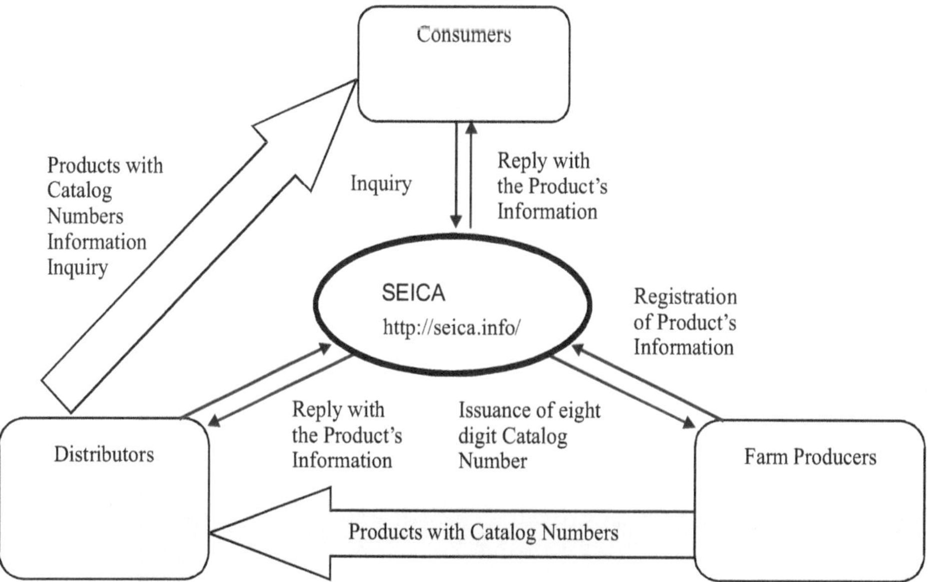

Figure 6 :SEICA network with Virtually Identified Produce System

- Web based traceability
- Free of charge, as it is a public database
- More than 1,700 items are covered
- XML Web Service implemented
- Mobile phone browsing is available

Registration of farmer's produce information in the SEICA database is accomplished by selecting from among 1,700 kinds of agricultural products and entering information about 140 items. Once registration is complete, a catalog of agricultural products is created automatically and an eight digit catalog number is issued. Once the eight digit number is attached to the agricultural product, wherever it ends up being sold, people interested in the product including consumers can access produce information using the eight digit catalog number. All services such as searching, browsing and registration on the SEICA system are operated by the public sector and are available free of charge (Figure 7).

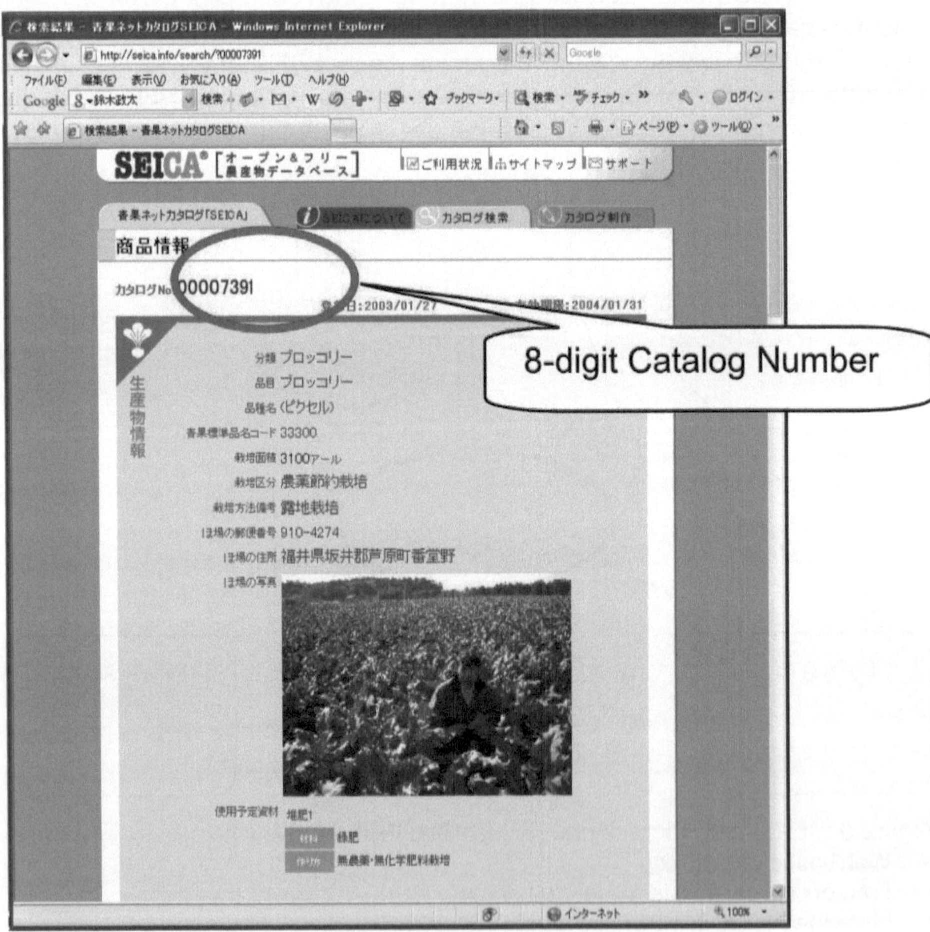

Figure 7 : SEICA search results by catalog number for broccoli
http://seica.info/search/?00007391

On SEICA, producers can register not only cultivation method, characteristics and shipping configuration but also voice and video clips, making it easy to communicate their commitment to their products directly to distributors and consumers.

SEICA provides a chance for producers, distributors and consumers to exchange information and producers can learn what consumers really think. These information exchanges promote production that meets consumer's needs, and it is expected that a new distribution of agricultural products will form.

SEICA, which is open to all people, is a useful reference for consumers in selecting safe food, allows distributors to provide more detailed product information and is useful for procuring products. The number of registered catalogs has steadily risen to nearly 10,000 as of December 2007 (Figure 8).

In the future, coordination with a point-of-sale (POS) system is a possibility (Shimakawa and Sugiyama, 2008).

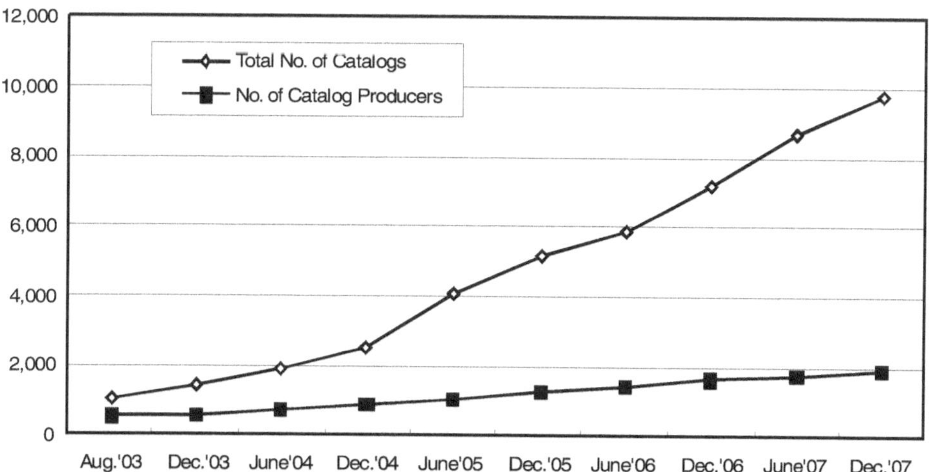

Figure 8 : Change in total number of catalogs and producers in SEICA

8. THE WAY FORWARD

The concrete future form of ICT available agriculture is shown in "IT strategy in the agricultural sector" in section 3.2. Although this form was revealed by MAFF in 2001, some aspects have already been brought about, while others are currently underway. Recently, ICT devices have become quite popular, so building a useful information system and enhancing the provision of necessary information has become very important.

From the standpoint of farm management, the amount of decision making information available from popularized ICT devices has significantly increased. However, there are many farmers who do not use ICT devices in their farming. Therefore, it is thought that the information gap between ICT using farmers and other farmers will increase.

Furthermore, ICT using farmers can get a lot of information from the Internet, but there is a lot of information which is difficult to evaluate because of different farming environments. In this case, farm advisors' opinions may be more important than ever in farmer's decisions. From this perspective, even though an ICT society comes about, the importance of public extension advisors will be maintained. Public extension advisors should improve their own abilities to be able to advise specialized professional farmers and make comprehensive decisions.

The nationwide information network system EK-SYSTEM for extension advisors has provided abundant information, such as 60,000 cases nationwide of extension activities. Yet many users of the system suggest improving and enhancing the capability to consult or exchange opinions with specialists to support specialized professional farmers. Furthermore, the e-learning system is a very effective training method for extension advisors, because they can learn in their spare time, so this system should be amplified to develop extension advisor's abilities.

References

Fukuda, K., 2004. A Study on the Characteristics and Usage of the New Extension Methods in Agricultural Extension Activities; Focusing on the "New Communication Technology" Based on Computer Systems, *Journal of Rural Community Studies*, No. 99, 2004, pp16-25.

Fukuda, K., 2005. Current Situations and Future Direction of Agricultural Extension Information Network System in Japan. *Seminar on Networking of the ATT&T*, 28th November - 1st December, 2005, Jakarta & Bogor, Indonesia.

Fukuda, K., 2009. Current Situation of and Prospect for Utilization of Mobile Phone Mail System in Extension Activities. *Spring Meeting of the Agricultural Extension Research Society of Japan*, 5th-6th March, 2009, Tokyo, Japan.

JADEA, 2008. EK-System News. *Monthly Journal of Agriculture and Extension*, 45(11), pp232-233.

Kojiya, H., and Ueda, S., 2001. Survey Report on Agriculture Information on the Internet Available to the Farmers who had Created Web Pages. *Research Report of Mie Prefecture Agricultural Technology Center*, 28, pp53-61.

MAFF, 2000. IT Strategy in Agriculture, Forestry and Fisheries in the 21st Century. *http://fgisf.ac.affrc.go.jp/forest/2001/MAFFITStrategy.pdf*, (Accessed on 31st January, 2009).

MAFF, 2006. Result of the survey on the use of personal computer and other devices in the farm, forestry and fisheries household. *http://www.maff.go.jp/j/finding/mind/pdf/20060308cyosa.pdf*, (Accessed on 31st January, 2009).

Shimakawa, Y., and Sugiyama, J., 2008. Design of Produce Item Number based on SEICA. *Agricultural Information Research*, 17(3), (2008), pp.51-58.

Sugiyama, J., 2008. Information Exchange Technology in the Agriculture and Food Industry. *Agricultural Information Research*, 17(3), (2008), pp.16-27.

Tsubota, T., and Kuniho, S., 2005. Leaf business which revitalizes a mountainous community. *http://www.nikkei.co.jp/digitalcore/local/18/index.html* (Accessed on 31st January, 2009).

Yamada, M., 1999. The Development of Computer Network Systems in Agricultural Extension Service in Japan and its Problems. *Agricultural Information Research,* 8(1), 1999, pp17-24.

Yamada, M., 2008. Japan. Country Chapter 7, In: R. Saravanan (Ed.) *Agricultural Extension: Worldwide Innovations*, New India Publishing Agency (NIPA), New Delhi.

Author Information

Mr. Masami Yamada

Technical Executive Officer
Dept. of Agriculture, Forestry and Fisheries
Fukui Prefectural Government
17-1, Ohte 3, Fukui city, Fukui Prefecture
Postal Code: 910-8580, JAPAN
Phone No. (O): +81(776)20-6049 (R): +81(776)36-9867
Mobile Phone No.: 090-2378-3925
Fax No.: +81(776)36-9867
e-mail: m-yamada@mitene.or.jp, m-yamada-1e@pref.fukui.lg.jp
URL: http://www.pref.fukui.jp/
Personal URL : http://www.mitene.or.jp/~m-yamada/

Mr. Masami Yamada has researched rice growth diagnosis, farm management *etc.*, in Fukui Agricultural Experiment Station for 18 years. After that, he established a prefectural rice growth diagnosis information system and expanded a prefectural extension information system as a subject matter specialist for 9 years. During the period, he chaired a committee of a nationwide extension information system for 4 years and published about 20 articles.

❑❑❑

Chapter 9

JORDAN

Esmat Alkaradsheh, Asmahan Farred Hattar,
Ashraf Saber Alhawamdeh & Samia Nadim Akroush

LIST OF ABBREVIATIONS

ICT	Information and Communication Technology
R&D	Research and Development
IT	Information Technology
JRV	Jordan Rift Valley
GDP	Gross Domestic Product
Ha	Hectare
MOA	Ministry of Agriculture
NCARE	National Centre for Agricultural Research and Extension
NCARTT	The National Centre for Agricultural Research and Technology Transfer
WB	World Bank
UNDP	United Nations Development Programme
FAO	Food and Agricultural Organization
USAID	United States Agency for International Development
GTZ	German Agency for Technical Cooperation

IDRC	International Development Research Center
JICA	Japan International Cooperation Agency
ICARDA	International Center for Agricultural Research in the Dry Area
ACSAD	Arab Center for the Studies of Arid Zones and Dry Lands
AARINNA	Association of Agricultural Research Institutions in the Near East and North Africa
AFESD	Arab Fund for Economic and Social Development
AOAD	Arab Organization for Agricultural Development
HCST	Higher Council for Science and Technology
ACC	Agriculture Credit Corporation
WFP	World Food Program
FFS	Farmers Field School
RSS	Royal Scientific Society
MOICT	Ministry of Information and Communications Technology
JTC	Jordan Telecommunication Corporation
MGM	Metro Goldwyn Mayer
JTC	Jordan Telecommunication Corporation
EU	European Union

1. GENERAL INTRODUCTION

Agriculture, as one of the most important economic sectors in Jordan, is under increasing pressure due to rising population, consumer affluence, pressure from globalization and a shrinking resource base. Previous production challenges in Jordan have been met largely through improvements in traditional agricultural inputs such as seeds and agrochemicals, but the present challenges need different solutions. Better management, which involves higher degrees of information and knowledge, is widely seen as a way to address these challenges. The potentials that ICT has to offer range from the sharing and re-use of data, research findings, lessons learned and best practices among R&D institutions to developing quick response mechanisms for agricultural and natural resources crises situations; from permitting informed decision making among the agricultural officials to sounder policy making among the legislators; from improving the extension delivery systems in the rural areas to bringing e-Commerce to the farmers.

Agricultural extension is about the application of information into knowledge and adoption, such as application of scientific research to agricultural practices through farmer training. ICT broadly defined as facilitation by electronic means the creation, storage, management and dissemination of information. Two types of ICT - old and new are practiced in extension service; the former refers to radio and TV which are low cost and require little skill to use while the latter forms of ICT include networked computers, satellite-sourced communication, wireless technology and the internet. Assume key factors explaining different ICT taken up by country are: per capita income; levels of education; the rural concentration; and dispersal of a country's population.

The development of ICT in Jordan is fast and the Jordan's IT capacity related to agro-information dissemination needs to be vastly improved at the local level to allow the effective horizontal transmission of information and communication between the farmers and institutions. However, technological challenges have still to be addressed such as the lack of trained communicators, poor ICT infrastructures high costs and limited access to ICTs and lack of relevant content.

2. BACKGROUND

The country of Jordan is located between 29° 11′ to 33° 22′ north, and 34° 19′ to 39° 18′ east, with an area of approximately 90 km^2 with altitude ranges from less than - 400 m at the surface of the Dead Sea up to the 1750 m of Jebel Rum. Jordan is bordered in the north by the Syrian Arab Republic, in the northeast by Iraq, in the southeast and south by Saudi Arabia, in the far southwest by the Gulf of Aqaba (northern shores of the Red Sea) and in the west by Israel and the West Bank.

The country is divided into four physiographic regions:

- The Jordan Rift Valley (JRV) along the western border of the country, with a total area of around 5 000 km^2, starts at Lake Tiberias in the north (212 m below sea level) and continues south through the Jordan Valley into the Dead Sea (417 m below sea level).
- The Highlands to the east of JRV, with a total area of around 5 000 km^2, run from north to south. They consist of ranges of mountains and plains at an altitude between 600 and 1 600 m above sea level and numerous side wadis sloping towards the JRV.
- The plains, with a total area of around 10 000 km^2, extend from north to south along the western borders of the Al-Badiah desert region.
- The Al-Badiah desert region in the east, with a total area of around 69 000 km^2, is an extension of the Arabian Desert.

Total population is about 5.7 million (2005), of which around 21 per cent is rural (Figure 1). The annual demographic growth is estimated at around 2.5 per cent during recent years not including fluctuations caused by international political events. Currently, about 90 per cent of the population is concentrated in the northwest quadrant of the country, where rainfall is highest and where most of the water resources are located.

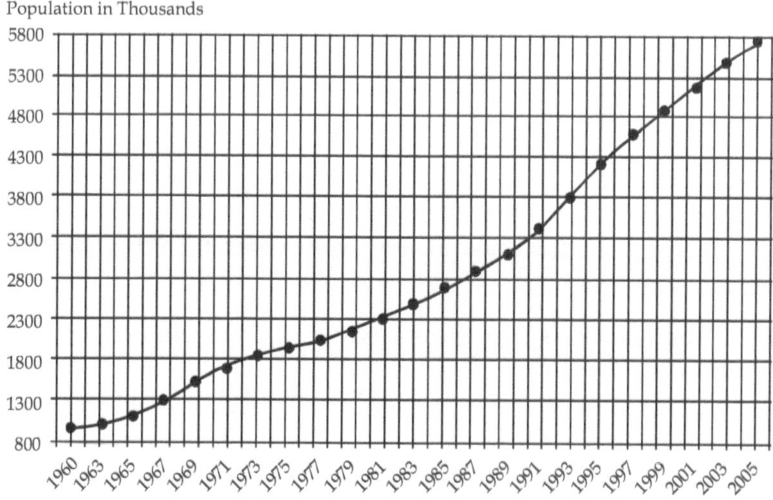

Figure 1 : Jordanian population growth

The climate varies from dry sub-humid Mediterranean in the north-west with rainfall of about 630 mm to desert conditions with less than 50 mm over distance of only 100 km. Rainfall varies considerably with location, mainly due to the country's topography and usually occurs between October and May. Over 91 per cent of the country receives less than 200 mm of rainfall per year. Average annual rainfall registered from 1937/38 to 2004/2005 is 94 mm but it is only 80 mm during the last ten years of this period (Directorate of Planning and Water Resources, 2005). The average for the period 1961-1990, given by IPCC, is 111 mm/year.

The climate of Jordan is semitropical in the JRV, Mediterranean in the Highlands and with continental influence in the eastern desert and plains region. Winter is the rainy season which is warm in the JRV, moderate to cool in the Highlands and extremely cold and dry in the desert land, whereas the summer is hot in the JRV, moderate in the Highlands and hot in the plains and the desert.

3. AGRICULTURE IN JORDAN

The Ministry of Agriculture is the apex body at the national level with the mandate to formulate policies and design and implement a number of programs that are aimed at bringing about a sustainable agricultural development in the country.

Agriculture is one of the important sectors in Jordan, socially and economically. The agricultural sector of Jordan employs 7 per cent of the labour force, and contributes 16 per cent to the country's exports. While primary agriculture accounts for only 3 per cent of the GDP, the entire food and agricultural sector (upstream and downstream linkages like agribusiness services and agro-industry) contributes about 30 per cent. In 2004 total agricultural production reached 2.13 million tons, of which 69 per cent were vegetables, 29.5 per cent fruits from fruit trees and 1.5 per cent field crops (cereals), which were consumed locally and exported to the markets of the neighboring countries.

Despite the country's insufficient water and cultivable land resources, the agricultural sector is considered as one of the most important economic pillars for integrated development in Jordan. The land suitable for cultivation is around 886 400 ha, or around 10 per cent of the total area of the country. In 2005, the total cultivated area was estimated at 270 000 ha, of which 184 000 ha consisted of annual crops and 86 000 ha of permanent crops.

However, it may happen that half of the rainfed land is left fallow in a year due to fluctuating and unevenly distributed annual rainfall. For instance, the harvested annual crops area was 168 435 ha in 2003 and 76 266 ha in 2004. Also, it is estimated that between 1975 and 2000, around 88 400 ha of good rainfed land was lost due to urban expansion. Data through the last three decades show an increase in irrigated land and in land planted with permanent crops, mainly in rainfed land of the Highlands.

Total population economically active in agriculture is estimated at 194 000 inhabitants, which is 9.8 per cent of economically active population in 2005, of which 70 per cent is female and 30 per cent is male. In JRV around 350 000 people are the main beneficiaries of irrigated agriculture and women form an important component of the labor force beside the foreign labors.

In spite of the low contribution of agriculture in the GDP, both rainfed and irrigated agriculture are vital socioeconomic activities in the country. They are the source of fresh vegetables all year, play an important role in the national economy and provide demographic stability in the rural communities.

Irrigated agriculture covered around 33 per cent of the cultivated area in 2004. Permanent crops represent 56 per cent of harvested irrigated area and 78 per cent of the harvested rainfed area. They consist of citrus, bananas, olives and vineyards. Main annual crops are vegetables, potatoes and cereals (wheat and barley).

Besides the climate (drought, fluctuating rainfall and hot winds) the main difficulties in rainfed agriculture are the fragmentation of farm holdings and the erosion of top soils in the steep slopes whereas the constraints in irrigated agriculture are the limited available water resources, overexploitation of groundwater, wastewater used in irrigation, silting of dams, and agricultural production marketing problems.

In general, the agricultural sector undergoes a strong competition from other sectors and it receives few national and international investments in comparison with other economic activities. Agriculture was not expected to remain the most important sector in the national economy, as it has been since the 1950s. The nature of subsequent socio-economic development, which led to the rapid development of other sectors of the economy, especially services and industry, has resulted in a continued decline in the contribution of agriculture to the GDP. The decline in the contribution of agriculture to the national economy could be attributed to the following main reasons:

- Failure of government policies to provide a suitable enabling environment to encourage the private sector to invest in agriculture.

- Failure of the private sector to establish agribusiness projects, with appropriate economic size and structures and to build strong organizations that would enhance farmers' and stakeholders' role in agricultural development and in participation in shaping and implementing agricultural development policies.
- Failure of public and private institutions to support agricultural development. Cooperatives and farmer associations failed to solve farmers' economic and social problems Moreover, universities and public institutions involved in research and technology transfer did not succeed in carrying out the duties assigned to them.

4. NATIONAL CENTRE FOR AGRICULTURAL RESEARCH AND EXTENSION (NCARE)

NCARE establishment goes back to the late 1950's, when the Department of Research and Extension had been created. In 1985, the research activities had been separated from the Department and enrolled under a new research centre "The National Centre for Agricultural Research and Technology Transfer, NCARTT". In 1993, NCARTT became a semi-autonomous institution. Until mid 2007, the responsibilities of the Ministry of Agriculture (MoA) and the National Centre for Agricultural Research and Technological Transfer (NCARTT) to provide technical assistance (extension services) to farmers overlapped. Since July 2007, the mission of NCARTT was amended and extended to include the extension services previously attributed to the MoA and NCARTT became the National Centre for Agricultural Research and Extension (NCARE). As the expansion of the mission of NCARE implies stronger focus on extension services, the Centre will have to re-orient its professional and financial means to eventually shift from purely fundamental and academic research activities to applied research activities to be able to answer technical constraints encountered by local farmers. NCARE built fruitful cooperation relationships with many national, regional and international institutions such as WB, EC, UNDP, FAO, USAID, GTZ, IDRC, JICA, ICARDA, ACSAD, AARINNA, AFESD, AOAD, HCST and many others. NCARE operates according to a special by-law as a semi-autonomous institution with administrative and financial independence. The centre receives its core funding from the government. Other direct fund is received from national and international donors to finance the implementation of proposed projects.

NCARE consists of a main headquarter in Baqa'a and seven regional centres located in Dair Alla, Ramtha, Mafraq, Mshaggar, Rabba, Tafilleh, and Shobbak., NCARE operates 12 research stations representing different agro-ecological conditions and 11 extension units and 12 subunits throughout Jordan. NCARE is mandated to conduct and coordinate applied agricultural research and extension to assist the transfer and adoption of appropriate and improved technologies, facilitate continuous professional development, and recognize the significant professional contributions of its members. Among the responsibilities of NCARE are:

- Organizing agricultural research and extension plans for sustainable development,
- Adopting the latest agricultural techniques for local conditions,
- Conducting economic studies aimed at improving farmers' organization,
- Capacity building of researchers and extension agents in collaboration with local, national and international partners and provides training and cross-learning opportunities in order to better addressing agricultural development problems.

5. AGRICULTURAL EXTENSION IN JORDAN

Presently agricultural extension activity in Jordan is being carried out through public institution (NCARE) and private sectors (NGO, GO, input supplier/ dealer, farmer organization, *etc.*).

In Jordan, the public extension is challenged with major bottlenecks that limit its reliability. Among the major challenges are the following:

- Insufficient logistical facilities,
- Low skills of the extension staff,
- Unclear job descriptions,
- Lack of working methodologies for on-farm follow-up, and
- Low motivation of the extension agents

These factors result in low presence of public extension agents on the field. Farmers instead obtain agricultural information and technical services from personal experience and observation or from the private sector (*e.g.* seeds companies, commercial vendors of equipment and materials). The private sector's role as a form of extension service has lately been increasing in importance, since it has been active in providing most of the new agricultural technologies and technical services. It must be noted though, that this private sector form of extension has the tendency to favor capitalistic intensive enterprises and to induce over-use of agricultural inputs.

As mentioned earlier, due to the overlap in providing the extension services to farmers between MoA and NCARTT, the mission of NCARTT was amended and extended to include the extension services in 2007 and NCARTT became the National Centre for Agricultural Research and Extension (NCARE). NCARE has started its mission in extension service with limited number, 69 extension agents (as shown in table 1), which summaries the characteristics of the extension agents at NCARE. A capacity building plans have been prepared to the extension agents to cover, beside the technical issues, communication skills training, problem solving and time management.

Table 1 : Characteristics of the extension agents at NCARE.

Characteristic		No.	Per cent
Sex	Male	48	69.6
	Female	21	30.4
Degree	Diploma	3	4.3
	B.Sc.	58	84.1
	M.Sc.	8	11.6
Specialization	Plant Production	17	24.6
	Plant Protection	17	24.6
	Soil & Irrigation	5	7.2
	Horticulture	8	11.6
	Animal Production	4	5.8
	Economics & Extension	14	20.3
	General Agriculture	2	2.9
	Field Crops	2	2.9
Date of birth	1965 - 1969	18	26.1
	1970 - 1974	13	18.8
	1975 - 1979	19	27.5
	1980 - 1984	8	11.6
	1985 -	11	15.9
	TOTAL	69	100
Experience in extension	< 5 years	40	58
	5 - 10 years	18	26.1
	10 - 15 years	9	13
	> 15 years	23	2.9

Away from NCARE, there are also some NGOs/private sectors that implement extension according to the integrated rural development approach.

Among those: The Agriculture Credit Corporation (ACC), Agricultural faculties in Jordanian universities and institutes of agricultural official, farmers organizations, Independent advisers, Ministry of Agriculture / Projects Department, Non-Governmental Organizations (NGOs) (*i.e.* The Jordanian Hashemite, Noor Al Hussein Foundation, Near East Foundation), International organization through their implemented projects (World Food Program (WFP), IFA, FAO, CARE International).

6. APPROACHES OF AGRICULTURAL EXTENSION IN JORDAN

Generally, the main features of the approach of the agricultural extension in Jordan are programmed or un-programmed extension services. The programmed extension services implement specialized extension programmes with specific goals in all production areas, which can be measured and planned in a decentralized manner by extension units and followed up by the central extension in NCARE, which implements up to 45 extension programmes per year. The un-programmed extension services use extension tools to meet the immediate needs of extension campaigns. Tools include field visits, TV and radio, publications, CDs, DVDs *etc.*

It is believed that extension agents are expected to become catalysts, mobilizing farmers to experiment on an identified need/solution, recognizing local innovations and helping to assess and encourage them. The approaches used to disseminate agricultural information to the farmers are:

6.1. Individual or Face-To-Face Method

The extension agent meets the farmer at home or on the farm and discusses issues of mutual interest, giving the farmer both information and advice. This individual contact between the extension agent and the farmer can take a number of forms: farm visits are the most common form of personal contact between the agent and the farmer and because they take up so much of the agent's time, it is important to be clear about the purpose of such visits and to plan them carefully. Just as the extension agent visits the farmer, so he can expect that from time to time the farmer will visit him at his office. Such a visit is often a reflection of the interest which the agent may have aroused among the local farmers. The more confidence local farmers have in the extension agent, the more likely they are to visit him. Such office visits are less time consuming for the extension worker, and offer some of the advantages of a farm visit. Although this approach seems efficient, but it is limited by the above mentioned challenges, mainly, the availability of logistics such as transportation and motivation.

6.2. Group Methods

The use of groups in extension has become more common over the past decade in Jordan, and a number of new ideas have emerged about how groups may be used most effectively. For example, the Regional Integrated Pest Programme in the Near East Project is based upon group methods, which is the approach of the Farmer Field Schools (FFS). The group method offers the possibility of greater extension coverage, and is therefore more cost-effective. Using the group method, the extension worker can reach more farmers and in this way make contact with many more farmers who have had no previous contact with extension activities. However, as veritable machinery, which is sustaining and relatively cost-effective, the farmer-group approach is dependent on sufficient mobilization at the grassroots and in social units in order to achieve the desired objectives of the approach.

6.2.1. Extension Programmers

The age-old practice of extension-farmer contact on a one-to-one basis, though very effective, is expensive and unsustainable as the sole means of reaching farmers with agricultural technology. New methods emphasize the passing on of agricultural technology to farmers in organized groups (farmer groups). A farmer group is a collection of farmers interacting with one another towards achieving a common goal. Usually, the interaction between the members of the group is more than with those outside the group. Membership of a group varies, and it is advantageous to have a small number of people forming it. A group size of between 15 and 20 is ideal and manageable in order to provide a face-to-face interaction, better communication and the free flow of information. The programmed extension services implement specialized extension programs with specific goals in all production areas, which can be measured. Planned in a decentralized manner by extension units and followed up by the central extension in NCARE, it implements up to 45 extension programs per year. The un-programmed extension services use extension tools to meet the immediate needs of extension campaigns. Tools include field visits, TV and radio, publications, CDs, DVDs *etc.*, examples on the extension programs are:

- Mushroom cultivation.
- Processing soap from olive oil.
- Home garden cultivation.
- Raising sheep and poultry -(family business).
- Farm Management of sheep.

- Bee-keeping.
- Food processing.
- Cultivation of aromatic and medicinal plants.

6.2.2. Farmer Field Schools (FFS)

Farmer Field Schools are schools without walls where groups of farmers meet periodically with facilitators during the crop or animal cycle. It is a participatory method of technology development and dissemination, based on adult learning principles and experimental learning. It reflects the four elements of experiential learning cycle, namely: concrete experience, observation and reflection, generalization and abstract conceptualization, and active experimentation.

The operation of the extension delivery approach is that developmental organizations partnered with extension personnel to identify or form farmer groups based on particular topics. This provides an opportunity for each participant to teach others what they have learned. At the end of the FFS cycle, certain farmers are chosen by the group to be farmer facilitators. They can then lead their own Farmer Field School the following season. The extension officer's role has evolved from that of a primary knowledge source to that of a facilitator of knowledge creation.

One important issue in FFS is that of sustainability without outside funding. It is a participatory approach, which facilitates farmers' demand for knowledge and offers opportunity for the end users to choose, test and adapt technologies according to their needs. Through participation in FFS, farmers develop skills that allow them to continually analyze their own situation and adapt to changing circumstances.

In Jordan this approach had been introduced in 2004 by the Food and Agricultural Organization (FAO) and funded by the Italian government through the Regional Integrated Pest Management programme in the Near East GTFS/REM/070/ITA. This program in Jordan trained more than 1500 farmers (20 per cent females) and succeeded to establish the farmer to farmer extension approach between farmers.

6.2.3. Demonstration Method

It is a another method which farmers like to see how a new idea works, and also what effect it can have on increasing their crop production. Both purposes can be achieved by means of a farm demonstration. A good, practical demonstration is an invaluable method in extension work. Method demonstrations basically show farmers how to do something. In the method demonstration, the farmer is shown the technology by step by step.

6.2.4. Field Days

Field days are usually opportunities to hold method or result demonstrations on a slightly larger scale and are usually run in a more informal and less highly structured manner. The purpose is often to introduce a new idea and to stimulate the interest of as many farmers as possible. Experimental stations or other government centres may be used for field days, but it is more usual and profitable for them to be held on a local farm and more impact if the farmer plays a part in running it and explaining the purpose.

6.2.5. Farmers Visits

Farmers like to visit farms in other districts to see how they work, what they grow and what kinds of problems the farmers there are facing. A tour is a series of field demonstrations on different farms, or at different centres, and can often attract a lot of interest from local farmers. The tour should give local farmers a chance to see how other farmers cultivate their land and to exchange ideas and experiences with them. It is important, therefore, that the area to be visited be in some way similar agriculturally to that of the visiting farmers.

6.3. Mass Media

Mass media, known as the channels of communication which can expose large numbers of people to the same information at the same time, are widely practiced in Jordan to convey different types of agricultural information. They include media which convey information by sound (radio, audio cassettes); moving pictures (television, film, video); and print (posters, newspapers, leaflets). The attraction of mass media to extension services in Jordan is mainly due to the high speed and low cost with which information can be communicated to people over a wide area. However, mass media cannot do all the jobs of an extension agent, and cannot offer personal advice and support, teach practical skills, or answer questions immediately.

6.3.1. Radio

Radio is now common feature among Jordanian rural communities where information can reach households directly and instantly throughout the country. Urgent news or warnings can be communicated far more quickly than through posters, extension agents or newspapers. Radio in Jordan is used for:

- Spreading awareness of new ideas to large numbers of people and can be used to publicize extension activities.
- Enable one community or group to share experiences, and
- Many extension agents had the opportunity to speak over the radio.

Radio currently broadcast three programs:

- "Green Land" for 40 minutes in the morning every day except Friday,
- "Wealthy Land" for 10 minutes, once in a week.
- "Land perfume" for 10 minutes, daily.

6.3.2. Television

Jordanian television since 1969, combines vision with sound and like radio, can also be an instant medium, transmitting information directly to a mass audience. Currently the television broadcasted extension clips produced by NCARE for (3-5) minutes and mostly these programs provide guidance and extension advice to farmers. Also the Jordanian TV has two regular programs related to agriculture:

- "My Home" for 40 minutes, once in a week.
- "Environment and Agriculture" for 15 minutes, once in a week

6.3.3. Films

The main advantage of film as a mass medium for extension is that it is visual; the audience can see as well as hear the information it contains. It is easier to hold an audience's attention when they have something to look at. It also makes it possible to explain things that are difficult to describe in words, for example, the color and shape of an insect pest or the correct way to transplant seedlings. During the last 10 years, extension produced 105 films in different topics, the timing for each film vary between 6-10 minutes and used to deliver these film *via* different way such as local TV and copied the films in (VHS&CD) and distribute to beneficiaries upon request and for the field staff to be used during training and implementing extension programs.

6.3.4. Campaign

In an extension campaign, several media are used in a coordinated way and over a limited period of time in order to achieve a particular extension objective. The advantage of campaign is that the media can support and reinforce one another.

6.3.5. Publications

One of the most important methods used in the extension is the publication. In Jordan more than 700 titles had been published and distributed to more than 20000 farmers, specialist, students *etc.*, this publication can give a brief description to the case which talks about with some guide pictures which can help farmers to solve their problems and find the solutions.

6.3.6. Exhibitions

More than 50 exhibitions conducted in Jordan through the extension department in different areas in Jordan and for different topics, one of the most important are the olive exhibitions which conducted yearly and more than 50 farmers participate in this exhibition. More than 5,000 visitors used to come to this exhibition every year. More than 200,000 US$ are the income of the farmers who participate in this exhibition and also many contracts done through it for local and international market directly from the site of the exhibition. Also now the geographical indication approach has been considered for many exhibitions and festivals. For example, there is an exhibition for Jordan valley products and another one for pomegranate crop.

6.3.7. Newspaper

Messages are published by official newspapers (Alrai, Addustour, the Arabs Alyom) almost daily. These messages include technical advice to farmers in all areas of agricultural production.

7. INFORMATION AND COMMUNICATION TECHNOLOGIES (ICTs)

7.1. Importance of Introducing ICT in Agricultural Extension

As indicated, agriculture is one of the most important economic sectors in Jordan, yet this sector is under increasing pressure due to population growth, pressure from globalization and shrinking resource base. Previous production challenges in Jordan have been met largely through improvements in traditional agricultural inputs such as seeds and agrochemicals, but the present challenges need different solutions. Better management, which involves higher degrees of information and knowledge, is widely seen as a way to address these challenges. Agricultural extension is about the application of information into knowledge and adoption, such as the application of scientific research to agricultural practices through farmer training. ICT broadly defined as facilitation by electronic means of creation, storage, management and dissemination of information.

Effective communication is considered the central of knowledge and information dissemination essential for empowering farmer communities. Traditional media and emerging Information and Communication Technologies (ICTs) have high potential in the diffusion of information to farmer communities. ICT is the product of the convergence of digital technologies encompassing computers, telecommunications, audio-video and publishing and can greatly enhance people's ability to share experiences and knowledge in support of agriculture development. It covers mobile phones, personal computers, the Internet, email, imaging technology, digital audio-video and digital broadcasts, even cable television. Products of ICT include databases, websites and multimedia packages, which may be in the form of CD-ROMs. Basically, ICT facilitates two elements critical in the Research-Extension-Farmer-Market Interface and technology transfer process: information access and networking. The storage and retrieval of research results facilitates information access while telecommunications facilitates networking.

7.2. The Information and Communication Technology (ICT) in Jordan

The Information Technology Centre (ITC) was founded in 1972, two years after the establishment of the Royal Scientific Society (RSS). It was the first to introduce computerization in Jordan. In 1992, the centre expanded its operations to cover industrial studies and conduct long term specialized training courses in the field of Information Technology at the regional level in cooperation with Japan International Cooperation Agency (JICA).

ITC is committed to maintaining well-performed applied research and specialized technical consultation and services based on approved national and international standards, procedures and methods of software development and studies, ITC has been ISO 9001 certified since, January 2001 and has also adopted RUP (Rational Unified Process) in project implementation. At the same time RSS has concentrated on building capacity and professional expertise in important fields like presentation skills, Database Design and Administration and Software Development (.NET, Oracle, Java, PHP, *etc.,*). The main objectives of ITC include:

- Transfer of accumulated knowledge, practical experiences in the Information and Computer Technology.
- Contribute to the socio–economic development in Jordan.

Fields of expertise:

- Software development,
- Establishment of integrated management information systems,
- Computerized requirement analysis studies,
- Supervision of computer centres,
- Database design and technical consultations,
- Computer Based Training (CBT) software,
- Specialized training in the field of computer and related subjects,
- Socio-economic studies,
- On-line information services from international databases, and
- Maintenance of PCs.

In 2002, the Ministry of Information and Communication Technology (MoI&CT) has been established as the heart of the Kingdom's transformation into the knowledge based economy that contributes to the socio-economic development of the Kingdom. MoI&CT is the policy maker of information technology, telecommunications and fosters ICT initiatives that promote the adoption of ICT by all segments of the society and stimulate local and foreign investments. The government's commitment to ICT as a catalyst to development, particularly through e-learning and e-government, is creating business opportunities. Jordan's ICT industry is thriving and has become a major contributor to Jordan's economy, boasting one of the more developed and advanced ICT sectors in the developing countries. Jordan's network readiness has been improving; ranking 44 out of 104 countries in 2004.The industry enjoys strong support from the government and His Majesty King Abdullah II, progressive regulatory and policy reform and is undergoing an ambitious privatization plan. It is growing locally, respected regionally and fuelled by a committed and talented pool of ICT professionals. Highly qualified human resources, the availability of world-class infrastructure, and the success of Jordanian IT companies contribute to positioning Jordan as a major regional IT hub. Jordan's dynamic and growing ICT environment presents attractive opportunities for investors and ICT businesses. Scientific research and technology is critical to the future of the ICT sector of Jordan. Government recognizes the key role it plays in providing an enabling environment for innovation and research and in building the human capital that is required for the future knowledge economy. Since early 2000 there have been important changes in the system of innovation in the country, specifically in the ICT sector. Jordanian institutions and their programs are better aligned with national needs. Innovation programs, particularly technology incubators, were established in Jordan to encourage innovative solution development in partnership with private sector. These have begun to strengthen technology diffusion and harness the entrepreneurship of the Science and Technology community.

Jordan's contribution to global research in the ICT sector has not been significant so far. This will require strengthening Jordan's linkages with global research networks, and developing Centres of Excellence in specific areas of R&D. In addition, the protection of intellectual property, and the support of products development from idea inception to business commercialization has to be facilitated. In this regard, it is critical that Government develops a strategic view on future scientific and technological research for the sector and address the above-mentioned issues effectively to allow better governance, more effective resource allocation and utilization and better outcomes in the short, medium and long term. Accordingly, the Ministry of Information and Communication Technology (MoI&CT) embarked on the task of drafting a Research and Development Strategy for the ICT sector.

ICT usually restricted to digital technologies. The former refers to newspapers, radio and TV which are low cost and require little skill to use. The latter forms of ICT include networked computers, satellite-sourced communication, wireless technology and the internet. Assumed key factors explaining different ICTs taken up by the country are: per capita income; levels of education; the rural concentration; and dispersal of the country's population.

The communication infrastructure in Jordan are reliable and modern. This is the Jordan Telecommunication Corporation (JTC) which is developing and expanding this line network. The telecommunication is not only the line network but also the cellular phones network, pagers data transmission and satellite networks. There is a fast growing demand for all these services. The development of internet in Jordan is also very fast. Most companies are now connected to the internet network.

7.3. National ICT Strategy

ICT initiatives in Jordan face challenges that do not arise mainly from hardware and software engineering issues, but from the nature of the existing social and administrative regimes, which cannot be easily re-engineered to accommodate the evolving network digital environment of the state model wished for by ICT optimists (Mofleh *et al.*, 2008). The development of the systems, networks and infrastructure necessary to make ICT initiatives work requires the re-alignment not only the government but also the social and cultural frameworks of the country.

In order to further strengthen its growing Information and Communication Technology (ICT) sector, Jordan has begun implementation of a new policy to enhance "e-participation", outlining a set of aggressive targets to boost internet usage and access across the country by 2011. The new ICT National Strategy 2007-2011, issued in late 2007, significantly revises the kingdom's initial 1999 ICT policy, which was known as the Reach Initiative. The new policy outlines a number of objectives for the country to reach within the next three years, including fostering a $3bn ICT sector, encouraging the development of 35,000 jobs and pushing the internet penetration rate towards 50 per cent. The strategy was devised with extensive input from the private sector, including the Information Technology Association of Jordan (Int@j), an industry lobbying group.

Jordan has traditionally been one of the most developed knowledge economies in the region, with statistics pointing towards an annual growth rate of 50 per cent in recent years. When the Reach Initiative was initially proposed eight years ago, the ICT sector was expanding rapidly, with a number of innovative start-ups such as Maktoob.com and d1g, as well as a labour force that was in increasing demand. To capitalize on this, the initiative proposed the creation of 30,000 ICT-related jobs, $550m worth of exports and $150m of FDI, all by 2004. However, by 2006, Jordan was still below the Reach targets; in some ICT sectors, the kingdom had even fallen behind its neighbours. According to a UN global e-governance index, between 2002 and 2005 Jordan ranked below the UAE, Egypt and Lebanon in terms of its e-government projects and its overall online engagement. As of late 2007, in spite of the Reach efforts, Jordan's ICT-related workforce remained at 16,000, while government statistics indicated that internet penetration hovered around 11 per cent. Similarly, PC ownership was only 7.1 per cent, while ADSL penetration was less than 1 per cent. Jordan's ICT sector also suffers from something of a brain drain, as high Gulf salaries lure away talented staff. Although an estimated 5000 ICT students graduate each year, skilled labour can be difficult to come by.

The 2007-2011 strategy directly addresses these and other weaknesses; including the high cost of broadband access, legal and regulatory challenges, and the lack of diversified ICT demand (the government is the major buyer of ICT services in Jordan). The new policy will provide a more detailed approach to facilitating government support and encouraging greater investment. The new policy looks to boost the sector's current 10 per cent GDP contribution by focusing primarily on connectivity, research and development, labour and education, regulation and the investment climate.

Although the policy is still quite new, the private sector is responding to the strategy with enthusiasm. Foreign companies are also showing interest, with industry heavyweight (*i.e.* Cisco Systems having selected Jordan as a training centre for its Middle East and North Africa operations). Niche sectors with ICT services are also becoming increasingly visible, with the Jordan-based animation and multimedia company Rubicon recently announcing a production and licensing partnership with Hollywood's MGM studios.

7.3.1. e-Government in Jordan: Vision

e-Government in Jordan is dedicated to delivering services to people across society, irrespective of location, economic status, education or ICT ability. Jordan remains committed to its long-term national vision for e-Government as to create a society where electronic government is a contributor to the electronic and social development of Jordan (Al-Omari, 2006). The e-Government will empower and benefit all in the society through access to government information, public-private partnerships and improved public services, communications and transactions with the government. It represents an opportunity to make a major contribution to economic development through assisting Jordanian business in reducing their operating costs by providing immediate access to government information and procurement. Jordan can achieve the vision of providing government services to all by enhancing infrastructure, developing skills, modernizing laws and working in partnership with the private sector. As part of its efforts to transform its society, economy and government, Jordan is pursuing a national e-Government strategy that aims to:

- Deliver high-quality services to consumers, businesses and organizations;
- Improve government performance and efficiency;
- Enhance Jordan's competitiveness;
- Ensure public sector transparency and accountability;
- Reduce costs and increase ease of interaction with government;
- Promote development of Jordan's ICT sector;
- Develop skills within the public sector;
- Boost e-commerce activities; and
- Improve information security.

To meet its goals for improving delivery of services and government efficiency, the e-Government Program has, over the past five years, initiated major projects including:

- Completion of two phases of the **Secure Government Network** (SGN) which provides connectivity, Internet and e-mail services to 18 government entities.
- Preparation of five **"fast track" vertical e-services** for government departments already launched.
- Establishment of a state-of-the-art **Operations Centre** that hosts the SGN and will eventually host other e-Government components and services.
- Launch of **e-Government Information Portal**—first phase of the e-Government Portal that will offer information and directory services about the Government of Jordan. Ultimately, the portal will be a "one stop shop" for user interaction with all Government of Jordan entities.
- Launch of the first phase of the **Contact Centre** that supports the 18 government entities connected to the SGN.
- Completion of comprehensive **information security roadmap** for the Government of Jordan, produced as part of the Security Assessment Project.
- Provision of nearly 7,900 **training** opportunities including programs on ICT literacy, project management, vendor certified training programs and IT professionals.
- Definition and prioritization of the e-Government **Portfolio of cross-governmental services** for the Second Wave of the e-Government Program.

Jordan varies in its preparedness to undertake e-Government services to businesses, citizens and within government institutions themselves (Al-Omari, 2006).

7.3.2. e-Readiness and preparedness for e-government in Jordan

Infrastructure

- FLAG network brings worldwide high-bandwidth connection to Jordan.
- Four major government networks exist.
- Lack of local "feeder" networks.

- High cost of telecommunications and internet access.
- Poor internal government interface and lack of intranet.
- Few municipalities outside Amman are computerized.
- Jordan Telecommunication Corporation (JTC) will provide nationwide infrastructure.

Back office management

- Lack of client-centered philosophy in many government's offices.
- No common data standards.
- Limited IT expertise in government institutions.
- Not all data kept in electronic format.

Policy and Legal:

- Electronic information sharing allowed, but not actively encouraged under law.
- Legal barriers to recognition of electronic documents and signature.
- No legislation allowing for electronic payment.
- Legal obstacles in intra-ministry submission documents.

Community and Education:

- Low internet penetration in Jordanian homes.
- Growing network of internet cafes.
- No government-established online access points.
- Lack of trust in electronic payment.
- Plans to teach computer skills in schools.

Major building blocks for e-government in Jordan

- *Identifying e-services applications*: This will identify and plan the government services to electronically offer to citizens and businesses. The sequence of rolling out these services will depend on the readiness of the particular ministries and institutions and the importance of the particular service to citizens and businesses. Within the eservices programs, eight specific services have been chosen as "Fast-Track Projects".

- *Developing technology infrastructure*: Without sufficient technical preparation and planning, Jordan's e-Government initiatives will risk duplication of effort and installing physical systems that do not connect efficiently and securely. The Infrastructure Program will consist of three major components; Access Media for e-Government Services, Delivery Networks and Technical Policies and Standards.
- *Developing legal and regulatory framework*: This discusses legal and regulatory measures that Jordan must take before successfully implementing its e-Government program.
- *Reforming education and develop skills*: Major national education and training initiatives are required both to provide the resources for e-government development and to provide Jordan's IT sector with the skills necessary to become a leading contributor to economic growth and job creation.
- *Designing management and organizational framework*: The proposal outlined must embody the concept of an effective public-private partnership, but has the strong central links to effect change within the government.

Challenges facing the e-government initiative in Jordan

- *Low level of Internet penetration*: The low level of Internet penetration in Jordan, 0.7 per cent of population in terms of account subscribers and 1.9 per cent in terms of users, is a significant barrier to the introduction of e-Governence, citizen-oriented services. The relatively high cost of Internet access and telecommunications services exacerbates this problem.
- *Infrastructure constraints*: There are significant ICT barriers in Jordan today. They include high cost of telecommunications services and lack of an adequate civilian telecommunications "backbone" network nationwide.
- *Digital Divide*: A digital divide exists in terms of geography (most users are clustered in Amman, Irbid and Zarqa), age (users tend to be young), skills (significant portions of the population lack computer skills), gender and income. In the business sector, the inequalities in terms of ICT are even more pronounced; most small and medium size businesses that account for the vast majority of Jordanian enterprises lack computer skills and ICT.
- *Privacy versus security concerns*: There are legitimate concerns about the citizens' rights to privacy versus the state's national security concerns.

On the one hand, there are concerns that the government can know too much about people and could use that information inappropriately. On the other hand, there are governmental concerns the unfettered access to information could undermine national security and therefore social stability.

- *Limited Information Technology (IT) skills*: There is a fundamental lack of computer literacy in Jordan that would limit the participation of citizens, businesses and government institutions in e-Governence.
- *Limited public sector reform efforts*: The e-Government is not a substitute for public sector reform. The limited state of public sector reform efforts in Jordan mean that e-Government initiatives need to be tightly focused on specific service areas.
- *Lack of an enabling legal framework*: The legal and regulatory framework to enable a full range of e-Government services does not currently exist in Jordan. e-commerce is deterred by the lack of adequate legislation and implementation capacity.
- *Lack of awareness*: The general population and the government and business sectors currently have very limited idea about what e-Government is and how to benefit from it.

8. MAIN CONCLUSIONS AND RECOMMENDATIONS

- The problems and some of the needs of improved Information and Communication Systems have been recognized and individual attempts are made to solve part of the problems. Collaboration and a comprehensive and more detailed policy and action plan could speed up the process and attract effective support and greatly advance the major concern of the government to have appropriate communication and information systems in place for accession to the EU in 2007.
- Non-electronic information (informal and print) still plays the most important role in agriculture extension and both electronic and non-electronic communication need to be improved through policy and specific investments.
- Jordan's IT capacity related to agro-information dissemination needs to be vastly improved at a local level; to allow the effective horizontal transmission of information and communication between the farmers and institutions.

- At a local level, rural farmers are widely dispersed with little or no access to computers and with very limited hardware and IT skills. This situation could be improved if farmers joined specific associations or co-operatives where information from the institutions and on markets and research could be made accessible.
- New demands for agricultural extension open the door for examining how communication can be addressed more effectively using new ICTs. However, technological challenges have still to be addressed such as the lack of trained communicators, poor ICT infrastructure, high costs and limited access to ICTs and lack of relevant content. The revolution in digital communications are fast transforming the way production is organised, the way governments provide services to their citizens and how people all over the world interact with each other. There are some unique attributes offered by ICTs, such as reducing isolation, facilitating dialogue, participation and fostering interactive networks. For the first time those using ICTs can be producers of their own information, not just passive recipients. However, this growth in networking calls into question our understanding of information and knowledge and raises questions of ownership of knowledge.
- The agricultural sector has lagged behind in exploring and tapping the potentials that information and communication technology has to offer. These potentials range from the sharing and re-use of data, research findings, lessons learned and best practices among research and extension institutions to developing quick response mechanisms for agricultural and natural resources crises situations; from improving the extension delivery systems in the rural areas to bringing e-Commerce to the farmers.
- Traditional media and emerging Information and Communication Technologies (ICTs) have played a major role in the diffusion of information to rural communities and have more potential. Currently, many scholars and international agencies argue that ICTs have the potential to increase agricultural productivity through communicating knowledge and information to rural agricultural communities, providing capacity building, accessing markets and credit, restructuring of extension and scaling up inter-linkages of development interventions.
- Communication for Development including ICTs provides a potential tool for building partnerships and enhancing coordination among extension, research, education, farmers and other actors in rural development. Examine how extension and research can be made more effective using appropriate communication for development tools and approaches, including ICTs.

- Information and Communications Technologies (ICTs) can play a key role in agricultural development. The use of radio for agricultural extension and propagate use of new technologies for improved agriculture formed the basis for the massive increases in farm productivity. The role of television in agricultural development, though not as successful as the radio for a variety of reasons, was also of significant importance. With the advent of use of new ICT, especially computers, the Internet and cellular telephony, there is an ongoing transformation of agriculture through innovation that is largely enabled through information sharing and exchange between agricultural communities through ICT enabled information systems
- The use of Information and Communications Technologies (ICTs) in agricultural research for development had initially been in analysing and managing research data and later in managing scientific and technical literature.

References

Al-Omari. H. 2006. e-Government Architecture in Jordan: A Comparative Analysis. *J. Computer Sci.*, 2 (11): 846-852.

Mofleh, S., M. Wanous and P. Strachan. 2008. Developing countries and ICT initiatives: Lessons Learnt from Jordan's experience. *EJISDC*, 34(5): 1-17.

Authors Information

Dr. Esmat AlKaradsheh

Researcher
Water, Soil and Environment Research Directorate
National Centre for Agricultural Research and Extension (NCARE)
Ministry of Agriculture (MoA)
P.O.Box 639, Baqa' 19381
Amman, JORDAN
Phone No. (O): +962 6 4725071 Ext. 357; (R): +962 5 3245389
Mobile Phone No.: +962777713935
Fax No.: +962 6 4726099
e-mail: esmatk@ncare.gov.jo, esmatk@yahoo.com
URL: www.ncare.gov.jo

Dr. Esmat AlKaradsheh, Ph.D. in Water Management and Environment, University of Kassel, Germany 2003. Worked as teaching and research assistance in the UoJ, Faculty of Agriculture and private companies. In 1993, started work as researcher at the National Centre for Agricultural Research and Extension (NCARE), in Water and Environmental Management Research Directorate. During this period, conducted many field studies and research in irrigation management and presented papers in many regional and international conferences and member in some national committees as well as member in evaluation missions.

Er. Asmahan Farred Hattar
Agricultural Engineer
Director
Department of Information and Farm Management
National Centre for Agricultural Research and Extension (NCARE)
Ministry of Agriculture (MoA)
Baq'a (19381), P.O.Box 639, Amman, JORDAN
Phone No. (O): 00962474996; (R): 0096264725071
Mobile Phone No.: 00962777402545
Fax No. :009624724996
e-mail: asmahan_hat2001@yahoo.com, asmahan@ncare.gov.go
URL: www.ncare.gov.go

Er. Asmahan Farred Hattar is having 17 years of solid experience in the field of Agricultural extension. Certified trainer from IFAD & INVENT IN Development Project Management. Participated as a co-author in many studies at NCARE.

Mr. Ashraf Saber Alhawamdeh
National IPM/FFS Project Coordinator
Extension Department
National Centre for Agricultural Research and Extension (NCARE)
Ministry of Agriculture (MoA)
P.O.Box 152, Jerash 26110, JORDAN
Phone No. (O): 0096264726206
Mobile Phone No.: 00962777757904
Fax No.: 0096264724996
e-mail: ash_agri@yahoo.com
URL: www.ncare.gov.jo, www.moa.gov.jo

Mr. Ashraf Saber Alhawamdeh started working in Ministry of Agriculture in 2002 as a mycologist officer and in 2006 became the National IPM/FFS project in Jordan. His expertise areas are in plant diseases diagnostic, plant protection techniques, isolation and identification of plant parasitic fungi from soil and plant samples, Integrated pest management of vegetables and fruit trees and a good experience in the participatory extension methods especially the farmer field schools approach.

Dr. Samia Nadim Akroush

Director of Socio Economic Directorate

Socio-Economic Research Directorate

National Centre for Agricultural Research and Extension (NCARE)

Ministry of Agriculture (MoA)

Baq'a (19381). P.O.Box 639

Amman, JORDAN

Phone No. (O):0096264725071 : Ext.234; (R): 0096264710366

Mobile Phone No.: 00962777841700

Fax No.:0096264726099

e-mail: samia_akroush@yahoo.com , samia@ncare.gov.jo

URL: www.ncare.gov.jo

Dr. Samia Nadim Akroush is a Socio-Economist researcher at the National Centre for Agricultural Research and Extension (NCARE). Started working at NCARE since, 1998. He is having 10 years of wide range of experience in designing, conducting, computer coding & analysis using statistical packages and writing final technical reports for the Socio-economic studies. Participated in many training courses. He has participated 23 regional conferences, meetings and workshops, 5 published papers and 6 reports in addition to participation as co author in many internal studies at NCARE.

❑❑❑

Chapter 10

NIGERIA

Olufemi Martins Adesope, Moses Okwusi
and Ike Nwachukwu

1. INTRODUCTION

In the past, conventional ICTs that have been applied to agricultural extension services include the use of television, radio, *etc.*, and these media have been criticized as archetypical, passive, top-down in nature; hence they have not created meaningful impact on agricultural information delivery (Akpabio, 2007). As a result of this, extension roles have undergone several transformation, over the years. One of the objectives of agricultural extension is to create conducive platform for empowerment so that food security can be guaranteed. In order to guarantee food security there is need to utilize faster and more efficient techniques of project identification, selection, implementation and evaluàtion in the agricultural extension process (Braimoh and Oladele, 2000). The inadequacies of interpersonal communication strategy in development gave credence to the use of media support in development process (Yahaya, 2000). In recent times, further improvements have been made on the use of media support through digital media. This is what makes contemporary ICT more relevant.

Modern ICTs emerged in Nigeria at a time when there was need for change in the agricultural extension delivery system to suit the contemporary situation. The conventional forms of communication used by extension agents include written (bulletin, leaflets, newspapers, articles, personal letters, circular, *etc.*); spoken (general meeting, special meeting, farm and home visit, official calls, telephone calls and radio); visual (result demonstration, exhibits, posters, motion pictures, charts, slide) and spoken/visual (method demonstration meeting, result demonstration meeting, meeting involving motion pictures, meeting involving charts). Torimiro *et al.,*(2000) noted the non-usage of telephone calls in extension work, giving reasons as preference for the locally accessible and affordable forms of communication they can employ.

Studies had reported low utilization of modern ICTs among public sector extension officers even though they had access to a wider range of ICTs than the private sector extension officers (Akpabio, 2007). In a similar way, Adesope *et al.,* (2007c) found that extensionists had higher ICT proficiency requirements than researchers, obviously because they required ICT for their day-to-day activities. The importance of ICTs in contemporary Nigerian agricultural extension system can be noted because it provides large volume of information which can be accessible within a short time, and it is also a repository of information.

Aderinto *et al.,* (2008) in their study covering Ondo State, reported that extension personnel principally use traditional ICTs. Radio and television were always used, while newspaper and telephone were sometimes used. However, majority indicated that fax, internet, CD ROM, websites were sometimes used or even never used, making the researchers to conclude that majority of extension personnel in Ondo State were yet to be modern ICT compliant. Aboh (2008) in a study on frequency of ICT tools usage by agricultural extension agents found that out of 9 ICT tools identified for use, only 2; computer and mobile phones were frequently used, while CD-ROM, television, internet, web, chatting, e-mail, search engines were less frequently used. Mobile phones were affordable and portable for use. As a personal item it can easily be used to make calls on farmers and co-researchers or extensionists. This is however, not so with the other complex ICTs.

Diffusion of technology is a function of adoption. There is no doubt that the level of ICT adoption is commendable in the entire Nigerian polity. The situation in the rural areas is also improving but it can be better. Telecommunication as a means of sharing information goes beyond connection between person to person or person to groups, but engenders a chain reaction effect on the development process. According to Yekini and Hussein (2008), a new set of challenges to agricultural production and management activities in developing countries is emerging. These set of challenges include:

i. The exponential increase in the demand for food and fibre.
ii. Issues regarding the continuous occurrence of new pests and diseases.
iii. Issues of cost of farm and domestic energy requirements.
iv. Value added concerns in exported agricultural products.
v. The phyto-sanitary requirements for exported agricultural products.
vi. The realities of challenges inherent in World Trade Organization's policy on trade in agricultural products.

Those among other issues, they observed, have brought the burden of globalization to bear on the agricultural practices in developing countries. Information management therefore is being considered more seriously than before.

The use of Information and Communication Technologies in the Nigerian agricultural sector is becoming more appealing to the major stakeholders in recent times. According to Adesope *et al.*, (2007c), Agricultural Extension the world over is facing a new challenge with the advent of Information and Communication Technology (ICT). For effective extension service, therefore, the need to have adequate access to agricultural information by the extension personnel cannot be overemphasized. It has been reported that the main focus of ICT in agriculture is meeting the farmers' needs for information (Meera *et al.*, 2004). This has implications for agricultural extension which emphasizes change in attitude, knowledge and skills brought about by adequate and appropriate educational process. The change brought about relates to what the end-users know, how they react to situations and in what they can do with their hands. Ani (2007) while noting the concern of extension education as an educational process, to help farmers to make a decent living, reiterated the fact that through learning and acceptance of new improved practices, the farmers may be able to produce more and live better. There is no doubt that the only way through which information on improved agricultural practices can be copiously delivered to end-users is through the application of information and communication technology in the contemporary situation.

2. AGRICULTURAL INFORMATION

Information has been identified as one of the resources required for the improvement of agricultural production (Aina, 1995). It is said to be a resource that must be acquired and used in order to make an informed decision. Those who possess appropriate and timely information will make a more rational decision than those without. The various agricultural information users' populations include policy makers and planners, researchers, extension staff, educators and students, agro-based industries and services staff and farmers. It is also observed that each of the sectors mentioned above contribute directly to the improvement of agriculture; hence relevant information provided to each category of these user populations will contribute positively to the development of agriculture in Africa. Information is the potential ingredient for the advancement of agriculture. Without information the agricultural sector may degenerate into extinction. Information provides the basis for scientific innovations in agriculture. Agricultural information has been defined as all published or unpublished knowledge on all aspects of agriculture.

3. INFORMATION TECHNOLOGY

The phrase "information technology" is used to encompass a range of "new" technologies and their applications, including all aspects of the use of computers, microelectronic devices, satellite and communications technology (Commonwealth Secretariat, 1991). These emerging technologies seem revolutionary because of the speed of their development but they can also be seen as the most recent developments in our evolving ability to record, access, manipulate and communicate information. In another dimension, information technology is the study or use of electronic equipment, especially computers, for storing, analyzing and distributing information of all kinds, including words, numbers and pictures (Oxford Advanced Learner's Dictionary, 1998).

Information Technology according to Zorkoczy (1982) can be looked at from the point of view of the science of electronic systems and to computer science to legitimize the products of their convergence. In another development, information technology is the application of computers and other technologies to the acquisition, organization, storage, retrieval and dissemination of information (Madu and Adeniran, 2000).

In the opinion of Vernon (2001) Information and Communication Technology (ICT) is a collective term covering all those technologies, both hardware and software, dedicated to the capture, storage, processing, transmission and presentation of information. Information Technology is a complexity of resources for effective communication. As a result it includes the following facilities: e-mail, electronic conferencing and computer bulletin boards. The essence of these facilities is to share ideas, exchange information, and develop partnership, which ultimately will result in a desired objective. In a simple term information technology, which can be viewed, as information and communication technology is the use of computers and the linkage between more than two computers. Babu *et al.,* (1997) observed that information system refers to a computer-based system, one that is designed to support the operations, management and decision functions of an organization. Accordingly, information systems in organizations provide information support for decision makers. It was further opined that information systems encompass transaction processing systems, management information systems, decision support systems and strategic information systems.

From a cursory standpoint, Information Technology comprises computer, telecommunication, microelectronics and storage facility. It now cuts across virtually all sectors of the economy. Madu and Adeniran (2000) explained that Information Technology has some important values that have made its use to be concrete. These include the fact that individual and company information-related needs have been met easily. Information Technology has helped to reduce the cost of technical developments in the microelectronics field. It is also easy to handle information in a variety of forms. As a result of all these we now live in what has been called "the information society" which is associated with rapid technological change (Jackson, 1984). The three expected consequences, therefore, include:

a. The refurbishing of manufacturing industry
b. A new range of high-technology industries, such as electronics, communication satellites, switching equipment and biotechnology
c. New leisure industries

Information Technology is gaining a wider recognition all over the world, even in Nigeria. But the inadequacies of basic infrastructure in Nigeria has made it a bit difficult to be fully appreciated. Nigerian rural areas are mostly affected as many are still lacking infrastructure. Prakash (2000) stated that modern technology tends to bypass local communities found in remote regions. However, recent technological advancements in communication have made it possible to introduce new information and communication technologies to rural areas. Evidence abounds that telecentres now dot many rural communities so that information are easily accessed in these rural areas. In Dakar, Uganda, Mali and Congo rural people can now get and send agriculture-related information, messages that will help them in their various agricultural endeavours (IDRC, 1999). Asiabaka (2002) reported that the use of modern Information Technology such as CD-ROM and the Internet could form the basis of teaching new technologies to rural farmers.

Banigo (2001) has rightly captured the almost incredible impact of ICT. It is on record that it took 38 years of Radio to reach 50 million people and 30 years of television to achieve the same result. Empirical evidence has shown that the same number of people adopted the internet in just four years. According to NITEL Internet Business Solution Sales Manual, the internet provides the ability not to only create and manipulate electronic information, but also to access and disseminate it on a global scale. The internet effectively opens up and extends the PC into the outside world, making all the knowledge and information stored in thousands of interconnected computers available to users whenever they want it.

Despite their centrality, PCs and internet are not the only important forces in this new, more distributed stage of the information revelation. Similar technologies are also becoming more powerful, flexible, user-friendly, low-cost and widely applied (Telephone, GSM, Faxes, modems, PC-Soft-ware, CD-Rom disks, Cable Television, cellular telephones, pagers, faxes, Hand Held Computers, Global Positioning System (GPS) *etc.,*).

According to AGREN (2004), the first ever global summit on the information society concluded with a declaration of principles and plan of action with the emphasis on coordinating action to improve equality of access to Information and Communication Technology (ICT). Representatives from 176 countries endorsed the declaration of principles expressing their commitment to build a people centered inclusive and development-oriented information society, where everyone could create, access, utilize and share information and knowledge.

The plan of action addressed agriculture from two perspectives. First, to provide more knowledge and information especially in rural areas through the systematic dissemination of information using ICTs on agriculture, animal husbandry, fisheries, forestry and food. Secondly, to maximize the use of ICTs as an instrument to improve production (quantity and quality) through public private partnership.

According to Phones 4 All (2002), the establishment of the National Information Technology Development Agency (NITDA) which took off on April 18th, 2001 under the leadership of Prof. Gabriel Ajayi was aimed at creating the awareness of ICT usage. This gave birth to the establishment of seven scaleable IT parks as follows- one in Abuja and one in each of the six geo-political zones.

In addition to that, various IT initiatives that emerged lately include, but not limited to the following: Akwa Ibom IT Development Project and the Lagos IT Resource Centre which is one of private sectors - led initiative to diffuse information technology availability and usage in the country. It is a project of a number of member organizations namely the Nigeria internet group.

Furthermore, there is the establishment of West African Institute for Information and Communications Technology (WAIICT) located at Asaba, in Delta State. Others are Bojugboju IT resource centre, the Zinox project IT and the computerized Nigeria initiative. Also, a major workshop on agricultural policy networking organized by CTA in November 2000, disclosed that information and communication had become crucial in facilitating dialogue between all those involved in policy making, researchers, extension workers, representative of farmers organizations, environmental bodies, trade organizations and the policy makers themselves. The workshop brought together CTA's partner organizations from all six ACP regions to discuss the development of strategies to support agricultural policy making.

Private sectors are not left in this information technology revolution. Nigerian university campuses are now well equipped with Information Communication Technology facilities. It is no longer strange to see lecturers and students doing their research and other academic works using various IT devices like e-mail and internet. One is happy to note that both the federal government of Nigeria and other international funding agencies are now interested in the general development of ICT in higher education in Nigeria. For example, Federal Ministry of Education has embarked on the establishment of National Virtual (Digital) Library Project. One of the objectives of this is to provide, in an equitable and cost effective manner, enhanced access to national and international library and information resources and for sharing locally-available resources with libraries all over the world using digital technology. A model Virtual (Digital) Library at National Universities Commission (NUC) will be the hub of the Virtual Library through the internet, CD-ROM and Wide Area Network (WAN). When this is fully materialized, it will definitely be a boost to the development of ICTs in Nigerian educational system. Some other international funding agencies like Carnegie Corporation also supports the establishment of scientific databases at the University of Ibadan, Oyo State, Nigeria. With similar grants for academic development, the Corporation which is presently in collaboration with Obafemi Awolowo University, Ile-Ife, Osun State, Nigeria and the university library, is well positioned to gain from the grant. For instance, through the various grants from these international granting agencies, Information Technology and Communication Unit (INTECT) is well established on the campus of Obafemi Awolowo University Ile-Ife and it is at present responsible for the Internet services in the university. In other words, it is a major ICT provider on the campus.

The popularity of ICTs and its use among rural dwellers all over the world has been cited in literatures by scholars worldwide. Rural communities can actively participate in globalization process, provided they have access to basic ICTs. Thioune (2003) explained that in agriculture, African farmers expect ICTs to facilitate access to high yielding activities at competitive prices, input supplier, credit institutions and information on how to improve their farming practices. But the benefits of knowledge and technology are not available to a large majority of the world's population and there remains a disparity to the extent to which access and skills are available. Not only do these differences affect developing nations and disadvantaged communities, but within societies also there are significant disparities.

4. INFORMATION TECHNOLOGY INITIATIVES IN AFRICA

In parts of Africa, Information Technology strategies have been experimented successfully. Munyua (2000a) gave the following examples:

- Successful experience with "audiovisual pedagogy" from Latin America has led FAO to apply the participatory video approach to agricultural development projects in Africa to empower rural communities. The methodology mediates between the needs of rural people, sources of information and expertise to respond to the expressed needs through the production and use of documentaries and training presentations.
- The Technical Centre for Agricultural and Rural Cooperation (CTA) works towards improving the dissemination of information for the benefit

of farmers through improved adoption of new technologies. CTA has a programme called Rural Radio Support Programme that supports rural radio and development of audiovisual aids in ACP (Africa, Caribbean and Pacific) countries. The programme was launched in 1990 and it distributes information packages consisting of taped interviews accompanied by a complete transcript and radio talks. Audio files are hosted by ONEWORLD (http://tv.oneworld.net/cta/) and can be downloaded and re-broadcasted.

- In Uganda, the National Agricultural Research Organization (NARO) and CAB International (CABI) are implementing a project titled *"Electronic Delivery of Agricultural Information to Rural Communities in Uganda"*. The project, funded by the International Development Research Centre (IDRC) under the Acacia Initiative, aims to improve access to agricultural information by rural communities through use of selected traditional media and modern ICTs such as radio, video, television, print media, e-mail, CD-ROMs and the Internet to increase agricultural production. The project will accelerate the delivery of appropriately packaged agricultural information using existing telecentres at Nabweru, Buwama and Nakaseke as information resource centres. Information needs and ICT preferences will be identified with special focus on women, community leaders, youth, frontline workers, NGOs and CBOs. Local content will be developed and building the capacity of the local communities will be a priority. The Evaluation and Learning Systems Approach (ELSA) developed by the Acacia initiative will be used to ensure continuous learning and evaluation.
- FAO has prepared a proposal for a National Agricultural and Rural Knowledge and Information System (NARKIS) and network for Uganda. The aim is to increase "farmers' access to information, knowledge and technology through an effective, efficient, sustainable and decentralized extension with increasing private sector involvement in line with government policy". The project hopes to assess user needs and facilitate agricultural and rural information flows and knowledge building, as well as capacity building. The programmes to be implemented include the building and management of an agricultural Web portal, packaging of multi-media publications, rural networking and ICT deployment and support services. The project will be managed through contract arrangements (Van Crowder and Fortier 2000).
- The Faculty of Agriculture of the University of Mauritius has developed a computer based information system, the Potato Extension and Training

Information System (PETIS). PETIS uses the Internet and will test whether rural communities can use the web to access information. The system, destined principally for the small-scale potato growers, is equipped with audio files that provide information in English. Illiterate users have an option that reads the summary of the content in Creole and Bhojpuri, and icons and pictures that enable most rural users to navigate easily the basic levels on the site. The system has been rated very successful and the research team is now exploring touch screens (Lukeeram *et al.*, 2000).

5. ROLE OF INFORMATION TECHNOLOGIES IN RURAL DEVELOPMENT

The modern information technologies have potentials for communities and rural development especially in developing countries. According to Munyua (2000b), the internet, one of the modern information technologies, is not a panacea for food security and rural development problems and it can open new communication channels that bring new information resources to rural communities. Traditional communication channels were noted to have been used successfully but these had been monologic and had not allowed for much interaction with users. Radio, for example, has been very effective for disseminating information to all types of audiences, but broadcasting times are sometimes not appropriate for the target audience. But, radio could be linked to the internet and a few initiatives have been started on this concept, such as the project Internet Radio in Sri Lanka (http://www7.itu.int/itudfg7/fg7/CaseLibrary/ShowSummary.asp?contrib=35). This enables users to access programmes on the web at a convenient time and send feedback through e-mail or chat. Broadcasters could then disseminate the latest information promptly. The following are some examples of areas where Information and Communication Technologies could play a catalytic role in developing rural areas:

a) Decision-making process - Sound decision-making is dependent upon availability of comprehensive, timely and up-to-date information. Food security problems facing developing countries demonstrate the need for informed researchers, planners, policy makers, development workers and farmers. Information is also needed to facilitate the development and implementation of food security policies. E-mail and the Internet could be used to transmit information to and from rural inaccessible areas.

b) Market outlook: Farmers could promote their products and handle simple transactions such as orders over the web while payment transactions for the goods can be handled off-line. This method has provided a cheaper and faster way to trade online than on paper-based medium, telephone or fax. Electronic-commerce could, therefore, enable entrepreneurs to access global market information and open up new regional and global markets that fetch better prices and increase farmers' earnings.

c) Empowering rural communities: ICTs can empower rural communities and give them "a voice" that permits them to contribute to the development process. With new ICTs, rural communities can acquire the capacity to improve their living conditions and become motivated through training and dialogue with others to a level where they make decisions for their own development.

   Improved policy formulation and strategies, however, require an educated and informed populace to reduce poverty, geometric population growth, environmental degradation and other factors that are most often the direct causes of hunger. New ICTs have the potential to penetrate under-serviced areas and enhance education through distance learning, facilitate development of relevant local content and faster delivery of information on technical assistance and basic human needs.

d) Targeting marginalized groups: Most rural poor people lack the power to access information. ICTs could benefit all stakeholders including the civil society, youth and women in particular. Other disadvantaged groups that could be targeted include the disabled and subsistence farmers.

e) Creating employment: Through the establishment of rural information centres, ICTs can generate employment opportunities in rural areas by engaging telecentre managers, subject matter specialists, information managers, translators and information technology technicians. Such centres help bridge the gap between urban and rural communities and reduce rural-urban migration problem.

6. TYPES OF INFORMATION TECHNOLOGIES AND LEVEL OF INTEGRATION IN RURAL AREAS

Information Technology comprises the various computer related mechanisms that aid communication technology from one or more computers to others, facilitated by man. Below are contemporary Information and Communication Technologies.

a) *Internet*: This is the short form of international network. It is the global network of computers, which provides the basic electronic infrastructure over which all Internet activity takes place. The history of the Internet, according to Prevost (1995), is long and unfortunately not all that interesting. During the World War II, a United States destroyer used a peaceful tuber - the potato, to make war. It was observed that through a strange twist of fate Soviet citizens used an implement of war the Internet - to wage peace and procure potatoes. After the United States dropped the first atomic bombs on Japan and ended the war, the devastating effect the nuclear war had on the communication infrastructure made the United States military engineers take the destruction of Hiroshima and Nagasaki to heart and they constructed a communication network over which military, research and educational institutions could communicate. The Internet was born in 1969 (then called ARPAnet) and it quickly became the communication method of choice amongst many people around the globe.

   The Internet should be thought of as the infrastructure for communication and by using that infrastructure in different ways, various services can be obtained (CTA, 1997). The Internet works by linking computer to a set of connections using a modem (device linking a computer system, *e.g.* a telephone line) a large amount of information can be obtained. The internet has not been integrated into the rural environment because of the low literacy level of the people. Apart from that the poor electricity situation does not give room to internet usage.

b) *Electronic mail*: This is also known as e-mail. With this, text messages can be sent from one individual to another or many individuals over the internet (CTA, 1997). Puliatti (1997) noted that the electronic mail works by sending a message from one person (normally typed into a computer) to another person through a telephone and is stored until the receiver asks for it. It is used to contact another person who desires particular information from another person or organization. Sotade (2001) stated that the e-mail is the postal system of the Internet. Message sent and received could contain sounds and pictures. A message may be sent or copied simultaneously to more than one user (Guiness Book of Knowledge, 1997). The e-mail software controls the way the message is processed. The software includes Microsoft Internet Explorer or Netscape Navigator. Each user of e-mail must have a unique address which takes the general form of name @ provider. The name portion consists of one's name (or whatever one would like to be called) using lower-case letters and with no spaces (though full-stops and hyphens can be used) the provider part is set by the IP (Internet protocol). If the provider is hotmail dot com and the user name is Marty Lufem, the e-mail address may be marty.lufem@hotmail.com. The e-mail has the advantage of being able to contain a very large volume of message. It is a faster information transfer system than facsimile (fax) or telex. One can send and receive messages to millions of people all over the world. Another advantage of the e-mail is that whoever is to receive the mail does not need to be present at the time his or her mail is being sent/received, reply of mails is also very easy.

c) *Newsgroups*: The network of news groups is collectively known as USENET. Newsgroups are discussion groups classified by any of the vast number of topics. Each news site is locally run and is organized by a news administration (Guiness Book of Knowledge, 1997).

   According to Sotade (2001) news groups are also called Bulletin boards and they work like e-mail. However, instead of writing messages to individual users, participants in a news group post their messages to a news server. On the news server, messages called articles are stored, most of the times in hierarchical directories. Users participate in news group services by reading messages and posting responses to the client software that can view a bulletin board service called a newsreader. Usenet is like a notice board that anyone, anywhere, can look at and to which any one can pin his or her own notice. It is a way of asking questions and receiving answers from people who are interested in the same subject.

   CTA (1997) explains that Usenet groups are Internet discussion locations. Questions, opinions or comments can be posted to relevant groups. Any one with Internet access can then read and respond as they wish. Examples of Usenet groups include *alt. agriculture. fruit, bionet. agroforestry, sci. agriculture.*

d) *Mailing lists*: These are specialized groups of people who use private e-mail to discuss specific topics of interest (Prevost, 1995). They are also used by public and private Internet groups to keep members informed.

e) *File archives*: Organizations such as universities and businesses offer large archives of highly useful texts, graphics and application files.

f) *Electronic magazine (e-zine):* Prevost (1995) opined that this is the Internet's answer to the office photocopier. Accordingly, e-zines offer a cheap and effective way for anyone to make "public" his ideas and spew them out into the electronic other. They range from the slick and professional to the crude and practically unreadable, but they all offer something of value to patient readers.

g) *World Wide Web (www)*: The Guinness Book of Records (1996) explained that the fastest growing part of the Internet is the World Wide Web. This is simply known as the web. It is a vast collection of information pages according to CTA (1997). Web pages can contain text, pictures, graphics, sound and moving images, among other things. They often contain links, which can take one to other sites.

   Guinness Book of Records, (1996) recorded that though the web was first implemented at the CERN Physics research laboratory in 1990, it became popular only in 1993 when the first web browser, *Mosaic,* was released by America's Natural Centre for Supercomputing Applications (NCSA). There were 623 web sites by the end of 1993, 10,022 by the end of 1994, and at least 100,000 by January 1996. The basic document of the World Wide Web is the page. Pages are written in an evolving language called HTML (Hypertext Markup Language). An HTML page contains text along with tags embedded commands that supply information about the page's structure, appearance and global community. Typically, web addresses start with http//: www, followed by specific site address whose sections are separated by a full stop. These site addresses reveal the country of origin and also reveal whether the source is academic (.edu or .ac), a commercial firm (.co or .com), government organization (.org or .gov). The commonest way to move around the web is by clicking text or pictures called links or hyperlinks, which have addresses coded into them. The mouse pointer changes to a hand shape when it passes from one page to the other and you can then browse (surf) the web and search the Internet for words you have specified. The use of websites is becoming interestingly useful to the agricultural industry. Qamar (2000) reported that more than 30 per cent of extension staff in Estonia use the Internet, and there are such programmes as "virtual gardens" and "virtual farms" on the World Wide Web. The Web according to Vernon (2001) is rapidly becoming more widespread across the internet, replacing the older methods of access and interaction. It is sometimes used as a synonym for the Internet.

h) *Chat room* : This can be used to connect people around the world. In the chat room, a real-time typed conversation is conducted with one or more persons. Osuagwu (2001) observed that once connected the user is brought into contact with a large number of users from around the world. All messages typed within a chat room appear almost simultaneously on the computer screens of all participants for the chat room.

i) *Compact Disk-Read Only Memory (CD-ROM):* This is a compact disk, which contains texts, sounds and picture that are coded digitally. User populations who value the technology for utilization access them easily. They are useful for the distribution of programmes that have been stored in them. Reports have shown that database producers and vendors are distributing their products on CD-ROM for the benefit of users (Ehikhamenor, 1993). It contains a large volume of data based information that can be printed out anytime it is required. Nelson and Farrington (1994) reported that CD-ROM systems are used primarily for the distribution of bibliographies and abstracts and also for full texts. The systems can be interrogated by using keywords to limit searches, and they provide useful method for disseminating information collection by larger libraries, *e.g.* the National Agricultural Library (USA), CAB International, the Technical Centre for Agricultural and Rural Cooperation (CTA) and the Consultative Group on International Agricultural Research (CGIAR). It was also reported that in France the Sesame CD-ROM is used widely as a resource for agricultural workers.

j) *The mobile phone*: This is a pocket-sized information and communication technology that has recently entered into the information technology market. It came into Nigeria fully in the last quarter of the year 2000. The technology is described by many in different appellations. Salihu (2000) referred to it as palm assistant, because of its handy nature. Qamar (2000) noted that the use of cellular phones is being used for a rural development project in Bangladesh.

k) *Video conferencing*: A communicator can from one location see, talk, and listen to an audience in one or more remote locations. It is applicable, according to Zorkcoczy (1982) for two or multi-way audio and television link-ups.

l) *The Essential Electronic Agricultural Library (TEEAL):* In contemporary information technology parlance, such terms as "library without walls", virtual library", are common. The Essential Electronic Agricultural Library (TEEAL) is described as a library in a box and is accessible through a CD ROM. It was made possible by the support of the Rockefeller foundation and Dr. Robert W. Herdt, the Vice President of the foundation. TEEAL is one of the recent advances in the information technology industry and is being used in the developing nations. The package makes 140 of the best scientific journals affordable to the developing world (Cornell University, Undated).

The TEEAL software has a profound effect on the agricultural system, according to Elisio Ponce, The Director of Philippine Bureau of Agricultural Research. It provides access to the world' s most important scientific literature, which would otherwise be unaffordable even to key research institutions. The software provides the following benefits:

a. It can help a researcher prepare scientific manuscripts.

b. It can be of immense advantage for postgraduate research.

c. It could help a field worker to obtain up-to-date research materials to facilitate his work.

d. It could help agricultural librarians to manage agricultural information effectively.

e. It could help agricultural personnel at both supervisory and managerial levels for decision-making.

The Electronic library has a demonstration disk from which the essential information can be accessible. According to Cornell University (Undated) the demonstration disk contains a full-text periodical library of agricultural sciences. The disk provides a hands-on experience to allow easy access to evaluation of the features of the TEEAL software, database and full-text images.

To gain access into the Compact Disk (CD), a user inserts it into the CD drive of the Personal Computer (PC) and the TEEAL software is loaded in the process. To perform a search for any required information select Search and then Basic Search from the toolbar, or click on the button containing the icon with the binoculars and red "1". Type search words into the keywords box of the TEEAL Search window. Click on Search (or press Enter). If for instance a user needs information on "weed control", simply type weed control into the TEEAL Search window and double-click on the citation list to see the full abstract and access the link to its full-text article.

This technology can be used by all cadres of personnel in the agricultural industry especially where a Personal Computer is available. It will make the search for information useful for the dissemination of agricultural ideas. With the disk, information access is guaranteed at all times.

Summary of ICT Types and Level of Integration in Nigerian Agricultural Extension System

Types of ICTs	Level of integration in Nigerian agricultural extension system
a) Internet	Very low
b) Electronic mail	Very low
c) Newsgroups	Very low
d) Mailing lists	Very low
g) World Wide Web (www)	Very low
h) Chat room	Very low
i) Compact Disk-Read Only Memory (CD-ROM)	Low
j) The mobile phone	Very high

7. ROLE OF COMMUNICATION TECHNOLOGY IN AGRICULTURAL EXTENSION

One of the basic functions of agricultural extension remains the dissemination of useful agricultural information from research to the end-users. The mass media have been greatly used in delivering this information. But the information boom of this century requires that worthwhile strategies of communication be developed and utilized. In order to meet up with the information explosion trend, agricultural extension needs to intensify efforts to acquire the necessary Information and Communication Technology facilities that are available. The categories of persons who should have access to Information Technologies for proper direction are the extension Managers and Supervisors. They know enough to influence decisions in their organizations. Waldron *et al.*,(1997) stated that an extension programme manager is the person that is vested with formal authority over an organization or one of its sub units. He or she has status that leads to various inter-personal relations and from this comes access to information. Information enables the manager to devise strategies, make decisions and implement action. VITA (1995) pointed out that modern communication techniques and facilities, which ensure uninterrupted access to large amount of agricultural information, have become necessary in hiding the distinction between urban and rural dwellers. These modern communication techniques are generally referred to as Information Technology (IT) facilities. Moll (1982) cited in Madu and Adeniran (2000) classified IT into two groups *viz* high and low. The low technology includes equipment and machines like typewriters, offset printing machines, microform readers, projectors, staplers, paper cutter, perforators, catalogue, radio, telephone, television sets *etc.*

The high information technology, on the other hand, consists of the various means of obtaining or transferring information using computer, telecommunications and microelectronics. Nzeakor (1991) declared that information technology could help rural communities to leapfrog from peasant farming to modern agriculture through:

a) Provision of equitable access to new techniques for improving agricultural production.
b) Reduced food storage losses through more efficient distribution network.
c) Efficient marketing of agricultural productions through information and telecommunication resources.
d) Establishment of information supplies for monitoring market performance and measuring market failure.
e) Development of information system to address food security issues.

Prakash (2000) posited that modern technology tends to bypass local communities found in remote regions. However, some recent technological advancements in communications have began to blur geographical distances and infra-structural bottlenecks. This was clearly emphasized in the Nakaseke Multi-purpose Telecentre in Uganda.

Mchombu (1992) observed that the provision of information to rural communities in the Malawi sub-region of Africa is a responsibility fragmented among several government ministries, non-governmental agencies and parastatals. With the exception of library services, most of the other agencies supply information to rural communities as a small package within a larger package or services and inputs. This is certainly the case with extension agencies in agricultural, health, community development and adult literacy.

Agricultural extension has a major role to play in the area of Information and Communication Technology. This is because, as an educational outlet that is dynamic in its dissemination delivery system, a more cosmopolitan outlook is needed. This explains why the modern technology gadgets are inevitable. Considering the fact that we are in the Information Technology age, extension needs to be in tune with the application of information technology to access and utilize the international network system. Ekemode (2000) observed that agriculture has moved through the various phases of development as society transforms into modernity. He therefore noted that the Information Technology era, involves the application of satellite technology for obtaining agricultural information through the Internet. Geographical Information Science is also another branch of InfoTech applicable to

Agriculture in the area of remote sensing like crop inventory, forestry inventory, yield prediction, land use classification and mapping, water forecasting, water resources management, ocean surge and coastal survey for fisheries and pollution.

Agricultural extension personnel are very important in the development of agriculture. They utilize strategic vital agricultural information for the individual and general improvement of the farmers, homemakers and youths. This could be in the area of their farming techniques, farm and family nutrition, health and community development. Agricultural extensionists serve as link between farmers and researchers. In order for them to effectively carry out their extension functions in the contemporary society, adequate information is inevitable. Information technology is a major ingredient in the communication of ideas. In a study, Ojiambo (1989) noted that extension personnel reported that they need information on types of insecticides, crop production management animal production management agricultural research results, marketing, credit and cooperative systems in the country, legislation on agricultural products, government policies on food and other agricultural productions.

8. FARMERS CAPABILITY IN THE UTILIZATION OF INFORMATION AND COMMUNICATION TECHNOLOGY FOR AGRICULTURE IN SOUTH-EASTERN NIGERIA

The media of information are as varied as the message themselves, ranging from personal visits and the use of mass media methods. But the limitations of these media in improving extension work are many (Adesope *et al.*, 2007a & b). But the unique method of feedback by the farmers to extension and research helps to modify messages and reorient research towards solving actual production constraints in the farm (Nwachukwu, 2003; Nwachukwu, 2005).

Today, it is important to know that a new paradigm of agricultural development is fast emerging. In both developing and developed countries, the overall development of rural areas in extending important service to citizens is being challenged; and traditional societies are being transformed into knowledge society all over the world. The report of Government of India (GoI) (2001) emphasized the necessity of developing the capacity to generate, absorb, disseminate and protect knowledge and exploit it as a powerful tool to derive societal transformation. Information and Communication Technologies (ICTs) are seen as an important means of achieving such a transformation, when used as a broad tool for providing local farming communities with scientific knowledge. ICTs herald the formation of knowledge society in the rural areas of the developing world. However, this can only be realized when knowledge, land and information are effectively harvested for overall agricultural and rural development. The advent of ICTs has narrowed the borders between nations to a thin line, making the world a global hamlet with nation encapsulated as one. Now it is possible for individual not to be physically present at the point of interaction of any business, with boundaries no longer a barrier in the current information flow. ICTs are revolutionary in nature; as a result, information seekers and communicators will need to follow the trends to make good use of the new tools. According to Ehikhamenor (1993), information revolution can be described as fundamental changes in the production and use of information occurring in the late 20th century. Human societies throughout history have had "information specialist" (from traditional healers to newspaper), and they have had "information technologies" (from cave painting to accountancy) but two interested developments, social and technological, underpin the diagnosis that an information revolution is now occurring. The tremendous advancement in computer technology, software engineering and telecommunications as well as the resultant radical changes and advances in the techniques of data and information processing, storage, retrieval and dissemination, in developed countries within the brief period between the mid-1990s and the late 2000s, is, no doubt, a revolution. Information processing has become increasingly visible and important in economic, social, and political life. The information-processing aspect of all work can be reshaped through information technology.

ICTs in general sense refer to the new technologies that have emerged from the integration of information technology (IT) and communication technology (CT). Akintunde (2004) stated that Information and Communication Technology (ICT) as terminology has overtaken information technology (IT) because of its appropriateness and its utilitarianism. Whereas, it was used to describe the new technology from the mid 1980s, to mid 1990s, focusing on computer; ICT, since the beginning of this decade, is a new concept that emphasizes the use of technology for development. ICT emphasizes the use of computer and other technologies such as telephone to process, transport and transfer voice and other data singularly or mixed with least interference or distortion of contents. The acronym ICT captures a multitude of equipment and services. These ranges from satellites communication system, telephone booths in rural areas, the Internet and electronic data bases, to e-commerce service *via* the World Wide Web (WWW). ICT encompasses an enormous variety of computer telecommunication and network hardware and accompanying software:

1. Digital Communications, specifically satellites, mobile telephone, wireless local loops and digital radio.
2. Computers and software including personal data assistants, e-mail, speech recognition technology, computer assisted distance learning, global positioning systems and of course the Internet.
3. Delivery mechanism such as public access rural information centres and tele-centres. Mobile internet vans, village kiosk and radio listening group (ISNAR, 2002).

The World Bank defines ICTs as "the set of activities which facilitate by electronic means processing, transmission and display of information" (Rodriguez and Wilson, 2000). ICTs "refer to technologies people use to share, distribute, gather information and to communicate through computers and computer networks". ICTs can be described as a complex varied set of goods, applications and services used for producing, distributing, processing, transforming information (including) telecoms, TV and radio broadcasting, hardware and software, computer services and electronic media. ICTs represent a cluster of associated technologies defined by their functional usage in information access and communication, of which one embodiment is the internet.

Regardless of the tool, the focus must be on the people who will use the technologies and the context rather than the technologies themselves. Some agricultural development services that can be provided in the developing world using ICT are:

- Online services for information, education and training, monitoring and consultation, diagnosis and monitoring, and transaction and processing.
- e-Commerce for direct linkages between local producers, traders, retailers and suppliers.
- The facilitation of interaction among researchers, extension (knowledge) workers and farmers.
- Question-and-answer services where experts respond to queries on specialized subjects. ICT services to block and district-level developmental officials for greater efficiency in delivering services for overall agricultural development.
- Up-to-date information, supplied to farmers as early as possible, about subjects such as packages of practices, market information, weather forecasting, input supplies, credit availability, *etc.*

- Creation of database with details of the resources of local villages and villagers, site-specific information systems, expert systems, *etc.*
- Provision of early warning systems about disease, pest problems, information regarding rural development programmes and crop insurances, post-harvest technology, *etc.*
- Facilitation of land records and online registration services.
- Improved marketing of milk and milk products.
- Services providing information to farmers regarding farm business and management.
- Increased efficiency and productivity of cooperative societies through the computer communication network and the latest database technology.
- Tele-education for farmers.
- Websites established by agricultural research institutes, making the latest information available to extension (knowledge) workers and obtaining their feedback (FAO, 1996).

In the same vein, CTA (2003) listed the following as a potential ICT application in extension:

- Enhance farmers' ability to collate demands.
- Collaborative learning.
- Exchange of time sensitive information. *e.g.* market prices, diseases outbreaks, *etc.*
- Make extension systems and structures more efficient.
- Engage farmers in assessing their own needs and solutions.
- Facilitating across multi-stakeholder brainstorming.
- Exploring alternative productive technologies.
- Facilitating access to market and credits.
- Training and Demonstration.
- Community learning.
- Search, select and compile information for individual clients.
- Early warning for disasters *etc.* and weather forecast.
- Peer to peer sharing and exchange among extension.

Thus the application of ICTs in agricultural extension as can be seen from above is almost limitless.

ICT is thought therefore to achieve information transfer more effectively than other communication methods in extension. Evidence relating to its impact in agricultural extension in some developing countries in Africa, Latin America and South Asia has been widely documented. In countries such as Mauritius, Egypt, India, Bangladesh, Chile, Mexico, Peru, South Africa and Uganda, ICT has played a major role in diffusing information to rural communities and shown even much unexpected potential (Munyua, 2000b; ADB, 2003). However, there are contemporary issues of concern to the relevance of ICT to agricultural extension in developing economies. These issues centre broadly on the pervasive poverty, illiteracy, poor infrastructural facilities and the constraint of manpower that are common in most developing countries.

The issue of ICT has received serious attention in Nigeria lately. The Federal government through the National Information Technology Development Agency which was established in 2001 is bringing useful ICT initiatives to enable Nigeria have a proper structure for ICT. But, it is also frustrating that the rural agricultural sector has not received adequate attention to warrant adequate utilization of ICT and this has obvious implications for rural farmers. In southeastern Nigeria, the ICT usage culture is near non existent as farmers have no direct touch with the ICT facilities except of course the use of mobile phones which is affordable and portable. To get a clearer picture of the situation this chapter investigates farmers' capability in the utilization of information and communication technology for agriculture in Southeastern Nigeria. The study also determines extent of the availability of ICTs to farmers in the study area.

8.1. Methodology

The study was conducted in southeast agro ecological zone of Nigeria. The choice was predicated upon the high agricultural potential of the large farming and trading population as well as abundant natural resources, where maximum utilization could be guaranteed through the effective use of ICTs. The South East agricultural zone of Nigeria lies between latitudes 4° 20′ and 7° 25N and longitudes 5° 25′ and 8° 51E. It occupies a land area of about 109,524 km^2, which represents about 11.86 per cent of the total land area of Nigeria. This area lies mainly on plains under 200m above sea level. It is bounded on the South by bight of Bonny, on the East by the Republic of Cameroon, on the West by river Niger and on the North by Benue State (Monanu, 1975). The zone has a population of about 1.892 million or 21.48

per cent of the total population of Nigeria (NPC, 1991). The rural population accounts for about 60 per cent with a population density of 173 people per square kilometer being one of the most thickly populated agricultural zones in Nigeria. There are nine states within the South East agricultural zone, namely, Abia, Akwa-Ibom, Anambra, Cross River, Enugu, Imo, Rivers, Bayelsa, and Ebonyi states. About 60-70 per cent of the inhabitants are engaged in agriculture. They are mainly engaged in crop farming except the reverie areas which are populated primarily by fishermen (Unamma *et al.*, 1985).

The climate of the South East agricultural zone is typically equatorial. It is influenced by the Equatorial East lies namely the tropical continental and tropical maritime air masses. Their interactions are responsible for the two distinct climatic seasons, the rainy and dry seasons. The rainy season starts generally from March-April and ends in October-November and the dry season last till March of the following year. The major ethnic groups are the Ibos, Anambras, Ijaws, Efiks, Ogonis and Ibibios. The main annual temperature ranges from 25°C in the South to 29°C in the Northern parts. The relative humidity ranges from 82-84 per cent in coastal areas to 70-80 per cent in the Northern parts. The relative humidity is highest during the rainy season and lowest during the harmathan (Monanu, 1975).

The major soil types in the Zone are the flovizol, cambiso, aerisol and nitosol. The soil types coupled with the climatic influences, have great implications on the type of vegetation and soil fertility status of the zone. The zone is characterized by three major vegetations. They are the humid forest (made up of salt water swamp, fresh water swamp and the rain forest), derived savanna and semi Montana. The major are Cross River, Niger, Anambra - Marnu and Imo. The other rivers include Otamiri, Qua-Ibo, Sambreiro, Orashi, forcados, Engenni, Igu, Asu, Ebonyi, Calabar, Great Kwa and the Akpa. The road network in the Zone is generally seasonal and consequently the movement of farm produce and extension deliveries are impaired (Ekwe, 2004).

The total arable land in the zone is about 5.8 million hectares which average 0.31 hectares per person. The forest reserve accounts for about 5.4 per cent, 4.5 per cent and 26.4 per cent of the total land areas of Cross River, Anambra and Akwa-Ibom States, respectively. The average farm household size is seven members with farm sizes ranging from 1 hectare to 1.5 hectare per farm family. The farm activities provide about 60 - 70 per cent of the household income and food crop production is the predominant agricultural activity except for the reverie areas. Capital cropping systems are the yam and cassava based cropping systems with the mixed cropping pattern dominating. The yam based cropping system consists of the yam as the hub crop and a number of minor crops. For example, yam, melon, maize, telfaira, okro, aerial yam followed by cassava, yam, melon, maize, telfaira, okro, aerial yam, cocoyam. The minor crops are introduced and harvested by the womenfolk (Ekwe, 2004). The zone accounts for 36 per cent, 53 per cent, 70 per cent and 17 per cent of the total national yam, cassava, cocoyam, rice maize production, respectively (NARP, 1995).

8.2. Sampling Procedure and Sample Size

The respondents involved in the study were randomly selected from three states in the study area since all the states are fully engaged in agricultural production. The states included Anambra, Abia and Akwa Ibom States, respectively. Three agricultural zones were randomly selected from the states apart from Abia State with only three zones. The selected zones were Aba, Umuahia and Ohafia in Abia State; Ikot Ekpene, Uyo and Eket in Akwa Ibom State and Awka, Onitsha and Anambra in Anambra State.

From the three agricultural zones, one block was randomly selected. Also, from each block one circle was randomly selected and thereafter, 30 farmers each were also selected randomly from a list of farmers complied by the ADP staff. This implies that 90 farmers were interviewed in each state giving a total of 270 farmers from the three states. A total of 270 respondents were drawn from 9 circles of the three states. Nine ADPs staff were recruited and thoroughly trained in the administration of well designed interview schedule; working under researcher's supervision. One set of structured interview schedule was developed and used for data collection. Frequencies and percentages were used for data analysis.

8.3. Results and Discussion

8.3.1. Extent of the availability of ICTs to farmers

Table 1 shows that 100 per cent of the respondents indicated that radio was available to them suggesting that most of the respondents interviewed have radio. 99.3 per cent of the respondents also indicated the availability of television to them. This could be a result of age-long introduction of television in Nigeria which had made it diffused into the society. Also, 98.5 per cent of the respondents indicated availability of G.S.M. In spite of the newness of G.S.M compared with radio and television the result suggests that G.S.M technology adoption is rapidly spreading and used all over the study area.

Table 1 : Distribution on the Availability of ICT Resources in the Study Area (n=270)

ICTs	States					
	Akwa Ibom		Anambra		Abia	
	Freq	per cent	Freq	per cent	Freq	per cent
CD-ROM	6	6.7	12	13.3	16	17.8
TV	90	100	89	98.9	89	98.9
Radio	90	100	90	100	90	100
Telephone	50	55.6	38	42.2	16	17.8
GSM	90	100	88	97.8	88	97.8
Computer	74	82.2	55	61.1	55	61.1
Internet	40	44.4	38	42.2	6	6.7
Fax	4	4.4	3	3.3	4	4.4

Source: Field Survey Data, 2007
*Multiple responses recorded

About 68 per cent of the respondents indicated that computer was available in the study area. From personal observation, it was observed that computer was found more in the urban centres than in the rural area. Use of Telephone, Internet, CD-ROM and Fax were (38.5 per cent), (31.1 per cent), (12.6 per cent) and (4.1 per cent), respectively as shown in table 1. The implication of this is that these ICTs were not available to majority of the respondents. Also from personal observation, these ICTs were seen in small quantity in urban areas of the study. Tiamiyu (2001), in his study on Information Technology for transforming the visions and efficiency of Nigerian public agencies, reported inadequate levels of availability of ICTs (Internet, CD-ROM and Telephone) components within Nigerian public agencies.

8.3.2. Capability for utilization of ICTs for Agricultural purposes

Table 2 shows that respondents rated the capability in the use of television (92.6 per cent), radio (91.6 per cent) and GSM (91.9 per cent) very high. The high rating of the use of television and radio is not surprising; this is because of the age - long introduction and utilization of radio and television in the society. Besides, television and radio are the earliest ICTs used by Agricultural Development Programmes for agricultural information exchange with the farmers (Arokoyo, 2005).

Table 2 : Distribution of respondents according capability for utilization of ICTs for Agricultural purposes (n=270)

ICTs	States						Total	
	Akwa Ibom		Anambra		Abia			
	Freq	per cent	Freq	per cent	Freq	per cent	Freq	per cent
Television	89	98.9	90	100	89	98.9	268	99.3
GSM	85	94.4	88	97.8	90	100	263	97.4
Fax	6	6.7	5	5.6	6	6.7	17	6.3
Radio	88	97.8	88	97.8	90	100	266	98.5
Computer	20	22.2	20	22.2	22	24.2	62	22.9
E-mail	22	24.2	20	22.2	20	22.2	62	22.9
CD-ROM	2	2.2	5	5.6	6	6.7	13	4.8

Source: Field Survey Data, 2008
*Multiple responses recorded

GSM is a common ICT tool which is also easy to use. This finding has shown that the usage of GSM is high. This is attributed to the availability of GSM, availability of phone booths both in urban and rural areas and competency of GSM operators. There is no doubt that GSM use needs less competence and GSM centres are almost at every nook and cranny of the society. Lawal-Adebowale and Adebayo (2006) are of the opinion that information and communication revolution at the time of the 21st century shows that ICTs such as telephone (fixed and mobile lines) and internet services have the potential of reversing the inherent setback in mass extension by specifically making it possible for the source to effect responses from the receivers. For instance, mobile phone service which is fast spreading, even to the rural areas of the developing countries, is making it possible for listeners or farmers in particular to respond to agricultural programmes aired both on radio and television. This development, not only make it possible for farmers and extension agents to interact with one another, it has greatly assisted in shortening and reducing the distance and cost of interaction between stakeholders in agricultural development effort and also helps to improve accessibility, quality and relevance of agricultural information made available to the farmers (Bertolini, 2004). The use of mobile phone in exchange of agricultural information, as indicated by Developments (2005) equally makes it possible for farmers to either identify or create markets for their farm inputs need and or sales of their farm products. Computer (22.9), e-mail (22.9 per cent), fax (6.3 per cent) and CD-ROM (4.8 per cent) respectively were rated low by respondents as shown in table 2 in their capability and utilization. Indian Council of Agricultural Research (2006) reported that the new communication technologies where micro-computers, e-mail, CD-ROM offer opportunities for interaction which enhances agricultural development.

8.4. Conclusion

Despite the positive accolade accorded to modern ICTs, its application in the agricultural sector is not encouraging. This study revealed that ICTs were not available to majority of the respondents. And unfortunately, respondents were not capable of using these ICTs for Agricultural purposes because of the newness and complicated nature of manipulating these ICTs. This suggests that respondents need aggressive awareness of the importance and need of internet use for agricultural information exchange and training in the use of ICTs.

8.5. Recommendations

From the above findings, the following recommendations were made:

- Efforts should be made by government (Federal, State and Local Government) to provide adequate ICT resources in both urban and rural areas.
- The Ministry of Information and Communication need to carry out a massive sensitization of the potentials of the ICTs for the speedy or timely dissemination of information to farmers.
- Intensive training in the use of ICTs should be organized by government to enable farmers know how to make use of varied ICT resources.
- Government should establish and manage kiosks, phone booths and cyber cafes in rural and urban areas for agricultural purposes.

9. CASE STUDIES IN NIGERIA

Over the past 25 years, developing countries have considerably increased ICT access, especially for telephone services. Developing countries accounted for more than 60 per cent of the world's telephone lines (fixed and mobile) in 2005, up from less than 20 per cent in 1980. During this period, population increased by half and real GDP more than doubled in these countries while the number of telephone subscribers rose 28 fold (ITU, 2003).

ICT is crucial to sustainable poverty reduction, because it makes a country's economy more efficient and globally competitive, improves health and education services and creates new sources of income and employment for poor people. In addition, ICT enhances social inclusion and promotes more effective, accountable, democratic government, especially when combined with effective freedom of information and expression (ITU, 2003). There is no gainsaying the fact that awareness level of ICT has increased dramatically as there is virtually no sector of the economy that ICT is not being applied. Even the rural economies have their share of the technological advancement. With the diffusion of the technology by service providers such as starrcomms, reltel, MTN and Glo, many rural communities are now connected to the outside world. For instance, in a recent international survey involving the authors in Saura, a remote village in Keffi local government area of Nasarawa State, North Central Nigeria, where the researchers thought was cut off from civilization received full service from three telecommunication networks. The wireless network has been greatly responsible for the improvement in communication between farmers and also information on where to send their agricultural products for better prices. This is with the use of mobile phones. Similar cases were found in remote communities such as Agbaduma, Ochiagbede and Ogbonicha in Kogi State, Nigeria.

The private sector has played significant roles in the information and communication for rural development initiative in Nigeria. For instance, the telecommunications service providers like MTN Nigeria, Starcomms, Glo and Zain have been able to ensure that ICTs reached the people. As a result the rural people are adapting the technology to their environment and then achieving improved livelihoods. Phone call centres dot the rural environment and young people who were until now unemployed now earn a living and are empowered. In the process they save money to build their capacity through sustainable training and personal development. The MTN Nigeria Empowerment Initiative, for instance, has been able to provide affordable capital, appropriate technology and capacity building resources needed to lift themselves out of poverty. Through this initiative, especially women have been given to help them reap the rewards of their own labor, through microfinance. The specific initiatives include: the rural telephone project (RTP) and the rural housing project.

The Rural Telephone Project (RTP) is a Micro Finance Scheme where rural and urban women, referred to as 'Phone Ladies' are loaned money through Micro Finance Institutions (MFIs). The objectives of the project are to provide first class GSM services in the rural and deep rural communities alleviate poverty and empower women. Interestingly the project has had a significant impact on the lives of the beneficiaries as well as their communities. Through the project, the beneficiaries have improved their income significantly to an average of N17,000 per month (85 USD). Also, their social status has increased because now they have a voice in the community compared to when they had little or no income and did not have a sense of belonging. In the same vein, there has been enhanced productivity in the community due to access to telecommunications services.

In June 2004, a project, called The Akede Project was launched in Ilorin, Kwara State, in North Central Nigeria and about 30 women involved in the project. That was a giant stride which apparently was the first of its kind in Nigeria. It involves training and equipping rural women with equipment needed for them to start a call centre in their community. The equipment includes: a handset, SIM pack, airtime, promotional materials (Umbrella, Banner, Wind vane, T-shirts and face caps), antennae and Battery pack. The total cost of the equipment and airtime provided is converted into a loan given to the women through a Micro Finance Institution (MFI).

The women make monthly repayments of between N1, 000 - N2, 000 (USD 5 - USD 10) on the loan over a 10-month period. The repayments are used to extend the project to other rural women. Beneficiaries are also expected to pay a refundable insurance fee of N5,000 (USD 25) at the rate of N125 (0.6 USD) weekly as well as saving at least N100 a week. At the completion ceremony, a number of the beneficiaries had been able to save about N14, 000 (USD 70) over the period. There were no defaults on the loan as 100 per cent repayment rate was recorded. This is very encouraging. As a result of the success recorded, the Rural Telephone Project has 100 phone ladies in over 20 communities. Apart from Kwara State, the project also is present in Akwa Ibom, Edo and Enugu States. The prospect of the project is to empower over 200 women in 70 communities and it is expected that within the next 5 years the project will be extended to over 3000 beneficiaries across Nigeria.

10. IMPLICATIONS FOR AGRICULTURAL EXTENSION

The public sector agricultural extension is the major outlet of extension services in Nigeria, but the impact of this sector leaves much to be desired as bureaucratic bottlenecks always associated with government establishments make it impossible for appropriate services to be delivered.

With the private sectors impacting on the lives of rural people, agricultural extension personnel need to make themselves relevant as stakeholders in the rural development process. Therefore, the federal government should partner with these telecommunications service providers like MTN Nigeria, Glo, Zain and Starcomms, reltel, etisalat *etc.* In the final analysis, it is to conclude that ICT has not been fully integrated into the rural agricultural sector in Nigeria. Despite the milestones by the Federal Government in other sectors like the health sector, the rural farmers cannot be said to have benefited meaningfully from the dividends of information and communication technology in Nigeria.

References

Aboh, C.L., 2008. Assessment of the frequency of ICT tools usage by Agricultural Extension agents in Imo State, Nigeria. *Journal of Agriculture and Social Research (JASR)*, 8 (2), 152-157.

ADB, 2003. A Strategic Approach to Information and Communication technology toward e-development in Asia and the pacific, Asian Development Bank (ADB), November 2003.

Aderinto, A. Adedoyin, S.F., Awotide, D.O. and Adamu, C.O., 2008. Use of information and communication technologies (ICTs) among extension personnel in Ondo State. *Nigerian Journal of Rural Sociology*, 8 (1), 66-70.

Adesope, O. M., Asiabaka, C.C., and Agumagu, A. C., 2007a. Effects of personal characteristics of extension managers and supervisors on information technology needs in the Niger Delta area of Nigeria. *International Journal of Education and Development using ICT*. 3(2),1.

Adesope, O.M., Asiabaka, C.C and Matthews-Njoku, E. C., 2007b. Problems affecting the use of information technologies as perceived by Extension Managers in Nigeria. In: A.C. Agumagu, *et al.*, (Eds). *Contemporary Issues in Agricultural Extension and Development Studies- Series one*. Omoku: Molsyfem United Services

Adesope, O.M. Agumagu, A.C. and Adebayo, E.L., 2007c. Extensionists and Researchers proficiency requirements in Information and Communication technologies in southeastern Nigeria. *Journal of Extension Systems*. 23 (1), 55-69

AGREN, 2004.Information and Communication Technology in Agricultural Development: A comparative Analysis of three projects from India. *Agricultural Research and Extension Network, Newsletter: No.49*. January, 04. P. 14.

Aina, L., 1995. *Agricultural Information in Africa*. Ibadan: Third world publishers

Akintunde, S.A., 2004. Libraries as tools for ICT Development. In: *Libraries: Tools for Educational Development. A Compendium of Papers Presented*. Nigeria Library Association. Akure. June 20- 25, pp. 11

Akpabio, I.A., 2007. Agricultural extension officials and utilization of Information and Communication Technologies in the Niger Delta, Nigeria. *Journal of Extension Systems*. 23 (1), 29-41.

Ani, A.O., 2007. *Agricultural Extension: A Pathway for Sustainable Agricultural Development*. Kaduna: Apani Publications.

Arokoyo, T., 2005. ICTs Application in Agricultural Extension Service Delivery. In: Adedoyin S. F. (Ed) *Agricultural Extension in Nigeria*. Ilorinf: Agricultural and Rural Management Training Institute, pp 245-251.

Asiabaka, C. C., 2002. *Agricultural Extension. A Handbook for Development Practitioners.* Omoku: Molsyfem United Services

Babu, A. R., Singh, Y. P. and Sachdeva, R. K., 1997. Establishing a management information system'. In: Swanson, B. E. *et al.*, (Eds) *Improving Agricultural Extension. A Reference Manual.* Rome: FAO

Banigo. E., 2001. How Nigeria can bridge digital dived in IT. *Guardian,* Tuesday April 3, 2001 PP.63.

Bertolini, R., 2004. *Making Information and Communication Technologies Work for Food Security in Africa.* International Food Policy Research Institute (IFPRI), 2020 African Conference Brief, 11, pp1-6.

Braimoh, A.K. and Oladele, O.I., 2000. Geographical information systems for agricultural extension and poverty alleviation in Nigeria. *Proceedings of the 6th Annual National Conference of the Agricultural Extension Society of Nigeria (AESON),* from 10 to 12 April, 2000.

Commonwealth Secretariat, 1991. *Microcomputers in Schools: Policy and Implementation Guidelines.* London: Commonwealth Secretariat, Educational programme.

Cornell University (Undated). *The Essential Electronic Agricultural Library. Library in a Box. Demonstration Disk.* USA: Cornell University in association with the Rockefeller Foundation.

CTA, 1997. Information for rural development: The ACP States identify their priorities'. Technical Center for Agricultural and Cooperation (CTA), No 67, February.

CTA, 2003. *Annual Report.* Special paper PP. 7.

Developments, 2005. Mobile Development: How wireless technology and open source software can liberate the poor. *The International Developments Magazine.* Issue 31, Third quarter, p 40.

Ehikhamenor, F.A., 1993. Information Technology and Scientific and Technological Information in Nigeria: Revolution or Evolution? *African Journal of Library, Archives and Information.* 3 (2):119

Ekemode, K.O., 2000. *Introductory Agriculture for Sustainable Development.* Lagos: Nigerian Association of Agricultural Educators.

Ekwe, K. C., 2004. Factors Associated with utilization of Garri processing Technology in South Eastern Nigeria. *Unpublished, Ph.D. Thesis,* MOUA Umudike, Nigeria.

FAO, 1996. Agricultural training: Report of an evaluation study. Rome. FAO.

Government of India (GoI), 2001. Report of prime minister's Task Force on India as knowledge superpower. Planning Commission, Government of India, New Delhi.

Guiness Book of Knowledge, 1997. *Guiness Book of Knowledge.* USA: Guiness Publishing Ltd.

Guiness Book of Records, 1996. *Guiness Book of Records.* USA: Guiness Pub Ltd

IDRC, 1999. International Center Development Research - *Reports.* 23(4).

Indian Council of Agricultural Research, 2006. *Handbook of Agriculture.* Indian Council of Agricultural Research, New Delhi, India

International Service for National Agriculture Research (ISNAR), 2002. Briefing paper 55. September. PP1.

International Telecommunication Union, 2003. *World Telecommunication Development Report.* Geneva: ITU.

Jackson, A.R.H., 1984. The information scientist as a professional: status and marketing'. In: Oppenheim, C. (Ed.). *Information Comes of Age.* London: Rossendale

Lawal-Adebowale, O. A. and K. Adebayo, 2006. Information and communication technologies: Pathways to effecting feedback in mass extension method of message delivery. *Proceedings of the 11th Annual Conference of Agricultural Extension Society of Nigeria (AESON),* 3rd – 6th April, 2006.

Lukeeram, I., Bheenick, K. J. and Travailleur, C., 2000. Combining audio files and graphics on the web to increase accessibility to agricultural information for the non-English speaking and the non-literate farmers in Mauritius: A practical example with PETIS (Potato Extension and Training Information System). *Paper presented to 2nd Global Knowledge conference,* March 7-10, Kualar Lumpar, Malaysia

Madu, E.C. and Adeniran, T.N., 2000. *Information technology uses and preservation of resources in libraries and information centers.* Oyo: Odumatt press and publishers.

Mchombu, K.J., 1992. Information needs for rural development. *Africa Journal of Library Archives and Information Science.* 2 (1), 17-32.

Meera, S. N., Jhamtani, A., and Rao, D. U. M., 2004. Information and communication technology in Agricultural development: a comparative analysis of three projects from India. *Agricultural Research and Extension Network. Network Paper No. 135,* 13pp

Moll, P., 1982. Information technology and third world – aid or aggravation? *Focus on International and Comparative Librarianship.* 13 (3).

Monanu, P.C., 1975. *Geographical Boundaries of Nigeria.* University of Ife press, 1(2), 3.

Munyua, H., 2000a. Information and communication technologies for rural development and food security: Lessons from field experiences in developing countries. Sustainable Development Department (SD), Food and Ariculture Organization of the United Nations.

Munyua, H., 2000b. Information and Communication Technologies for rural development food security: Lessons from field experiences in developing countries paper on the role of information and communication technologies in rural development and food security. *CAB International, Africa Regional Centre Special.*

NARP, 1995. National Agricultural Research Strategic Plan for 1996-2010. *Draft Report,* National Agricultural Research Projects (NARP) Ibadan, Nigeria.

NPC, 1991. Provisional Population Census Result. National Population Press, National Population Commission (NPC), Lagos Nigeria.

Nelson, J and Farrington, J., 1994. *Information exchange networking for agricultural development. A review of concepts and practices.* The Netherlands: Technical Centre for Agricultural and Rural cooperation (ACP-EEC).

Nwachukwu I., 2005. Extension Communication; In: I. Nwachukwu and C. G. Onuekwusi (Eds). *Agricultural Extension and Rural Sociology*. Snaap Press Ltd, Enugu.

Nwachukwu, I., 2003. *Agricultural communication. Principles and practice.* Umuahia: LambHouse Publishers.

Nzeakor, A.N., 1991. The new information technology in local government administration. Workshop for local government functionaries, Nsukka: University of Nigeria Computer communication center.

Ojiambo, J.B., 1989. Communication of Agricultural information among research scientists, extension personnel and farmers in Kenya. *PhD dissertation*, University of Pittsburgh.

Osuagwu, D. E., 2001. 'New technologies and services in internet business'. *Journal of Professional administration, July – September, 2001, 29-41.*

Oxford Advanced Learner's Dictionary, 1998. *Oxford Advanced Learner's Dictionary.* Oxford: Oxford University

Phone 4 All, 2002. Evolution of Telecommunication in Nigeria 1886-1999. IT Initiatives 1999-2002.

Prakash, S., 2000. Uganda: Information Technology and Rural Development. The Nakaseke Multipurpose Telecenter. *IK Notes No.27,* December, 2000.

Prevost, R., 1995. *Internet Insiders.* USA; Mc Graw-Hill.

Puliatti, E., 1997. Alternex: A computer-based network by and for NGOs. *Development Communication Report,* 67: 5 - 7.

Qamar, M. K., 2000. Agricultural extension at the turn of the millennium: trends and challenges. In: Qamar, M. K ., (Ed) *Human Resources. Agricultural and Rural Development.* Rome: Food and Agriculture Organization.

Rodriguez, F. and E. Wilson., 2000. Are poor countries losing the information Revolution. *Info dev Working Paper,* Washington D.C.: World Bank.

Salihu, T.Y., 2000. IT landmark in the next decade. *Microbyte Magazine.* 3(12), pp. 26 – 27.

Sotade, N., 2001. The Internet and you. *Gramon Trends.* p.22

Thioune, 2003. Women: The key to food security. In: Fruoung, H. with Williams G, Clark, J. and Couey, A.1993. (Eds.) *Gender Issues in Communications,* http://www.cprs.org//gender/bawit.cfp.

Tiamiyu, M., 2001. Information Technology for transforming the visions and efficiency of Nigerian Public agencies. *Working paper 37*, Development Policy Centre, Ibadan.

Torimiro, D.O., Adedoyin, S.F. and Alao, J.A., 2000. Forms of communication used for strengthening agricultural technologies dissemination in Ogun State, Nigeria. *Proceedings of the 6th annual national conference of the Agricultural Extension Society of Nigeria (AESON), from 10 to 12 April, 2000.*

Unamma, R.P.A, Odurukwe, S.O. Okere, H.E. Ene, L.S.O. and Okoh O.O., 1985. Farming System in Nigeria. Report of Benchmark survey of farming system of South East Agricultural Zone of Nigeria. NRCRI, Umudike Nigeria. PP 116.

Van Crowder, L. and Fortier, F., 2000. National Agricultural and rural Knowledge and Information System (NARKIS): A proposed component of the Uganda National Agricultural Advisory Service (NAADS). FAO, 22pp.

Vernon, R., 2001. *Knowing where you're going: Information systems for agricultural Research Management.* The Hague: International Service for National Agricultural Research.

VITA (Volunteers for Technical Assistance), 1995. Information center. Paper presented to the World Bank. USA.

Waldron, M. W., Vsanthakumar, J. and Arulraj, S., 1997. Improving the organization and management of Extension. In: Swanson, B. E. *et al.*, (Eds) *Improving Agricultural Extension. A reference manual. Rome: FAO.*

Yahaya, M.K., 2000. Prospects of integrated multi-media communication model in mobilizing farmers for the adoption of sustainable poverty alleviation programmes in Nigeria. *Proceedings of the 6th Annual National Conference of the Agricultural Extension Society of Nigeria (AESON), from 10 to 12 April, 2000.*

Yekinni, O.T. and Hussein, L.A., 2008. An assessment of the relevance of information and communication technologies (ICTs) to agricultural and rural development by research and extension personnel in South-Western Nigeria. *Nigerian Journal of Rural Sociology*. 8 (1), pp. 79-88.

Zorkoczy, P., 1982. *Information Technology: An Introduction.* 2nd Ed. London: Pitman Publishing.

Authors Information

Dr. Olufemi Martins ADESOPE

Lecturer

Agricultural Economics and Extension

University of Port-Harcourt

PMB 5323, Port Harcourt, Rivers State, NIGERIA

Postal Code: 500001

Phone No.: +234 703 095 8561

Email: omadesope@yahoo.co.uk

Dr. Olufemi Martins Adesope, is a Lecturer in Agricultural Extension at the Department of Agricultural Economics and Extension, University of Port-Harcourt, Rivers State, Nigeria. He has about 15 years teaching and research experience with over 100 publications (books, chapter in books, proceedings, technical notes). His areas of interest include rural development, strategic planning, farmer field schools, socio-economic impact assessment, research methods, Information and Communication Technologies (ICTs) in Agriculture. Dr. Adesope has competence in the use of Statistical Packages for the Social Sciences (SPSS), GenStat, E-Views and Microsoft Excel for statistical data analysis. He has been a consultant to individuals and groups on issues related to sustainable development and has established farms for individuals and groups and facilitated various development trainings and workshops.

Dr. Moses Okwusi

Head of Unit, Marketing information and ICT services
Department of Agricultural Extension Services
National Root Crops Research Institute, Umudike, NIGERIA
Postal Code: 440001
Mobile Phone No.: +234 803 798 2882
e-mail: mosesokwusi@yahoo.com

Dr. Moses Okwusi is an Agricultural Extensionist with the National Root Crops Research Institute, Umudike, Nigeria. He is widely travelled and has several journal articles to his credit. Dr Okwusi has research experience spanning over 10 years.

Dr. Ike Nwachukwu

Associate Professor
Director, Extension Services
Department of Agricultural Extension and Rural Sociology
Michael Okpara University of Agriculture, Umudike, NIGERIA
Postal Code: 440001
e-mail: nwaike3@yahoo.com

Dr. Ike Nwachukwu is an Associate Professor of Agricultural Extension and Rural Sociology at the Michael Okpara University of Agriculture, Umudike, Abia State, Nigeria. He was previously a Research Officer (Extension Communication) with the National Cereals Research Institute, Badeggi, Niger State, Nigeria. He is currently the Director of Extension Centre, Michael Okpara

University of Agriculture, Umudike, Abia State, Nigeria. He is the author of Agricultural Communication, Planning and evaluation of Agricultural and rural development projects and a co-editor of the book titled Agricultural Extension and Rural Sociology. He has several journal publications to his credit.

□□□

Chapter 11
RWANDA

Charles Karemangingo, Edward Mutandwa,
Nathan Kanuma Taremwa and Frank Mugisha

1. AN OVERVIEW OF RWANDA'S SOCIO-ECONOMIC SETTING

Rwanda is a small landlocked country located in East Africa covering a total surface area of 26,338 km^2. It lies about 120 km south of the Equator and is bordered by Uganda in the North, Tanzania in the east, Burundi in the south and the Democratic Republic of Congo in the West. According to the national official statistics, the total population is approximately 8,200,000 people, with women accounting for 55 per cent of the people. The country's population density is 505 people per km^2, making it one of the most densely populated countries in Africa (National Institute of Statistics, 2002). Since the turn of the millennium, the annual growth rate of the population has been 2.9 per cent and it is believed that by the year 2020, the country will have an estimated population of 20 million, thus exerting pressure on the limited land resources. It is also important to note that most people (64 per cent), those living in abject poverty, reside in rural areas of the country (Government of Rwanda, 2008).

Rwanda's economy is largely predicated on agriculture, which accounts for at least 50 per cent of the Gross National Product (MINAGRI, 2004). The climatic conditions in the country are eminently tropical but tend to vary from region to region. However, agricultural production is largely subsistence with little commercialization for local market. Between the year 1994 and 2000, the country faced a food crisis due to the socio-economic problems and after effects of the war. As a result, the country heavily depended on food aid from donor agencies. Net cereal imports constituted 61 per cent with respect to national cereal consumption requirement in the country between 1961 and 1998 (Agriculture and Food-Rwanda, 2003). When discussing the country's socio-economic milieu, it is important to note that agriculture is a key economic activity sustaining the livelihoods of many people. Tourism also forms an important parallel economic activity generating significant foreign exchange for the country. It is estimated that this sub-sector contributes about 70 per cent of the country foreign currency, emanating from tourist visits to the country's three main national parks namely, Akagera, Virunga (Volcanoes) and Nyungwe national parks (MINECOFIN, 2002).

Rwanda is considered to be an emerging economy, whose national output has been increasing over the years. In 2008, the country experienced an one per cent growth in its national output. Inflation rate has generally been around 20 per cent in the last five years (2005-09). Despite these developments, Rwanda has been facing a multiplicity of socio-economic challenges which are *inter alia* diminishing agricultural productivity, narrow economic and export base, poor institutional capacity, public debt burden, extreme land fragmentation, diminishing land resources and per capita arable land, rapid population growth and high population density, low level human resource development, underdeveloped and under-funded social infrastructure and services, weak private sector and low capacity in World Trade Organization (WTO) negotiations on General Agreement on Trade in Services (GATS) (Government of Rwanda, 2008). These challenges are related to various socio-economic indicators such as low literacy rates and the HIV/AIDS pandemic. In terms of education, more males are educated (72 per cent) than females (52 per cent). When juxtaposed with other countries in Sub-Sahara, HIV/AIDS incidence in Rwanda is relatively low (3 per cent). However, HIV/AIDS poses serious developmental challenges to the country. The government of Rwanda has promulgated a number of policy initiatives to stave off the deleterious effects of HIV/AIDS on the socio-economic fabric of the country and also to ensure that women are an integral part of mainstream development. Subsequent sections of this paper will focus on the role of agriculture to Rwanda's economic landscape, the challenges and opportunities faced, the need for information and communication technologies, the ICT policy environment and characterization of innovations. These sub-sections are discussed in view of observations made above.

2. STRUCTURE OF THE AGRICULTURAL SECTOR

As already observed, Rwanda's economy is underpinned by the agricultural sector, which sustains the lives of at least 90 per cent of the population. Much of the agricultural activities are performed on small farms which are generally less than 0.5 hectares. Only three (3 per cent) of the national landholding belongs to large-scale farmers (MINECOFIN Report, 2004). Increasing population continuously creates pressure on the limited land resources. Access to land in Rwanda has been constrained by lack of a legal and regulatory framework for land use and ownership. For several decades, secure access meant acquiring and clearing more land. In addition, land distribution has also been a major constraint to securing access for most poor groups in Rwanda, as in other parts of sub-Saharan Africa (MINITERE, 2004).

The government thus focuses on land consolidation strategy, which seeks to increase agricultural productivity while simultaneously ensuring sustainable use of the environment. The country is conveniently sub-divided into four provinces namely North, South, East and Western provinces. The potential of agriculture depends on the prevailing climatic conditions and soil types. These conditions tend to vary across the different regions. In the northern and western provinces, annual rainfall received is high (1400-1800 mm) and the vegetation should predominantly be mountain forest. The eastern part of the country is characterized by drier conditions and vegetation is mainly tropical savanna. In the southern province, rainfall is also high (1000-1400 mm per annum) and it supports a wide array of crop and livestock sub-systems. The types of crops grown and livestock raised are determined by the government, through its Green Revolution Policy based on the agro-ecological regions. The main objectives of this policy are to buttress agricultural productivity and ensure household food self sufficiency (Government of Rwanda, 2008). Table 1 depicts the crops and livestock systems stipulated for the agricultural sector.

Table 1 : Main crops and livestock stipulated under the Green Revolution Policy

Province	Crops	Livestock
Eastern	Maize, short beans, bananas, sorghum, cassava	Cattle, goats, and rabbits
Southern	Soybeans, cassava, coffee, tea	Pigs, cattle, goats, sheep
Northern	Irish potatoes, wheat, climbing bean, pyrethrum, maize	Sheep, cattle, goats, pigs
Western	Coffee, Irish potatoes, Tea, Rice, Wheat	Cattle, goats, pigs and sheep

Source: MINAGRI, 2004

2.1. Contribution of Agriculture to Rwanda's Economic Vista

Like in most developing countries, agriculture has a multi-faceted role in Rwanda's development. Its strategic importance is underlined by the fact that the sector generally accounts for 46 per cent of real GDP. Agriculture also provides opportunities for job creation with 87 per cent of the rural population being employed. At least 80 per cent of the national exports emanate from the agricultural sector (MINAGRI, 2007). Given the essential role that agriculture plays in the achievement of household food security, the government of Rwanda has started to formulate policies and strategies in order to improve agricultural productivity.

Coffee and tea are the dominant cash crops in Rwanda's agriculture. Coffee is grown by approximately 500,000 smallholder farmers on a total area of 33,000 hectares (OCIR-Café, 2005). Despite the important role that the crop plays in the livelihoods of rural farmers, national coffee throughout has been declining since late 1980s. National production reached a peak of 43,000 tons in the 1986/87 coffee season. Between 2000 and 2004, national coffee production hovered between 16,000 and 25,000 tonnes with an average yield of 2 tons/ha (ibid). Between 1988 and 2004, the value of exported coffee has been between US$ 15 million and US$ 65 million per annum (OCIR-Cafe, 2005). Tea exports have also played a significant role in national agricultural export generation. In 2003, the value of tea exports was US$ 26.8 million (MINECOFIN, 2003). Pyrethrum is another crop that is increasingly becoming visible in terms of its contribution to national exports. It is mostly exported to the USA and different European destinations. Focus has been on value addition in order to establish the international market position of pyrethrum. However, output data for this crop is not easily available. A wide range of food crops are also important in Rwanda's agricultural landscape and these include beans, bananas, sorghum, Irish potatoes, sweet potatoes, cassava and maize. Figure 2 shows the variation of total area harvested for cereals (maize, wheat, sorghum and rice), pulses and oil seeds (beans, peas, groundnut and oilseeds), roots and tubers (sweet potatoe, Irish potatoe, cassava and yams) as well as fruits and vegetables.

In general, the total area harvested for cereals, fruits and vegetables has been increasing since 1998. This is attributable to a number of factors that include use of improved seeds, fertilisers, chemicals as well as access to agricultural extension services among rural farmers. For pulses, roots and tubers, the trend has generally been positive between 1997 and 2004. Plummeting output experienced between 2005 and 2007 is due to the increasing incidence of pests and diseases and also the erratic climatic conditions in some parts of the country.

Livestock production in Rwanda is important from a socio-economic and cultural perspective. In rural areas, livestock assets are considered as an important indicator of wealth. The predominant forms of livestock include cattle, goats, sheep, pigs, poultry and rabbits. The national statistical office estimated that as of 2007, the country had 991,697 cattle, 1,270,903 goats; 371,766 ovine; 211,918 pigs; 498,401 rabbits and 2,482,124 poultry (MINAGRI, 2007). The main livestock products include milk, meat, fish and eggs, which are mostly for household consumption. Generally, livestock numbers have been increasing through different policy initiatives such as the one cow for poor family program, whose main objective is to provide poor households with cows for each family (Government of Rwanda, 2008).

In general, agricultural productivity is hindered by a plethora of technical, socio-economic and institutional factors which include population pressure (which leads to environmental degradation), declining soil fertility, limited access to strategic resources such as agricultural credits, lack of improved seeds and erratic rainfall patterns. A number of programs have been enunciated to respond to the aforementioned production and environmental related constraints. These include the environmental protection program (bench terraces and ditches), Plans of Action (Imihigo), Watershed management program (Agasozi Kindatwa) and crop production programs on very steep slopes which seek to reduce soil erosion and promote sustainable crop production (Government of Rwanda, 2008). In view of these challenges, the government of Rwanda has been crafting various policies to improve agricultural productivity, ensuring food security and developing export-oriented agriculture.

3. AN OVERVIEW OF THE POLICY ENVIRONMENT AFFECTING AGRICULTURE IN RWANDA

Since the turn of the millennium, a number of policy initiatives have been enunciated by the government of Rwanda. However, all the policies designed by the government of Rwanda centre around the *Vision 2020* document describes the overall developmental direction of the country. Related policies embraced over the years include the National Poverty Reduction Strategy, the National Investment Strategy, the National Agricultural Policy, the National Decentralization Policy and the New Partnership for Africa Development (NEPAD). These policies affect economic agents such as households, the government and producers in the various sectors of the economy. The overall thrust of these policies is to ensure modernization of agriculture, promotion of public and private sector investment, good governance, political stability, macro-economic stability, effective institutions and involvement of local institutions and ultimately poverty reduction (Government of Rwanda, 2008).

The national agricultural policy seeks to ensure household food security and simultaneously stimulate agricultural exports. It is based on six major axioms, namely food security through the creation of an environment favourable to income generation; modern, professional, innovative and specialised agriculture; strict market oriented agriculture (domestic, sub-regional, regional and international); fair distribution of benefits from all products resulting from different stages of processing; integrated and diversified agriculture and development of agriculture that is friendly to the environment (MINAGRI, 2007). In order to implement the agricultural policy, a multi-pronged strategic framework was developed. It focuses on commodity-based intervention, selection of a limited number of priority crops, promotion of cash crops and fostering partnership between public and private sectors. However, a number of challenges have been militating against effective policy implementation including lack of coordination between extension and MINAGRI, lack of a clearly defined extension system, low levels of training among the population, dearth of financial resources and lack of a coordinated research-extension framework. Since the implementation of the national agricultural strategy depends on effective institutions, the next section will focus on the agricultural extension system. We outline the historical overview of agricultural extension in Rwanda and highlight the institutional setup in agricultural extension. In this section, an outline of the main strengths and weaknesses of the agricultural extension system in the country is made.

4. HISTORICAL EVOLUTION OF THE AGRICULTURAL EXTENSION SYSTEM IN RWANDA

To properly understand the historical evolution of the agricultural extension system in Rwanda, it is essential to conceptualize the term 'agricultural extension'. Adams (1994) defines agricultural extension as assistance to farmers which seeks to identify and analyze their production problems and to become aware of the opportunities for improvement. Van den Ban and Hawkins (1996) note that agricultural extension involves the communication of information to help people to form sound opinions and make good decisions. The overall objective of agricultural extension is to improve livelihoods of farmers through encouragement of local communities to use their own resources with minimum assistance from the government (Nzabamwita, 2005).

The earliest form of agricultural extension system was introduced in Rwanda by the then Belgian colonial administration around 1900. During that period the system was largely prescriptive and required farmers to produce crops for export to Belgium. Initially, two crops namely tea and coffee were introduced and production was monitored by expatriate extension agents from Belgium. To facilitate research in coffee and tea, research centres were established in Congo and the then Rwanda-Urundi. These were managed by the Agronomic National Institute of Agronomic Research of Congo (INEAC) until 1962, when the institution was re-named Rwanda Institute of Agricultural Science (ISAR). After attainment of political independence in 1962, the government of Rwanda adopted the system of controlled production which also included the use of fines for farmers who failed to achieve their targets (ibid). In 1975, G. Delepierre, an agronomist, developed agricultural regionalization based on soil types, climate and elevation. This regionalization was used as the basis for conducting and recommending crops by the government (Olson, 1994). Different models of extension approaches were used during this period and these included use of demonstration sites, training and visit approach, pilot zones and the group extension system. However, the system collapsed as a result of the war in 1994. After the war, the agricultural extension of districts has increasingly become decentralized with agricultural technicians being put under the supervision of mayors (ibid). Currently, the national extension system is predicated on participation in which the government encourages local communities to take an active role in the identification of needs and extension concerns. Important institutions in this respect include agencies of specialised services (for example NGOs) and farmers' cooperatives. The cooperative method is often argued to lower training costs of farmers in the short and long term (MINAGRI, 2007).

Figure 1 shows the directorates operating under the Ministry of Agriculture and in collaboration with other organizations that include non-governmental organizations and related ministries such as the Ministries of Land and Environmental Resources. According to this framework, the Ministry of Agriculture has the mandate to implement the National Agricultural Policy through the public institutions highlighted in Figure 1. The main areas of focus include animal husbandry, crop production, horticulture, adaptive agricultural research, and rural development reflecting the general objectives of the National Agricultural Policy.

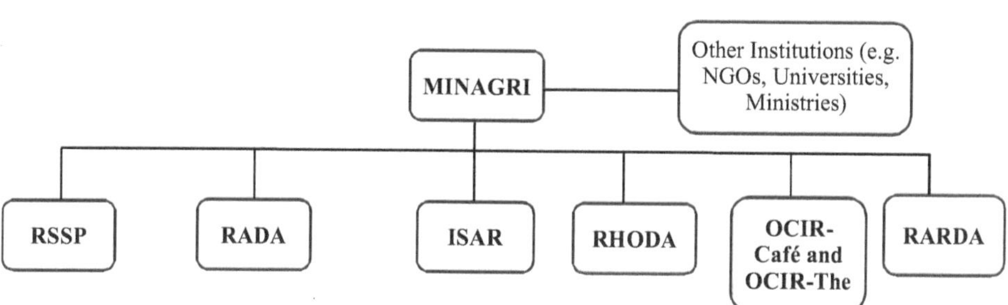

Figure 1 : Structure of agricultural extension system in Rwanda
Source : National Agricultural Policy Document, 2007

MINAGRI: Ministry of Agriculture, RSSP: Rural Sector Support Project, ISAR: Rwanda Institute of Agricultural Science, RADA: Rwanda Agricultural Development Authority, RARDA: Rwanda Animal Resources Development Authority, RHODA: Rwanda Horticultural Development Authority.

5. AGRICULTURAL EXTENSION POLICY IN RWANDA

The current national agricultural extension policy in Rwanda seeks to enhance farmers' incomes and ultimately the livelihoods of the rural poor communities. The main extension approaches highlighted in the policy are farmer-to-farmer extension and use of local extension agents. The role of extension agents is to act as a facilitator, catalyst, adviser and counsellor. In this policy, there is greater autonomy on farmers as they can choose to adopt appropriate technologies which are consistent with their local environmental conditions. Thus, extension agents introduce a basket of technologies rather than focus on uniform recommendations (MINAGRI, 2007). However, the system suffers from a number of constraints including lack of adequate extension staff at the district level, weak organization of producers, weak connection between research and extension and limited participation of local communities in mainstream decision making processes (MINAGRI, 2007). To overcome the aforementioned structural and institutional barriers to agricultural extension, the government of Rwanda, has in recent years, embraced the use of Information and Communication Technologies (ICT). In the next section, we outline the reasons why ICTs are increasingly being considered as an integral part of agricultural development process in Rwanda, and also characterize the ICT policy environment in terms of existing laws and statutes.

6. THE NEED FOR ICT IN AGRICULTURAL EXTENSION AND RURAL DEVELOPMENT

ICT is an acronym standing for Information and Communication Technologies and it is an umbrella term that includes any communication device or application, encompassing: radio, television, cellular phones, computer and network hardware and software, satellite systems and so on, as well as the various services and applications associated with them, such as videoconferencing and distance learning. ICTs are usually sector specific, for instance, ICT in education, in healthcare, *etc.* (BNet Business Network, 2007). The term 'ICTs' can be interpreted as including a wide range of media. However 'new ICTs' is used to denote "the use of computers and communication systems between computers" (CTA, 1999). Because of its increasing importance in socio-economic development of the global economy, researchers, development theorists and policy makers are integrating this concept in program formulation, design, monitoring and evaluation. ICTs open up new communication channels and information resources for rural users and therefore add value to all forms of assets (natural, social, financial, physical and human capital) in rural communities. Increased access to information on agricultural input and output markets, for example, can increase financial capital through gains in profit margins and improved accessibility to financial services in rural areas (Batchelor and Scott, 2001). Use of ICT has a multiplier effect on the efficiency of resource use, that is, land, capital and labor (human resources).

In Rwanda, the government recognized the importance of ICTs in socio-economic development of the country. In 1998, an ICT document entitled "An Integrated Framework for Socio-Economic and ICT Policy and Plan Development and Implementation for Rwanda" was developed after consultation with different primary and secondary stakeholders in the government, private sector and NGOs. Integration of ICT in agricultural development seen as a necessary condition for circumventing technical and structural barriers that include low agricultural productivity, poor access to agricultural technologies, dearth of the land resource, the geographical position of the country (landlocked) and low human resource development (Government of Rwanda, Policy on ICT, 2000).

6.1. The ICT Policy Environment in Rwanda

To promote general socio-economic development of the country, the government of Rwanda promulgated the Vision 2020 document. The long term thrust of this policy is to reduce poverty and therefore transform the country into a middle income economy (Government of Rwanda, Policy on ICT, 2000). In this broad vision, ICTs are considered as a means to an end, that is, to assist in the transition from a rural, subsistence oriented economy to a market based, information rich society. The adoption of ICT in Rwanda is facilitated by two strategies, National Policy on Science, Technology and Innovation and the National Information and Communications Infrastructure (NICI). The specific objectives of the National Policy on Science, Technology and Innovation are to promote knowledge acquisition, knowledge creation, knowledge transfer and encouraging innovation. The national ICT policy is based on 8 important pillars namely Human Resource Development Sub-Plan; ICTs in Education Sub-Plan, Facilitating Government Administration and Service Delivery Sub-Plan; Developing and Facilitating the Private Sector Sub-Plan; Deployment and Spread of ICTs in the Community Sub-Plan; ICT Infrastructure Development, Legal Regulatory, Institutional Provisions and Standard Sub-Plan; and Foreign Direct Investment Drive in ICTs Sub-Plan (ibid). The policy consists of four 5 year plan periods. The first five year plan period has been completed and the second plan period (NICI II) is ongoing 2006 - 2011 and it seeks to establish the requisite infrastructure necessary to facilitate enhanced usage of ICT in the country. This policy environment has provided a conducive environment for implementation of ICT innovations in the country. However, it can be observed that the implementation of ICT is still in its infancy.

7. ICT INNOVATIONS USED IN AGRICULTURAL EXTENSION AND RURAL DEVELOPMENT OF RWANDA

In order to ensure that the benefits of ICT trickle down to local communities, a number of specific initiatives in ICT have been developed by the public and private sectors. In this section, an outline of some of these developments is made.

7.1. Rwanda Info-Bus Project

In Rwanda, the majority of people who live in rural areas generally lack access to information about input and output markets and improved technologies, which results in low agricultural productivity. The ICT bus project is a response to the information needs of a cross-section of people in the Rwandese society and is being implemented under the e-Rwanda national project. The ICT Bus Project (a mobile computer laboratory) is an initiative being implemented by the Rwanda Information Technology Authority (RITA), a public institution, responsible for developing ICT infrastructure in the country. Under this concept, the ICT bus acts as a mobile tele-centre to assist rural communities to be integrated into the global information system. It targets farmers, traders, students, women, youth groups, entrepreneurs and other rural based Rwandans (RITA, 2007). The main advantage of this ICT concept is that it reduces the transactions costs of accessing information in rural communities. It also provides training to teachers, who will, in turn, assist their spheres of influence. It will also be used to promote innovation among farmers and business entrepreneurs in rural areas of the country. However, successful use of the ICT Bus project hinges on several necessary and sufficient conditions. Firstly, there is need for improved awareness among extension workers and greater capabilities among development institutions such as education and research organizations. Extension agents must be empowered with ICTs, in order to obtain advantages of these technologies during providing services to their clients. Secondly, there is need to establish an efficient road and communication network to be able to navigate into remote rural areas of the country. Third, for the rural population to use the ICT bus concept, they must be trained on how to use computers. Thus, programs targeting children, youths, women and men must be developed. Technical know-how, continuous training and exchange of information are maintained by systematic realization of the opportunities afforded by advanced information technologies (Koutsouris, 2006; Stienen *et al.*, 2007).

Figure 2 : ICT Info-bus
(Source: RITA, 2007)

Figure 2 indicates the ICT Bus with several state-of-the art computers, where people in different parts of the country can have access to information that affects their general livelihoods.

7.2. Access to HIV/AIDS and Health Information through Cell Phone Facilities

Over the years, HIV/AIDS has increasingly become one of the main socio-economic challenges facing the Rwandan economy. As part of the Vision 2020, the government has encouraged the establishment of projects which raise awareness about the transmission of HIV/AIDS and other health related issues among people. The HIV/AIDS and health information forum is based on voice and data communication networks, whereby people with various concerns, phone and ask questions on issues related to HIV/AIDS or general health. This program was enunciated by Voxiva, an international non-governmental organization in collaboration with TracNet in Rwanda in 2001 (Gahamanyi, 2003). The program has been able to take advantage of the high tele-density in the country, where most people have access to cell phones and radios. By providing relevant health information, local communities are more inclined to make informed decisions. A healthy people are also able to participate more effectively in mainstream development process.

7.3. Use of Video-Conferencing and Multi-Media Platforms

Under this method, audio and video facilities are used to enhance communication and information sharing between the recipients and the sources of information. In Rwanda, this technique has been used particularly in healthcare and education. In the education sector, the Kigali Institute of Science and Technology African Virtual Center has been offering university degrees in conjunction with an Australian university. Thus many students have managed to have access to high quality educational facilities. In the agricultural sector, this method seeks to enhance accessibility to input and output information which assists farmers in decision making about markets.

7.4. BCS Tuvugane Community Payphones

The word "Tuvugane" is a vernacular word from Kinyarwanda, the national language, implying to communicate. The BCS Tuvugane community payphones were developed by a private company called Business Communication Services (BCS). BCS sells payphones to private individuals and works in partnership with MTN. There is potential in using this platform since the tariff structure is lower tariff structure than ordinary mobiles. This could be an advantage to communities particularly those in rural areas, if specific programs could use this ICT platform.

7.5. National Radio and Television Programs on Agriculture

These are programs that are broadcast on national radio and television to deal with specific issues affecting livestock and crop production systems to reach out to a large number of farmers in the country. In Rwanda, both national radio and television programs related to agriculture are run under the auspices of the Ministry of Agriculture and Animal Resources (MINAGRI) in collaboration with public institutions such as RADA, RARDA and RHODA. Topics are demand driven, that is, they are suggested by farmers and other stakeholders. During the programs, technocrats from the aforementioned institutions present expected practices on a given topic. This program is used as a means of diffusing technological innovations among farmers located in different parts of the country.

7.6. Rwanda e-Soko Project

The e-Rwanda Project is a World Bank funded project being implemented by the Rwanda Information Technology Authority (RITA). The targeted clientele are farmers and the main objective is to empower rural farmers with requisite information on agricultural markets. This project will be implemented by MINAGRI. Under this concept, farmers in Rwanda will use ordinary cell phones to query up-to-date price databases run by the Ministry of Agriculture and receive relevant and updated information on market prices for agricultural goods. Furthermore, farmers with access to the Internet will be able to obtain this information from the MINAGRI website. Additional information on the website will include weather forecasts, pest control tips and advice on growing different crops (MINAGRI, 2007).

The field staff employed by MINAGRI to collect this information will be equipped with smart phones to collect and enter the pricing information in the databases. In case the field officers are out of network signal range, they should still be able to store the data in the phones and later synchronize it with the databases when they get to areas within network coverage.

8. IMPACT OF ICT INNOVATIONS IN RWANDA

Although a number of innovations have been developed over the past few years in Rwanda, most of them are still in the implementation phase. The ICT programs are still in their experimental phases. Examples of such ICT initiatives are the e-Rwanda project under the auspices of the Rwanda Information Technology Authority (RITA) and the Swedish Program for Information and Communication Technology in Developing Regions (SPIDER) project. However, according to various stakeholders there are isolated cases where farmers are adopting improved agricultural technologies. Various ICT methods are being used in different areas which include Gicumbi, Rulindo, Gakenke, Burera, Musanze and Nyabihu Districts. The most common methods used to disseminate technologies are telephones, the Internet, Short Message Services (SMS), radios and publications or bulletins (RITA, 2009). In Gasabo and Rulindo Districts of the country, some farmers have adopted the kitchen garden concept as a result of efforts of a non-governmental organization called VI-LIFE.

Some farmers have also indicated benefitting from information on organic farming techniques. Another group of tomato growers had found new blight resistant varieties as well as directions on organic field management and had set up their own research plots in Gasabo and Rulindo Districts. In Gicumbi District in the Northern Province, land consolidation is now a reality. Under this concept, farmers are mobilized to grow certain crops over large tracts of land thereby increasing outputs of crops of national importance. Lack of technical know-how, training and the generally high level of illiteracy are often cited as some of the challenges facing the effective implementation of ICT programs in Rwanda.

9. THE WAY FORWARD

In the foregoing section, it has been observed that the government of Rwanda has promulgated an auspicious policy environment to meet the socio-economic endeavors of the country. In ICT, there are specific provisions which have encouraged the development of innovations in agricultural extension. Such initiatives are largely driven by the public sector, along with the private sector and these include the e-Rwanda project, Info-bus concept, radio and television programs focusing on agriculture and HIV/AIDS related issues. Moreover, most of the ICT related agricultural initiatives are still in their experimental stage. As such, there is no concrete evidence on whether these programs have had a significant impact on rural populations. There is some preliminary evidence that some farmers have started to adopt environmentally benign technologies using advice obtained from different ICT forums. Examples include the kitchen garden, a concept where farmers use organic manure to provide plant nutrients for their crops, and rely less on chemical pesticides. Technical and structural constraints such as the dearth of ICT trained agricultural extension personnel, lack of accessibility to certain areas as a result of poor road and communication networks and lack of technical knowledge general computer use remain a challenge. In view of these observations, a number of recommendations are provided in order to fully capture the benefits from ICT in agricultural extension:

- There is need to consolidating infrastructure including roads, communication networks and electricity particularly in rural areas of the country. These are pre-requisites for efficient use of ICT in agriculture
- Sensitization of village leaders and local communities on the use and importance of ICT in agricultural extension. This can be done through workshops organized at sector, community and village levels. Such training programs could focus on important demographic characteristics such as age and gender of participants. Local leadership also needs to be involved to break cultural barriers in rural communities.
- ICT-led initiatives must be based on participation and dialogue rather than using the prescriptive approach to agricultural extension. This allows rural development planners to identify local knowledge held by the communities.
- Training of extension workers, who will act as the critical mass for further training and adoption of environmentally benign agricultural technologies.

- Financing of ICT-based initiatives focusing on infrastructure and applied research in ICT (for example characterization of local knowledge on ICT, best practices which are consistent with the socio-economic and institutional setting, ICT needs assessment for rural communities).

References

Adams, M.E., 1994. *Agricultural Extension in Developing Countries.* Longman, 6th edition, Scientific and Technical, Essex, UK

Agriculture and Food-Rwanda, 2003. Country Profiles, EarthTrends, Country Profiles, online at http://earthtrends.wri.org

Batchelor, S., and Scott, N., 2001. The role of ICTs in the development of sustainable livelihoods: A set of tables, Sustainable Initiatives, Version 2, 2001

BNet Business Network, 2007. Available online at http://dictionary.bnet.com/definition/ICT. Html

CTA, 1999. Information and communication technologies: A remarkable revolution, In: *SPORE*, No 79, February 1999, p45

Gahamanyi, J., 2003. How ICT will have an impact on poverty reduction in Rwanda, February 2003, Rwanda

Government of Rwanda, 2008. Country Publications, Kigali, Republic of Rwanda

Government of Rwanda, Policy on ICT, 2000. An Integrated Framework for Socio-Economic and ICT Policy and Plan Development and Implementation for Rwanda, Kigali, Rwanda

Koutsouris, A., 2006. ICTs and Rural Development beyond the Hype. *Journal of Extension Systems.* 22(1), pp:46-62

MINAGRI, 2004. The Agricultural Policy, Kigali, Rwanda

MINAGRI, 2007. Strategic Plan for Agricultural Transformation in Rwanda, Document prepared by GECAD, Kigali, Rwanda

MINITERE, 2004. Draft of National Land Policy, Republic of Rwanda, 52 p.

MINIECOFIN, 2002. Recensement General de la Population et de l'Habitat, Gouvernement du Rwanda Publications.

MINIECOFIN Report, 2004. Vision 2020, Draft no. 4, Kigali, Rwanda, p63

MINIECOFIN, 2003. In: Strategic Plan for Agricultural Transformation in Rwanda, Document prepared by GECAD, Kigali, Ministry of finance and Economic Planning. Rwanda

National Institute of Statistics, 2002. Statistical bulletin, Kigali, Rwanda

Nzabamwita, Haremimana, P., 2005. Extension agent as mainstay of rural development projects: The case of rural sector support project in Ruhengeri Province, Rwanda, A research project submitted to Larenstein University Profession Education, Deventer, The Netherlands.

OCIR-Café, 2005. Situation actuelle du café : RWANDA, Available online http://www.ocir.vrsrv.com

Olson, J.M., 1994. Farming systems of Rwanda: Echoes of historic land divisions reflected in current land use, Working Paper 2, Department of Geography and the Centre for Advanced Study of International Development, Michigan State University, East Lansing, USA.

RITA, 2007. An overview of ICT policies and projects in Rwanda, Government of Rwanda Publications, Rwanda.

RITA, 2009. The inventory of agri-actors using ICT Application in agricultural extension system in Rwanda, Republic of Rwanda Publications, RITA/RBD.

Stienen, J.W., Bruinsma and Neuman, F., 2007. How ICT can make a difference in agricultural livelihoods, International Institute for Communication and Development (IICD), *The Commonwealth Ministers' Reference Book.*

Van den Ban, AW and Hawkins, H.S., 1996. *Agricultural Extension*, 2nd edition, Blackwell Sciences Ltd, London, UK.

Authors Information

Dr. Charles Karemangingo

Senior Lecturer, Soil and Agricultural Engineering
Higher Institute of Agriculture and Animal Husbandry
BP 210, Musanze, RWANDA
Postal Code: 250
Mobile Phone No.: .(250)55106907
e-mail: ckareman@rogers.com
URL: www.isae.ac.rw

Dr. Charles Karemangingo is currently the Rector and Senior Lecturer with the Higher Institute of Agriculture and Animal Husbandry, Busogo, Rwanda. He has extensive knowledge in soil science, particularly in soil fertility management. Dr. Karemangingo has conducted adaptive field studies, written factsheets for extension and reviewed scientific papers and books. He has managed the Nutrient Management Software for New Brunswick, Canada before he moved to Rwanda in 2007.

Mr. Edward Mutandwa

Lecturer, Rural Development and Agribusiness
Higher Institute of Agriculture and Animal Husbandry
BP 210, Musanze, RWANDA
Postal Code: 250

Mobile Phone No.: .+250(0783813992)
e-mail: edwardmutandwa@yahoo.com
Alternate e-mail: emutandwa@buse.ac.zw
URL: www.isae.ac.rw

Mr. Edward holds an M.Sc. in Agricultural Economics from the University of Zimbabwe. He has been a lecturer in the Department of Agriculture at Bindura University of Science Education for four years. Currently, he is a lecturer in the Department of Rural Development and Agribusiness at ISAE-Busogo. He has published more than 20 articles in internationally referred journals and his research interests are in natural resources management, use of indigenous knowledge systems and economic analysis of rural projects.

Mr. Nathan Kanuma Taremwa
Lecturer, Rural Development and Agribusiness
Higher Institute of Agriculture and Animal Husbandry
BP 210, Musanze, RWANDA
Postal Code: 250
Mobile Phone No.: .+250(0788504820)
e-mail: taremwa2001@yahoo.com
URL: www.isae.ac.rw

Mr. Nathan Kanuma, holds an M.Sc. in Rural Development from National University of Rwanda. He has been a lecturer and Head of Department of Rural Development and Agribusiness at ISAE-Busogo for the past five years. His research interests are in the areas of biodiversity conservation and protected areas management, water and environmental management, agricultural credit management in rural areas and rural development.

Mr. Frank Mugisha
Teaching Assistant, Rural Development and Agribusiness
Higher Institute of Agriculture and Animal Husbandry
BP 210, Musanze, RWANDA
Postal Code: 250
Mobile Phone No.: .+250(0788740266)
e-mail: mugishafrank2000@yahoo.co.uk
URL: www.isae.ac.rw

Mr. Frank Mugisha holds a Bachelors degree in Development Studies from Makerere University in Uganda. He has worked as a sector youth leader in economic development in Nyarujenje/Rwezamenyo in Kigali. Currently, he is a lecturer in the Department of Rural Development and Agribusiness at ISAE-Busogo. His research interests are in the areas of rural development, human rights and environment.

□□□

Chapter 12

SRI LANKA

Rohan Wijekoon and M.f.m. Rizwan

1. OVERVIEW OF THE NATIONAL AGRICULTURE SCENARIO - NATIONAL EXTENSION SYSTEMS, SCOPE AND IMPORTANCE

Like in many developing countries, Agriculture Sector being the backbone of the Sri Lankan economy, contributes significantly (11.3 per cent in the year 2007) to the Gross Domestic Products (GDP) and the economic growth by 7.7 per cent (Central Bank of Sri Lanka, 2007). As to the situation in 2007, 80 per cent of country's population of 20.01 million is rural and 31.3 per cent of the total Sri Lankan work force is employed in Agriculture sector. Nearly 14.6 per cent of the population still depends on the agriculture as the main livelihood. The country's GDP has been growing at 6.8 per cent per annum (in 2007), though that of agriculture is growing at only 3.3 per cent.

Country as a whole is constituted by 25 administrative districts all of which is encompassed into nine provinces. Of these, nearly four fifths are predominantly agricultural districts with over 60 per cent of the total population living in them. Majority of the population in these districts constitutes the farming community and they, by and large, are the ones who keep the nation feeding. Apart from plantation agriculture which looks after large scale plantation level crops like tea, rubber, coconut along with other crops like cinnamon, pepper, coffee, cloves, nutmeg *etc.*, grown extensively for the export market, almost all other food crops are grown by farmers, majority of who own land plots less than 2 hectares. On the one hand, this makes agriculture mainstay of a major sector of the population and on the other hand they are spread nearly everywhere in the country rather thinly on the ground.

Agricultural extension service being the livewire to the agricultural development of the country, over several decades in the past Sri Lankan Agricultural Extension Service of the Department of Agriculture (DoA) has rendered the yeoman service to the Agricultural development of the country handsomely (Wijekoon and Newton, 1998). Up until 1985, *Krushikarma Vyapthi Seva Niladari* (KVSN: Agricultural Extension Service Officer) was the grass-root level extension officer who provided a face to face extension service by maintaining an extensive yet strong as well as direct linkage with and facilitating the smooth transfer of technology and information to the farming community. With the Central Government's decision at this stage to absorb this entire cadre to the Central Administrative System, Agricultural Extension System in the country suffered a serious setback. It took few years afterwards for the system to re-establish some linkages and got things moving, though at a lower profile.

2. NEED FOR ICTs IN AGRICULTURAL EXTENSION

Major as well as drastic political and administrative changes brought about at this juncture such as trade liberalization, withdrawal of agricultural extension workers from extension activities, devolution of powers to provincial councils contributed significantly to the inevitable land-slide and deterioration in the entire agricultural extension system in the country (Wijekoon, 2005). This created many voids almost everywhere in the system hampering regular information flow from knowledge repositories (Research Station) to farmers and back to researchers, from farmers to planners, from executors to farming community, from the trader to the farmers and back and so on and so forth. The weak inter-organizational/sectoral as well as intra-sectoral linkages among extension, research, training and service sectors like marketing network, financial organizations *etc.*, left the farming community destitute while the effectiveness of research, training and extension too crumpled badly.

It was at this juncture that the Ministry of Agriculture and the Department of Agriculture (DoA) in Sri Lanka having already realised the magnitude of the need of an alternative to cope up with the demand and cry for information from people in the system, took steps to capitalise on the new innovations and developments in the Cyber world, which were already getting closer to the farming community at a speed. Therefore, as an ICT initiative, for the first time in Sri Lanka "Cyber extension" mechanism was implemented by the Audio Visual Centre (AVC) of DoA in 2004 as an appropriate information exchange mechanism which seemed affordable and convenient to rural farmers in satisfying their information needs.

Cyber extension is an agricultural information exchange mechanism over cyber space, being the imaginary space behind the interconnected computer networks through telecommunication means. It utilizes the power of networks, computer communications and interactive multimedia to facilitate information sharing mechanism.

Cyber world getting exposed to the whole world intrinsically carries basic elements required for a communication system to be developed for agricultural information to be channelled through. People at all levels try and get access to these facilities on their own, basically for entertainment purpose. All possibilities are there. Therefore, to capitalise on this situation and creep into farmer dwellings with a material they, in fact, need agricultural information.

3. IT SCENARIO IN THE COUNTRY IN THE PAST AND ITs PENETRATION IN RURAL AREAS

In the year 2003 e-government index of Sri Lanka was only 0.92, which was well below the global mean e-government index of 1.62 implying that the e-government capacity of Sri Lanka was poor (Kumarawadu, 2003). By this time, country had a very small number of Personal Computers (PCs) and that amounted to a mere 56 out of every 10,000 people and only six out of every thousand people of the population were enjoying facilities of an on-line connection. A web-based research (Kumarawadu, 2003) covering government institutes revealed that 30 per cent of the ministries in the country did not have web sites or inaccessible even if they had as they were inactive. Thirty eight per cent of the ministries by this time were in the infant stage and information available in web pages not being updated was often stale in content and the number of pages was also limited to a few web pages. Only about 17 per cent of ministries offered interactive web contents, where users had the access to regularly updated information and could communicate through e-mail to download government documents through the Internet.

In the same study, it was found that 99 per cent of e-mails were not responded by the web masters. In the year 2003, the progress of e–government solutions was very slow and very little information was accessible to users through the internet. After considering the poor tele-density (ratio of telecommunication facility to people) and the poor e-government solutions in Sri Lanka (See Figures: 1, 2 and 3) web based ICT initiative was not introduced quickly with the cyber extension project in the year 2004.

4. e-EXTENSION IN THE PAST

The traditional extension system that was in operation for many decades in the country has a foundation which needs not changed very much in accommodating the new system. The infra-structural setting it has built on over years should conveniently cater to the needs of the system to be introduced. Great advantage one can see in this regard is the possibility for the cyber extension system that is going to be introduced to merge with the existing traditional system with least amount of structural alterations.

Considering the limitations in the original online cyber extension mechanism, which depends on telecommunication facility, the project was planned to implement giving time for the telecom system to get rooted in the society first. This was the time when almost all were set for a major change in terms of infrastructural development for the enhancement of the required telecommunication facility for people. With such developments round the corner, cyber extension program was thus planned and implemented in two phases;

Phase I An alternative approach to Cyber extension: Digital / Wireless extension strategies - (year 2005/2006)

Phase II Real cyber extension with internet and telecommunication facility - (year 2007)

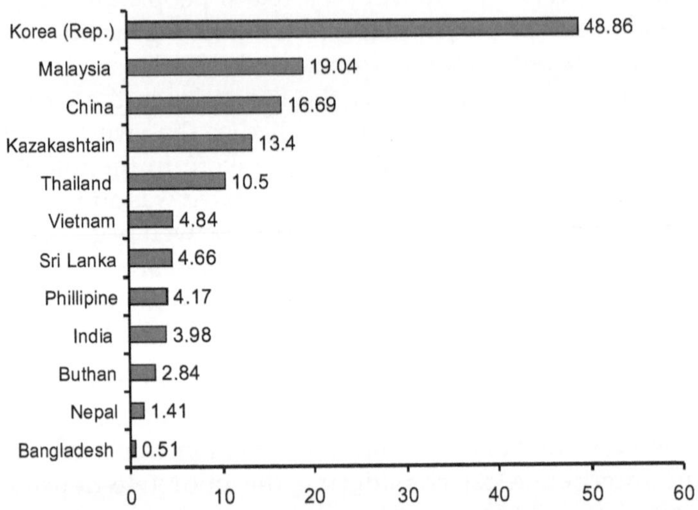

Figure 1 : Main Telephone Lines per 100 Persons - 2002 Inhabitants

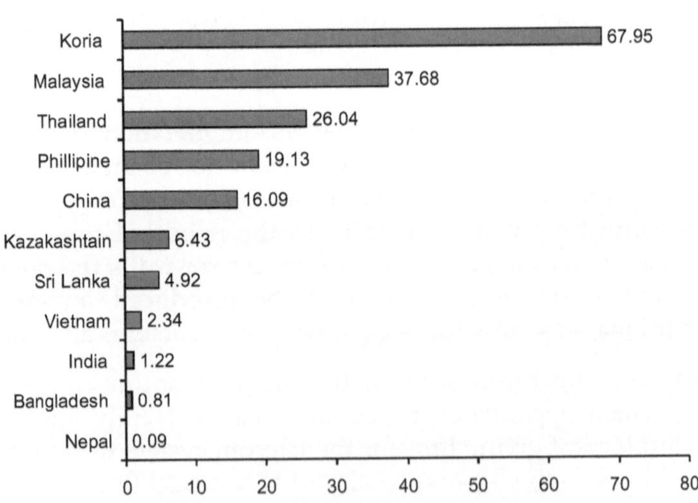

Figure 2 : Mobile Cellular Subscribers Inhabitants

Source – 2nd International Seminar on Internet Resource Management, Sri Lanka - 2004

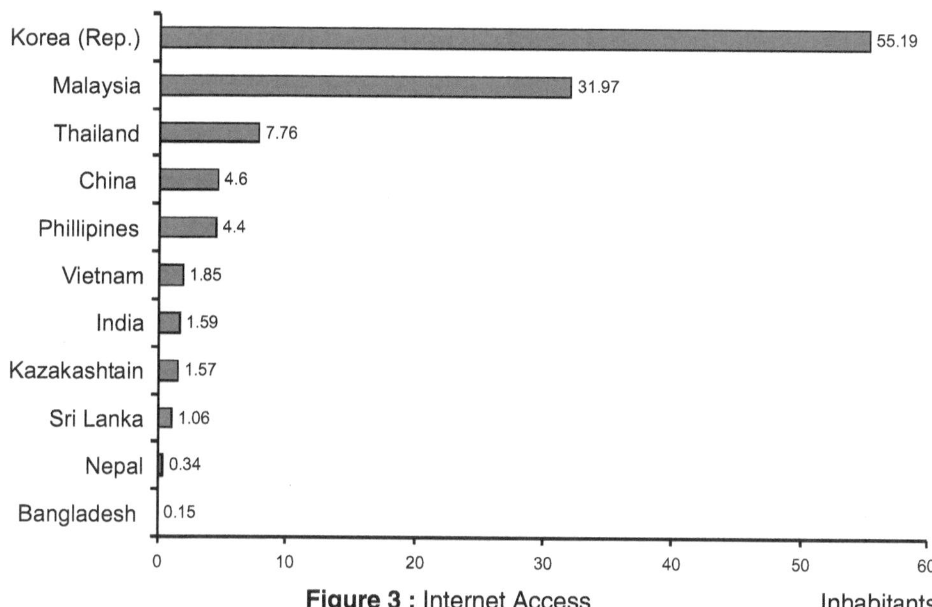

Figure 3 : Internet Access

Source – 2nd International Seminar on Internet Resource Management, Sri Lanka - 2004

4.1. Phase I- An Alternative Approach to Cyber Extension: Digital/ Wireless Extension Strategies

In the absence of well established basic telecom facility throughout the country, Digital/ Wireless Extension Strategies were considered as an alternative approach to Cyber Extension.

The first ever project towards this was established during the period from 2004-2006 and 45 Cyber Extension Units (CEU) were installed at Agriculture Instructors' offices, *Govijana Kendra* (Agrarian Service Centres) in 20 administrative districts in the country. Each Cyber unit was equipped with a technically high capacity computer complimented with other facilities like scanner, laser printer, digital camera and uninterrupted power supply unit (UPS) along with required office facilities to function.

The Cyber Unit was managed by a trained Agriculture Instructor (AI), and continuous computer training programs have been conducted at the Audio-Visual Centre (AVC of the Department of Agriculture) to improve computer skills of the AI. In this phase, multimedia e-learning strategies were implemented effectively by using Interactive Multimedia CD- ROMs (IMM CD-ROMs) produced by the AVC. Five digital extension strategies were implemented in the phase I.

4.1.1. Digital Extension Strategy 1

Use of IMM CD-ROMs as Crop Based Information Material

e-Learning strategies have been introduced to cyber units by using IMM CD-ROMs on agricultural subjects. Objective of using IMM CDs as ICT strategy was to assist farmers enhance their knowledge and skills on relevant crops and crop technologies in a community friendly, aesthetically pleasant and enjoyable learning environment. Extension workers were able to use these IMM CD-ROMs carrying most up- to-date technologies in agriculture as teaching tools (audio-visual aid) to deliver technological messages effectively to the farming community. Farmers either with a basic skill to operate a system or with the assistance of another person (or Agriculture Instructor) at the same time could use the same CD-ROMs as a self-learning package for upgrading his/her relevant knowledge skills. Farmers trained on use of information technologies in agriculture, this way would be well expected to become 'e-farmer' of the day. The extension workers in the mean time would become 'e-extensionists' who could lead the march forward at field level. The AVC of the DoA has already facilitated this revolution by designing, developing and producing forty three crop-based and technology-based IMM CDs and distributing them to established Cyber extension units.

In beating the human resource scarcity for agricultural development, which takes place through a series of activities over time like generating technology through research, converting them to extension messages by trainers and getting those messages over to the farming community along with many other facilities such as seeds and planting materials, *etc.*, e-technologies can be utilized very advantageously. Many individuals' efforts required to develop a technical skill in a group of farmers can now be compressed into an enclosure of a CD-ROM and this novelty device can be utilized by a trainer in turn more conveniently to simulate a training session. This will no doubt require lot of efforts from specialists from different fields to start with, to produce a material, but later become most convenient way to rapidly multiply and deliver them at farmer's door quickly.

Each CD-ROM was developed in a collaborative effort with the assistance of a senior research officer and a research assistant of the subject for the technical content, a training specialist for the process and a team of multimedia specialists of the audio visual centre of the DoA.

In addition to technical information being organized into several chapters on each crop, a separate chapter was devoted for compiling database on research papers and articles published by local authors in local and foreign journals. Each CD is further enriched with a separate link for video films, which were produced pertaining to each crop and telecast through *Mihikatha Dinuwo* and *Govibimata Arunalu* (weekly television programs of the DoA) by the Audio Visual centre.

Interface and Navigational Design of CD-ROMs

CD-ROMs were produced for extension workers as well as farmers, where a few of the potential users were expected to have had much previous exposure to such technologies. Considering the low computer literacy of extension workers as well as farmers, a familiar concept for the interface was used; an electronic book which was similar to reading an analogue book. Page turn is facilitated by two icons; 'Next' (to go next page) and 'Previous' (to go previous page). The contents were organized into chapters, topics and sub-topics. Most of the pages contain variety of multimedia presentations (video, sound / voice-overs, animations, graphics and text). All media are interactive and users may review and or skip section, as they desire. Each page is printable and farmers will be able to get a print. (See Figures: 4 & 5 for the interface design of an IMM CD-ROM)

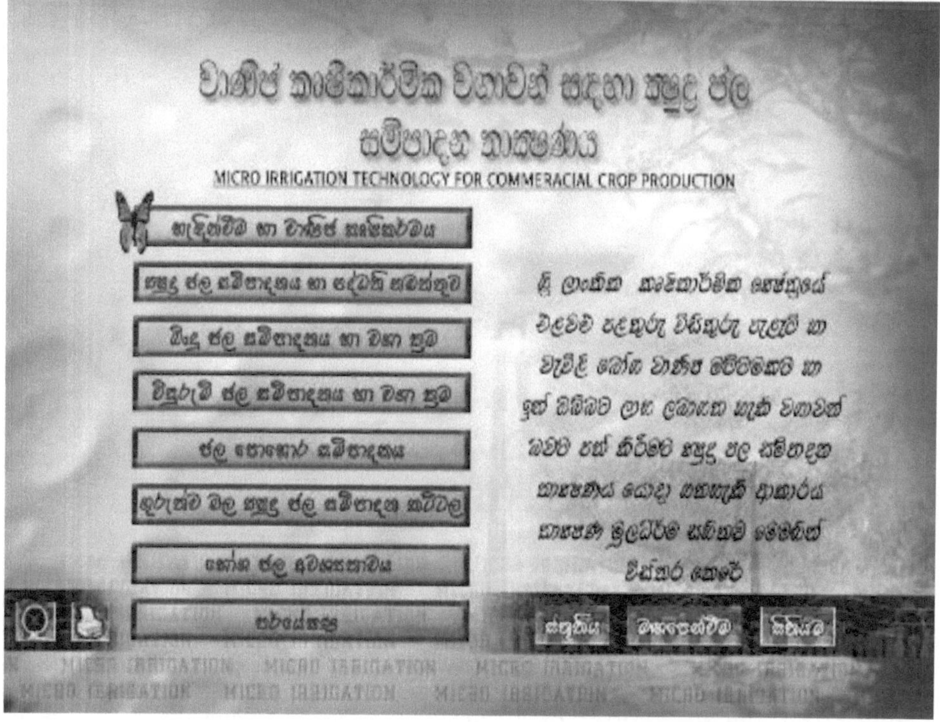

Figure 4 : Content Page of Micro-irrigation IMM CD

Figure 5 : Typical Page of Micro-irrigation IMM CD

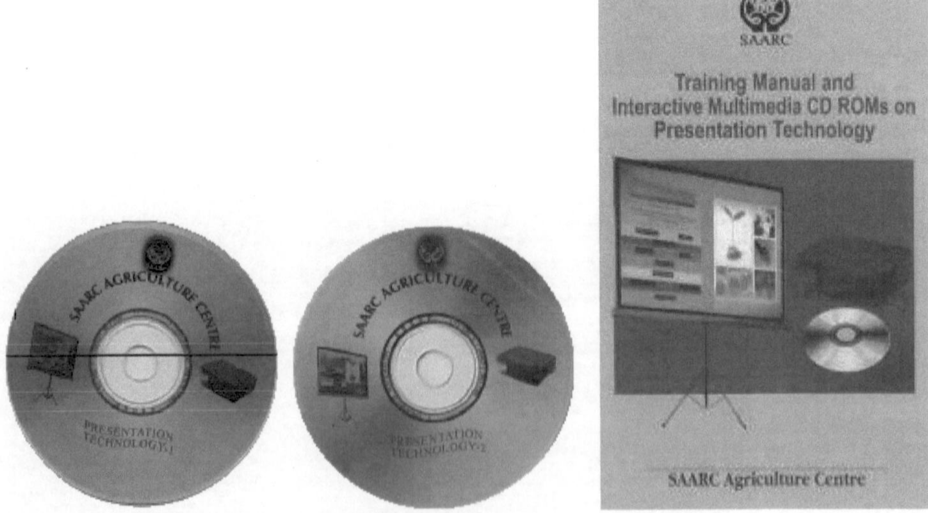

Figure 6 : Cover page of the IMM CD of Presentation Technology which was reprinted by the SAARC Agriculture Information Centre

4.1.2. Digital Extension Strategy 2- Use of Interactive Multimedia to Develop Low Cost Audio Visual Aids

Two CD-ROMs were designed, developed, produced and introduced to improve presentation skills of Agriculture Instructors and distribute them to all Cyber units. As a result, with the assistance of these materials extension workers were able to produce low cost audio visual aids at their doorstep with locally available material (Hi-tech to improve low cost instructional media) (Wijekoon, 1999 & Wijekoon and Newton, 1999). These two CDs with an instructional manual titled "presentation technology" were acknowledged and reprinted by the SAARC Agriculture Information Centre to distribute to other SAARC countries (see figure 6) (Wijekoon, 2006).

4.1.3. Digital Extension Strategy 3- New Dimension for Web Based (Internet) Delivery Mechanism with CD- ROMs

The DoA website (www.agridept.gov.lk), presents most of the agriculture related technical information as well as agricultural statistics, news, recently published books, *etc.*, (This web site was awarded as the best government website in the year 2008 by the Information and Communication Technology Agency (ICTA) which is the apex Institute for ICT in Sri Lanka). The Cyber extension project established a mechanism to distribute DoA web to Cyber units, in a CD-ROM. Until the day dawns for each Cyber Extension Units to have fully geared web-facilities for people around to make use of, this step will certainly help everybody concerned to update knowledge bases and quench individual info-thirsts.

4.1.4. Digital Extension Strategy 4- Develop Digital Training Material (Audio Visual Aids) for Extension and Training

In certain circumstances IMM CDs produced by the National Centre would not give all the answers to their location specific problems. Therefore, extensionist attached to the Agrarian Service Centres, is in a position to produce a location specific simple extension material like a 'Power Point' presentations, desktop publications in the form of simple handouts, leaflets, pamphlets, *etc.*, by using the facility of Cyber extension unit for their extension training *in situ.*

The staff at Cyber centres is expected to compile a visual database on all aspects related to agriculture in the areas like farmer innovations, local problems, key attractions, *etc.*, in each cropping season (*yala* and *maha*) and thereby researchers, trainers and whoever else is interested in them would be able to make use of at a later date. Such collections would make it convenient and suffice the need of material collection and transportation of them over geographical distances as well, for further investigations, training purposes, *etc.*

4.1.5. Digital Extension Strategy 5- Distance Learning Mechanism for Agriculture Research and Production Assistants (ARPAs)

Agriculture Research and Production Assistants (a set of field level facilitators recruited by the government who have no knowledge in Agriculture) attached to Agrarian Service Centres with no professional in-depth training in agriculture, would not form an effective and powerful force for bringing about a desired change in the farming community at a significant level. On the other hand, training of large number (9000) of such a cadre effectively within a reasonable period of time is not practical due to the limited facility of existing face to face training mechanism. Therefore, as an alternative and a stop gap arrangement, most of the Cyber units have taken initiatives to train ARPAs through IMM CDs.

4.2. Phase II - Real Cyber Extension with Connectivity

4.2.1. Expanded Use of Online Resources

In this phase, considering the development in telecommunication facility and rapidly grown dynamic e-government situation in Sri Lanka, (See Figures: 7, 8 & 9) original on-line Cyber extension was introduced for maximum utilization of connectivity. Cyber units were increased up to 85 in the year 2007/2008. CDMA (Code Defined Multiple Access) telephone and internet facility too were provided with the Cyber units of the project, at the same time.

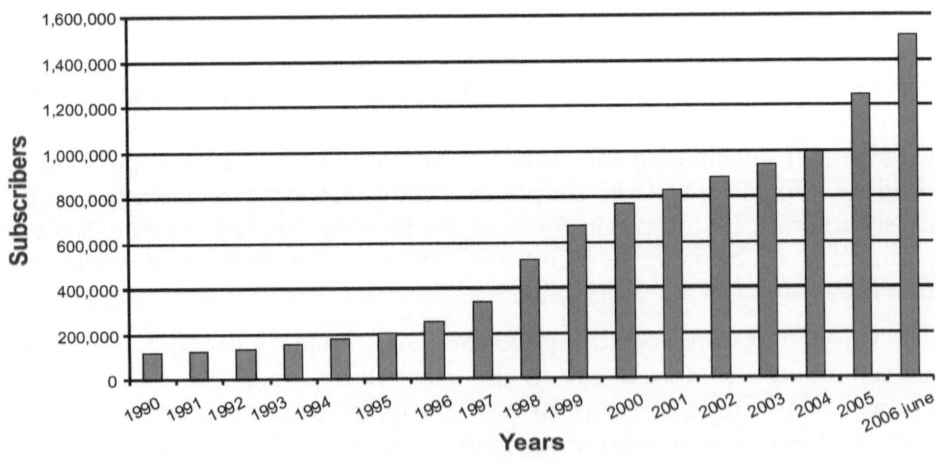

Figure 7 : Growth of Fixed Access Telephone Subscription - 2006
Source - http://www.tre.gov.lk/statistic.htm

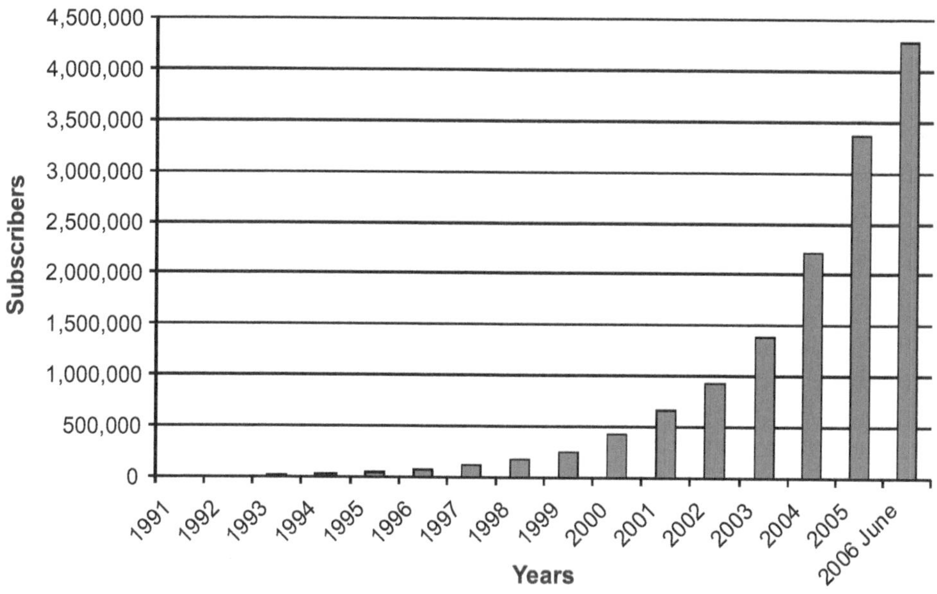

Figure 8 : Growth of Mobile Phone Subscription - 2006
Source - http://www.tre.gov.lk/statistic.htm

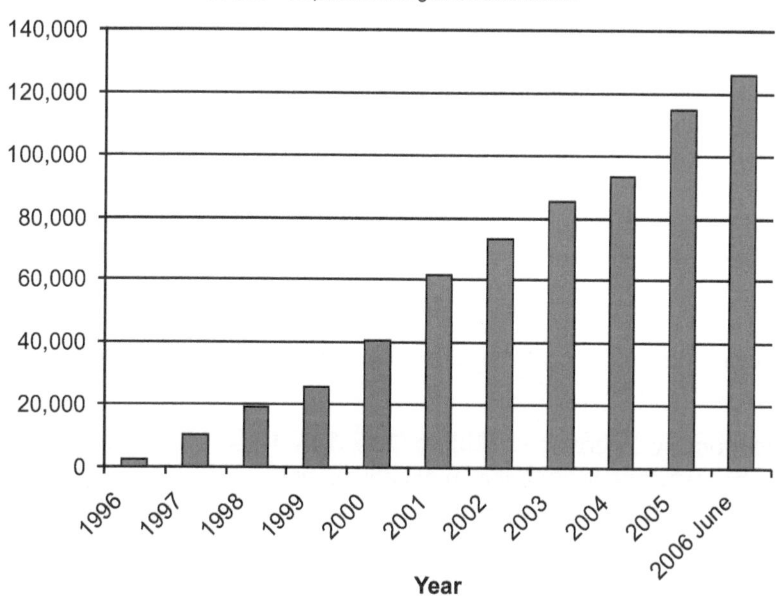

Figure 9 :

Using facilities at Cyber units, users could gain access to a wide range of information related to agriculture by browsing local and international websites of organizations such as DoA, International Rice Research Institute, Sri Lanka Rice Knowledge Bank *etc.*

Technical assistance is sought through e-mail queries with visual attachments. For instance, if there is any new problem such as emergence of a new pest or diseases, it can be visualized either by digital camera or by scanning the live specimen with the scanner and send to the Subject Matter Specialist (SMS) or the researcher at the Station far away for necessary action.

4.2.2. Farmer Database for e-Marketing

Avoidance of numerous obstacles and barriers being faced by farmers in marketing agricultural products is an essential task to be fulfilled to ensure the success of Sri Lankan agricultural development. Sri Lankan farmers currently face a wide range of problems in marketing their products. More often than not, low prices offered for products during harvest times is a frequent bitter and unpleasant experience which a great majority of farmers are a subjected to. Marketing problem of tomato and big onion farmers during July/August, marketing difficulties faced by lime and orange cultivators (Jan/Feb) as well as pumpkin and potato farmers are a few examples.

One of the attributes to this type of problems happened to be the lack of guidance and hence difficulties arising on the part of the grower in arriving at decisions pertaining to time, extent, crops and cultivars to be grown. As a result of such eventualities, large numbers of farmers grow same type of crops within the same time, without any planning, leads to further chaos during harvesting time. Very low prices are offered for their products by intermediaries who capitalize cunningly on the situation. If market forces are better coordinated, producers are better informed and the entire system operate on an integrated mode, whole sale traders will be in a position to get information about farmers. Then, they can go directly to them without being involved with intermediaries and purchase their products offering a better price to the producer.

Additionally, there are problems faced by those who attempt to market agricultural produce to overseas markets as well. Main reasons to this are the lack of reliable and accurate information about buyers who need products of certain specific standards at particular times in definite quantities and farmers who grow crops to supply overseas markets on a continuous basis.

Even policy makers are facing difficulties in arriving at correct decisions on imports and exports of agricultural products, due to absence of correct information. As such, farmers become victims of unfortunate eventualities like importing similar products from other countries when they harvest their crops. This situation may become worse, in the event information relevant to what products, when available in what quantities, *etc.*, are not available at national level.

A strong need has arisen for continuously updating statistics and information on farmers and farmer products as indispensable matter required in finding solutions to all the above problems. It will be easier to find solutions to above mentioned problems and overcoming difficulties by establishing modern information and communication technologies and implementing such strategies for updating statistics at national level.

Therefore, a farmer database of each cyber unit (name of the farmer, type of crop, extent, expected yield *etc.,*) was introduced to the network of information repository of the DoA web site in early 2007. (http://www.agridept.gov.lk/agmis.php). In addition, daily price information from the Dedicated Economic Zone (main vegetable wholesale market) at *Dambulla* is collected by the Cyber unit established at this centre in the year 2007, and it appears now in the DoA website. This service can be extended to other places as well.

5. OTHER ICT INITIATIVES FOR AGRICULTURE EXTENSION

5.1. Toll Free Agriculture Advisory Service

Recently a Toll Free Agriculture Advisory Service with a dedicated hot line number (1920) was established at the Audio Visual Centre. This service no sooner commenced, its activities became very popular among farming community and general public in Sri Lanka and daily on an average about 350 inquiries are received by the Toll Free Agriculture Advisory Service. A data base comprising Frequently Asked Questionnaires (FAQs) is being developed every week to analyse farmers burning problems of the week, and then weekly television programmes are designed to address their issues.

5.2. Information Dissemination through Agro-Technology Park

As a quick information dissemination strategy, the first Agro-Technology Park was established by the Audio Visual Centre in the year 2005 at *Gannoruwa* Agriculture Complex (Main Agriculture Complex in Sri Lanka) and the second was established at *Hambantota* in the year 2007. A wide range of Information and Communication strategies (live field demonstrations, self explaining field instructional boards, facilitators for each demonstration sites, instructional leaflets, IMM CDs, VCDs, DVDs *etc.,*) are introduced here for disseminating agricultural information to all those who visit these sites daily in thousands.

Each park is provided with a one-stop-shop which consists of a Cyber Agriculture Information Unit, Seeds and Planting Material Sales Centres, Books/CD sales centres, *etc.*, to cater to needs of visitors (Details of Agro Technology Park are given in the DoA website www.agridept.gov.lk).

The main objective of the Agro Technology park is to demonstrate current recommendations in agriculture of the Department of Agriculture, Department of Export Agriculture, Department of Animal Production and Health, Forest Department, Department of Indigenous Medicine and Faculties of Agriculture to educate farmers, school children, students in professional education in agriculture and general public in an aesthetically pleasing environment.

Figure 10 : Entrance of Agro Technology Park at Gannoruwa

5.3. Cyber Agriculture Wikipedia

A Wikipedia (www.goviya.lk) is the latest addition to DoA website and this site is now being developed. This is a web based agricultural information repository and it will serve as the single online 'meeting site' for farmers and agriculture expertise to discuss and share agricultural information and experience using blogs and vlogs therein. At the initial stage a blog will be established to open discussion for 21 crops identified by the government for national food production drive (Api wawamu - Rata nagamu).

6. IMPACT OF ICTs IN AGRICULTURAL TECHNOLOGY/ INFORMATION DISSEMINATION, TECHNOLOGY ADOPTION IN AGRICULTURAL PRODUCTION

Continuous monitoring and evaluation was done for last 4 years by using the following four methods.

6.1. Results of the Progress Reports of the Cyber Units

Every cyber extension unit should submit a monthly progress report to the Audio Visual Centre. Performance indicators were used in the evaluation and performance of Cyber unit was calculated by evaluating the targets and achievement. Performance indicators are given below.

- Number of training programmes conducted for farmers by using IMM CDs as a teaching tool.
- Number of self learning sessions for farmers and officers.
- Number of farmers trained at the Cyber unit.
- Number of power point presentations produced for farmer trainings.
- Number of inquiries made by farmers for which solutions are available at the Cyber unit.
- Number of e-mails sent to researchers and response.
- Number of leaflets produced.
- Number of CD burning (IMM CDs are duplicated and distributed free of charge at the Cyber unit).
- Number of flip charts produced with the use of IMM CD on presentation technology.
- Number of information provided to others (School Teachers, Students, Entrepreneurs).

Since majority of performance indicators are measurable and physically available at the Cyber Unit (eg. PowerPoint presentations, Leaflets, Journals *etc.,*) no one could fabricate figures and present faulty unaccountable reports. In addition the digital camera of the unit was used to keep visual records, such as training programmes conducted by the Cyber unit.

6.2. Results of the Observation of Cyber Evaluation Team

Officers of the AVC visit all the Cyber units regularly.

The evaluation is based on the following criteria.

1. Verifications on Cyber units functioning schedules (Cyber evaluation team visit the unit without prior notice).

2. Evaluation of activities of the unit by going through the recordings in the digital photo library of the unit (Every activity of the Cyber unit must be visualized through digital photos).
3. Ascertain Computer skills of the Cyber Staff.
4. Evaluation of training material/Instructional media material produced by the Cyber unit.
5. Assess orderliness, cleanliness, setting of the work environment for work-time efficiency of the Cyber unit.
6. Arrangements to keep open the Cyber unit for all five week days.

6.3. Results of the Questionnaire Survey

A questionnaire survey was conducted using structured questionnaires with users of the six Cyber extension units (Rassagala, Vavinna, Hanguranketha, Yatiwawala, Marassana, Urapola). Questions were based on the eight criteria and analyzis was done by using Statistical Package for Social Science (SPSS). Individuals for the sample were selected by observing the digital photographs of Cyber training programs as participants for training programs appear in the photo. Two Research studies (undergraduate final for dissertation and M.Sc. thesis) were done to evaluate the impact.

Criteria used for questionnaire survey

1. Method of awareness of Cyber extension unit.
2. Frequency of visiting the Cyber unit.
3. Objective of visiting the Cyber unit.
4. Quality, clarity and appropriateness of information provided by the Cyber unit.
5. Participation for training at Cyber unit and the quality of training due to new digital instructional media material (IMM CDs, Power Point, Visual database *etc.).*
6. User interface and navigation scheme of IMM CDs.
7. Perception of information presented in CDs.
8. Computer skills and interest of Cyber staff.

6.4. Field Impact Survey

One field impact survey was conducted with farmers under the purview of *Ambalantota* Cyber unit.

A special attention was given to *Wileyaya* (52 Acres, 18 farmers) in *Ridiyagama*, as rice yield of farmers in this village was very poor for several years due to low knowledge in latest innovation such as use of straw, paddy husk charcoal *etc.*,

Therefore, Cyber unit staff conducted several training programs at the Cyber unit in the year 2005 and the yield was gradually increased from the following season. This Cyber unit has also taken initiative to publish a quarterly magazine on local Agriculture ("Denimanya") issues by using the Cyber unit facility.

7. INDIRECT BENEFIT AFTER ESTABLISHING CYBER EXTENSION

7.1. Keeping the Rural Agriculture Extension Office Opened through the Entire Week

Customarily, the Agriculture Instructor's office of the Agrarian Service Centres is kept opened for public only on Wednesdays and this day is treated as the office day of the Agriculture Instructor (AI). Public can visit AI and consult him/her for any need in the office, during office hours, which stretch from 0830 hr till 1630 hr of the day

With the provision of Cyber facility to AI's office, arrangements were made to keep the Cyber unit and office opened for all five week days and Agriculture Research and Production Assistants (ARPAs) and field extension workers are required to manage the Cyber unit, while the AI is away. This arrangement while extending a novelty facility to the farming community, makes it possible for them to visit Agrarian Service Centre all five week days for their other agricultural needs, which otherwise was limited to only Wednesdays hitherto. However, we found that this arrangement is not working effectively in the field, as ARPAs are administrated by another government department.

7.2. Upgrading Rural Agriculture Extension Office to the Level of National Audio Visual Centre

As continuous training programs were conducted at AVC for Cyber Staff during past three years, majority of this staff has now gained skills to render them capable of producing latest digital instructional media materials. Cyber units spread in 20 administrative Districts have got the minimal facilities for the trained staff to practice what they learned. These units now working as satellite stations of the National Audio Visual Centre (AVC) at *Gannoruwa* can always link up with the National centre and dispatch a service very effectively and efficiently to the farming community. The entire staff is almost capable of producing, at least, power point multimedia presentations and print media materials, which are adequate to cater to initial demand. It is encouraging to note that a satisfactory percentage of the staff is at the moment using the flip chart IMM CD to produce flip charts and leaflets for their extension-training activities in the field.

7.3. Establishment of Cyber User Groups

With the establishment of Field Cyber Units, it was revealed that there was an enthusiasm in some areas to form Cyber Groups (user groups). A system can be developed to register these bodies as Cyber Clubs like Young Farmer Clubs of the DoA and much results can be accomplished through and with them in the field of Agricultural Extension for Agricultural Development in Sri Lanka. Cyber villages of *Biso Bandara* and *Mandalagiriya* of *Madirigiriya,* Cyber units of *Hathamuna* and *Siriketha* in *Hingurakgoda* are some of the villages, which have been given special attention to Cyber extension mechanism. Impact studies are being conducted in these Cyber Villages to assess the impact of Cyber Extension in the field and its contribution to agricultural development in the country.

7.4. Joint Ventures with Other Organizations

Having accomplished significant success in the field of Cyber Extension Mechanism of the DoA, joint ventures are now being considered with various other Rural Knowledge Centres (*Nanasala* of ICTA, *Vidhatha* of the Ministry of Science and Technology, Telecentre of *Sarvodaya*) especially to address the development issues and future needs. IMM CDs produced by the AVC (only multimedia agriculture content available in Sri Lanka) are being used in these centres at the moment. Some of the Cyber extension innovative projects being implemented with other organizations are given below;

a. Establishment of *Nanasala* at the Cyber unit at *Dambulla* wholesale market.

b. Two *Sarvodaya* telecentres (Anuradhapura and Nuwara Eliya) are now being operated as Agri Clinics.

c. The World Vision assisted Cyber unit at *Govijana kendraya* in *Kalpitiya.*

d. JICA has establish three Cyber units in their three project areas.

   (Kandurugasdamana, Kimbulvanaoya and Rajanganaya Left Bank Agrarian Services Centres)

8. ON GOING Ph.D STUDIES ON CYBER EXTENSION IN SRI LANKA

1. "Evaluation of the Existing agricultural extension services of the food crop sector in Sri Lanka", S.H.M. Hussain, Provincial Department of Agriculture, Eastern Province.

2. "Sustaining ICT project for Rural Development in Developing Nation an Investigation of Approaches and Strategies", Ranjula de Zoysa, University of Western Australia.

9. LESSONS

9.1. Lack of Awareness

Lack of awareness was identified as the main drawback for not using the Cyber facility of the *Govijana Kendraya.* Even though farmers have seen the computer facility in the office of the Agriculture Instructor they have thought that the facility is not for farmers and computer is for office work of the Agriculture Instructor. To overcome the problem, a name board and a poster were given to display at the office to inform people that computer facility is available for farmers (see Figure 11).

Figure 11 : A Cyber Extension Unit

Mass media strategies were not used as the project is confined to only 85 Cyber units out of 550 units in the island and therefore localized awareness programs were implemented.

The project management has launched a new communication awareness initiative known as Information days. This was a use of multiple media approach in one day at the Cyber unit and four such Information days were organized in the year 2007; one at *Marassana, Rassagala, Polonnaruwa and Buththala.* A Street drama, Muppet show, Mobile Cyber exhibition stall and Banners were used as media strategies, in addition to face-to-face farmer/ extension/research dialogues for specific agricultural technical issues in the area (see Figure 12).

Figure 12 : A Street Drama Group of the AVC demonstrating use of IMM CDs

Politicians, Administrators, Farmer leaders, School children and teachers in the area were invited for the information day. After this awareness program, number of farmers and especially, school children visiting the cyber unit were increased rapidly. It was observed that most of the school children visited the Cyber unit for completion of their Ordinary level and Advanced level project activities. In addition to the above program a mobile exhibition unit of the AVC participates for exhibitions in the country to create awareness (see Figure 13).

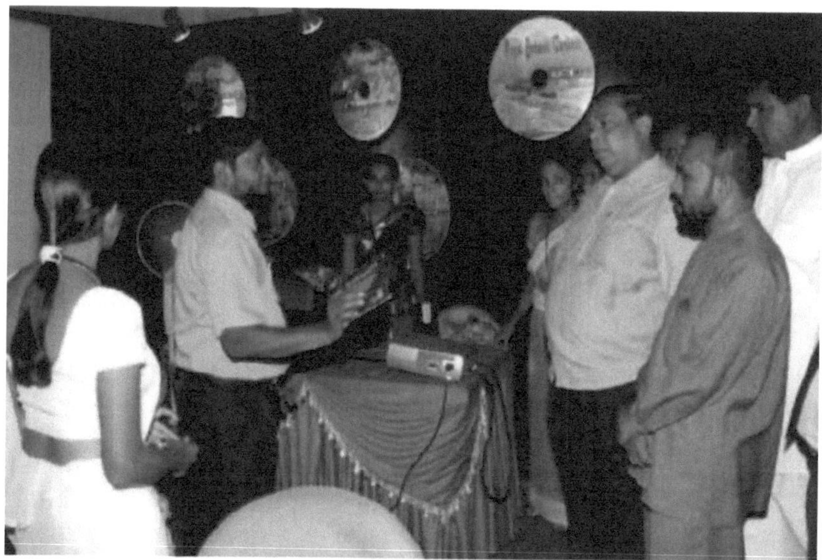

Figure 13 : Mobile Exhibition Unit to Promote Cyber Agriculture Extension Unit

9.2. Problem of Officers Who Manage the Cyber Unit

Since existing staff (Agriculture Instructors) of the DoA are utilized for Cyber extension, considerable effort is given to improve the computer skills of the Cyber Staff. It is evident that young officers are more interested and enthusiastic in gaining digital knowledge than older officers. It is also found that some of the Agricultural Instructors neglect to maintain the farmer database due to hectic work load.

Therefore, some of the older Agriculture Instructors of the Cyber unit were replaced with younger officers who were recruited recently. The digital touch to extension system has enhanced the practicality, assured effectiveness and sustained the credibility of the system and upheld the pride and dignity of the extension staff of the DoA. It was observed that new look given to Agriculture Instructors' office with additional facilities, as a part of the whole program, has generated extra-ordinary enthusiasm among visitors. So much so, most of the farmer societies of the area have shown keen interest in getting their Agriculture Instructors' offices modernized in the same way too.

9.3. Administration Problems

As usual with most government programs in the country, several administrative constraints and obstacles have been observed to occur in few Cyber units. Lack of administrative support and backing from higher officers (especially when their own offices don't have similar facilities as at Cyber units), insufficient financial resources for meeting day to day needs, transferring of trained Cyber staff, tendency to abuse and misuse of some of the digital facilities are a few of the problems encountered. Therefore, steps have been taken to appoint a senior officer at provincial and inter provincial level to co-ordinate Cyber extension activities with other inter-organizational as well as intra-organizational sectors.

9.4. Negative Attitude of Some Individuals at Senior Management Levels towards ICT Initiatives

Some individuals at this stage seem to feel that ICT is far too advance for developing countries like ours. Influence of such people on activities of this program will certainly have negative effects on the same. Therefore, need arises for considerable lobbying to bring about necessary changes in mindset. In this endeavour, political back-up that usually helps overcoming such problems would be a blessing. Thus, efforts have been taken to invite politicians and their participation in opening ceremonies of Cyber units. As a result some of the Provincial Ministers have taken steps to open new Cyber units, similar to DoA Units, in the Agrarian Service Centres in their provinces (31 units were established by the Eastern Provincial council).

10. THE WAY FORWARD

His Excellency, the President of Sri Lanka has taken initiative to establish cyber extension units in all Agrarian Service Centres (ASCs) after reading the value of farmer database. Therefore, a project proposal was submitted to the presidential secretariate to scale up the cyber extension project. After providing equipment to all ASCs (550) an island wide farmer census will be conducted to gather information of all the farmers in the island. This will be the first farmer census in Sri Lanka and each cyber unit will update this farmer database regularly.

As the farmer database of the pilot project is not maintained regularly due to the problem of officers (discussed elsewhere in this paper) a new officer will be assigned to each cyber unit as a cyber extension officer.

Since new recruitment is difficult due to financial barriers, one of the ARPA's will be assigned as cyber extension officer. Problem of paying telephone and internet charges by individual cyber units is identified as another problem. Therefore, island wide one network will be established with the assistance of Telecom and the AVC will bare the running cost on behalf of entire cyber units of the country.

References

Central Bank of Sri Lanka, 2007. *Annual Report, 2007*. Sri Lanka

Wijekoon, R.R.A., 2006. Interactive Multimedia for Technology Transfer in Agriculture. *Paper presented for conference on building digital collection for empowering Sri Lanka*. 1 - 3 November, 2006, Sri Lanka.

Wijekoon, R.R.A., 2005. Digital extension: An alternative strategy for Network extension strategy. *Proceedings of workshop on rural information network in Asia-Pacific: Innovative practices and future directions*. 14 - 16 December, 2005, Bangkok, Thailand.

Kumarawadu, P., 2003. Assessing and Benchmark our digital readiness for e-Government: Moving Towards e Sri Lanka. *Proceedings of the 22nd National IT conference*, Colombo, Sri Lanka.

Wijekoon, R.R.A., 1999. Strategies for the use of Interactive Multimedia to Training Extension Workers on Low Cost Audio Visual Aids in Developing countries, Annual Symposium of the Department of Agriculture, Sri Lanka.

Wijekoon, R.R.A. and Newton, S., 1998. Multimedia support for Extension Trainers in Developing Countries. In: *Proceeding of 7th International Conference on Computer for Agriculture*, Orland, Florida, 1998.

Wijekoon, R.R.A. and Newton, S., 1999. The use of Interactive Multimedia (CDROM) in Training Extension Workers on Flipchart Technology. In: Peiris S.E.(Ed). *Tropical Agricultural Research*, Vol.11, 1999. P.154-173.

Authors Information

Dr. Rohan Wijekoon
Assistant Director (Audio-Visual)
Audio-Visual Centre
Department of Agriculture
Gannoruwa, Peradeniya
Postal Code: 20400, SRI LANKA
Phone No. (O): 94-81-2388618; (R): 94-81-2210532
Mobile Phone No.: 94-71-5328281
Fax No.: 94-81-2388618
e-mail: avcentre@sltnet.lk
URL: www.agridept.gov.lk

Dr. Rohan Wijekoon, Obtained B.Sc (Agriculture) from the Faculty of Agriculture, University of Peradeniya, Sri Lanka and joined the Department of Agriculture as a Subject Matter Specialist in Audio Visual Media in year 1982. Obtained M.Sc. in Agriculture Extension from the University of Reading, United Kingdom in the year 1990. Obtained Ph.D in Digital Media Design from the University of Western Sydney in the year 1999. He is having 27 years of experience in instructional media production for technology transfer in Agriculture and works as a Director of the Audio Visual Centre of the Department of Agriculture which involves in wide spectrum of technology transfer activities in Agriculture (production of TV documentaries, VCD, DVD and IMM CD production, designing and development of DoA website and agriculture wikipedia, management of Agro Technology Parks, operation of Toll Free Advisory Service). He is also working as an external lecturer in ICT for Agriculture at the Post Graduate Institute of Agriculture, University of Peradeniya.

Mr. M.F.M. Rizwan

Programme Assistant (Agriculture)
Audio-Visual Centre
Department of Agriculture
Audio-Visual Centre, Gannoruwa, Peradeniya
Postal Code: 20400, SRI LANKA
Phone No. (O): 94-81-2388618
Mobile Phone No.: 94-71-4402554
Fax No.: 94-81-2388618
e-mail: avcentre@sltnet.lk; rizwan.doa@gmail.com
URL: www.agridept.gov.lk

Mr. M.F.M. Rizwan, completed B.Sc degree in Agriculture from the Faculty of Agriculture, University of Peradeniya, Sri Lanka and he started career as a Programme Assistant (Agriculture) at Audio Visual Centre of the Department of Agriculture. He had training on e-learning Content Development in Philippines in 2006. He has nearly 6 years of experience in various audio visual media production. Among them, his major contribution is for the production of 38 IMM CD-ROMs as an Executive Producer (Instructional System Designer), content developer and trainer for cyber agriculture, wikipedia and monitoring and evaluation of Cyber Extension.

❐❐❐

Chapter 13

SUB-SAHARAN AFRICA

Kristin Elizabeth Davis and Benjamin K Addom

1. INTRODUCTION

This chapter looks at ICTs in Sub-Saharan Africa (SSA), with a focus on several selected countries in the region. Exciting and positive developments are happening throughout SSA Africa in ICTs. For instance, in spite of extensive poverty, SSA has had the fastest growing mobile market in the world over the last five years. The mobile telephone in the region has morphed to serve uniquely African development needs, including those in rural areas. It is being used for market information, extension messages and as virtual currency. In Kenya and Uganda, mobile phone services provide cheap messages about crop price information *via* text messaging. In Tanzania, there are 'market spies', farmers who visit local markets and remain in contact with the village using mobile phones to negotiate for better prices.

In spite of the growing use of mobile telephony and other ICTs, Africa has lagged behind in harnessing ICT potential for extension and other rural development issues in comparison to countries such as China, India, and Chile. Sub-Saharan Africa has the lowest teledensity in the world (Bertolini, 2006a). Many challenges exist to ICT use in SSA, and there are concerns that many rural African communities are being further left behind through the 'digital divide'.

With the many challenges, how can ICT potential be harnessed to benefit rural communities in SSA? Is the ICT 'revolution' just hype? Or will ICTs be a key tool for bringing about sustainable development in Africa? Section 2 will give an overview of the national agricultural and extension scenario in the case study countries: Ethiopia, Ghana, Kenya, and Malawi. Section 3 will give an overview of the use of ICTs in agriculture in these countries, while Section 4 discusses the national ICT policy in the four countries. Section 5 discusses specific case studies of ICT use in the four countries. Section 6 covers certain issues surrounding ICT usage that affect agricultural extension in the region. Finally, section 7 discusses the impact of ICTs, lessons learned, conclusions, and the way forward for the region with regard to use of ICTs in extension.

2. OVERVIEW OF NATIONAL AGRICULTURE AND EXTENSION SCENARIOS IN SUB-SAHARAN AFRICA

Agriculture remains of great importance to livelihoods and economies in SSA. Most countries thus have public extension systems funded by government coffers to assist their farmers. In addition, many countries have increasing numbers of Non-Governmental Organizations (NGOs) and private sector providers of extension and other rural development services. This section will give an overview of the national agricultural scenario of Ethiopia, Ghana, Kenya, and Malawi. The literacy rates of the countries, disaggregated by gender, are also discussed, as this is of great relevance to the use of ICTs and other extension methods used.

Ethiopia has a population of at least 82 million. Agriculture accounts for about half of the country's gross domestic product (GDP) and employs about 80 per cent of the labour force. Horticulture has become an important agricultural sector in the past few years, creating 70,000 jobs (80 per cent to women) and earning the country 120 million US dollars in 2007. Other major agricultural exports include coffee, qat (*Catha edulis*), livestock and livestock products and oilseeds.

Ethiopia has had agricultural extension services since the 1950s, when a model similar to the United States Land Grant approach was used, where universities reached out to communities with research-based knowledge and through adult education. In 1993, Sasakawa Global 2000 (SG-2000) promoted the use of productivity-enhancing technologies and access to inputs and credit, coupled with training using demonstration plots. The SG-2000 pilots led to the current extension approach called the Participatory Demonstration Extension and Training System (PADETES) in 1995. PADETES used a similar approach to SG-2000 but extended the technology package to livestock, high value crops, post harvest technologies, and agroforestry. PADETES also used a menu-based approach rather than the former package approach.

In 2007, the Rural Capacity Building Project (RCBP) was initiated under the Ministry of Agriculture and Rural Development, focusing on capacity building of human resources in extension; supporting farmer training centres (FTCs) with physical infrastructure, and the implementation of institutional innovations such as decentralization and participatory financing mechanisms About 15-18,000 FTCs are planned (one in each local administrative area); 8,489 have been completed to date. Each FTC is to have three extension agents, called development agents (DAs). Part of the RDBP project is upgrading the skills of existing development agents, and hiring new ones to be able to cover the FTCs in each locality. About 36,000 DAs have been trained in the past few years (when this number is added to the existing DAs, it makes a total of 69,680 DAs). Each DA based at the FTC is expected to give modular training to 60 farmers every six months in his/her field of specialization (Ministry of Agriculture and Rural Development, 2007).

The use of ICTs in extension is very limited, although plans are in place to make better use of ICT tools for extension and other rural development efforts. More details on the use of ICTs in extension and agriculture is given in section 5.

Ethiopia has a literacy rate of 35.9 per cent among people aged 15 and older, with 50 per cent for men and 22.8 per cent for women, during the period 1995-2005 (UNESCO, 2007).

A much smaller country, **Ghana** has a population of about 23 million, with 64 per cent of the population being rural. It has a total area of 239,460 square kilometres (230,940 sq km of land, and 8,520 sq km of water). The country is well-endowed with rich natural resources such as gold, timber, industrial diamonds, bauxite, manganese, fish, rubber, hydropower, silver, salt, limestone and petroleum. Fifty-five per cent of the workforce (mostly smallholders) is involved in agriculture. Agriculture accounts for 36 per cent of the economy, with cocoa being the most important export crop.

In 2000, the agriculture sector employed 70 per cent of the country's labour force, contributed 39.6 per cent to GDP, and accounted for 57 per cent of foreign exchange earnings (Dzidonu, 2002). With more than 80 per cent of all agricultural production carried out on small land holdings of less than one hectare per farmer, the vast majority of small-scale subsistence farmers lack access to improved technologies for production, storage, processing, and market information. Ghana's agricultural growth continues to rise and fall with a rate of 2.4 per cent between 1987-1997, 3.6 per cent between 1997-2007, 6.0 per cent in 2006, and 4.6 per cent in 2007 (World Bank, 2008a).

Ghana's Ministry of Food and Agriculture (MoFA) is responsible for extension provision for the country, and MoFA is committed to the use of ICTs in extension delivery. Several approaches to extension have progressed in Ghana through the Transfer of Technology (ToT) model during the colonial period to the early sixties; the Training and Visit (T&V) model of the World Bank, which emerged in the nineties; and the current program of Farmer Field Schools (FFS). In addition, Ghana has extension provision through NGOs and private companies. In 1997 Ghana began decentralizing its extension system (Amezah and Dorman, 2004).

Extension services in Ghana are provided free of charge to farmers by extension workers, who are civil servants of MoFA. The system allows extension staff to go around visiting farmers on their farms to identify problems and to give advice on best practices in all areas of agriculture. However, the extension officer-farmer ratio is quite low (1:1500), resulting in the inability of the extension workers to visit many needy farmers. To a large extent, therefore, their impact on agricultural development is not being felt. Apart from the one-on-one visit by extension staff to farmers, MoFA and the research institutes are using few other avenues to transfer new technologies to the farmers.

According to the (UNESCO, 2007), the literacy rate between 1995-2005 for those aged 15 years and above was 57.9 per cent overall. When disaggregated by gender, men's literacy rate is 66.4 per cent and women's is 49.8 per cent.

Kenya, with a population of nearly 38 million, relies slightly less on agriculture, with contribution to GDP standing at about 23 per cent. Horticulture is a major income earner for the country, as are coffee and tea. Maize is the staple crop of smallholders.

Extension was established in Kenya in 1903 with the colonial Department of Agriculture (Ng'eno, 2004). Kenya has several extension systems and models. Kenya's National Agriculture and Livestock Extension Programme (NALEP) was started in 2000. The NALEP approach, supported in part by Sida, focuses on stakeholder inclusion, bottom-up planning, and group-based approach in focal areas. Many different NGOs and projects also provide extension, as do private companies such as British American Tobacco and East African Breweries. Farmer Field Schools, an extension and education approach based on adult learning principles with a focus on empowerment and marketing, have become an important extension approach in the country.

Kenya's literacy rate stands at 73.6 per cent for 1995-2005, with men's rate at 77.7 per cent and women's at 70.2 per cent of those aged 15 years and over (UNESCO, 2007).

Malawi has a population of about 14 million. Agriculture contributes 34 per cent of the economy, and 80 per cent of the labour force is involved in agriculture. Malawi is one of the poorest SSA countries, with two-thirds of the population living below the poverty line. Malawi mainly exports tobacco, sugar, cotton and tea. Maize is the staple crop of smallholders.

Major constraints to agricultural production and growth in Malawi include: high population growth; limited access to land; small size of the domestic market; low prices of crops; limited credit; lack of sufficient and effective extension advice; and persistent droughts, pests, and diseases (Dzidonu, 2003).

Malawi's extension policy focuses on pluralistic, demand-driven, and decentralized extension systems, with bottom-up participatory planning (Chinsing and Cabral, 2008; Munthali, 2004). The four pillars of district extension system are (a) organization of farmer demand, (b) facilitation of service provider response, (c) coordination and agricultural strategy development, and (d) funding. Initial attempts at public-private partnerships in extension provision have met with major problems (Matemba and Charman, n.d.).

The "one village one product" (OVOP) approach that is now taking roots in some areas as an extension model was initially a presidential directive as a means to enhancing rural economic growth (http://www.ovop.org.mw/). In this approach, villages focus on producing one specific product, and production is zoned. The "farmer to farmer initiative" where innovative farmers are identified is also important, as is the Farmer Field School (FFS) approach.

In Malawi, literacy for 1995-2005 is 64.1 per cent nationwide (for those aged 15 and over). Male literacy rate is 74.9 per cent, and female is 54 per cent (UNESCO, 2007).

As can be seen from the above descriptions, agriculture is of major importance to each country, and extension has heavy responsibility to assist smallholders to improve production and livelihoods. While Ghana, Kenya and Malawi have relatively high literacy rates, especially for men, Ethiopia's rates are troublingly low. Many issues, from policy all the way down to the human capital of smallholder farmers, will affect the way extension can do business and harness the benefits of ICTs for extension.

3. NEED FOR ICTs IN AGRICULTURAL EXTENSION AND RURAL DEVELOPMENT

There is increasing evidence that infrastructure such as telecommunications helps to reduce poverty and provide opportunities to people in developing countries (Torero and Chowdhury, 2005). Making information readily and cheaply available can enhance production, increase bargaining power and incomes, and ultimately, lead to poverty reduction and economic development. Furthermore, making information readily and cheaply available can enhance production by enabling farmers to plant and harvest at appropriate times (weather information), improve the functioning of credit markets (by facilitating social capital as through increased communications farmers develop and gain trust and reputations), facilitate access to more efficient technologies (through finances obtained through increased credit ratings/worthiness), and in the long run, transform production processes through more rapid and diffuse spread of technological innovations amongst a broader range of interacting agents - consumers (with different tastes and preferences leading to product differentiation or new product development), sellers (of inputs products and services), traders, and processors (Eggleston *et al.*, 2002).

ICTs are thus increasingly being used in many parts of Africa for the dissemination of market, price, weather, extension and political information. However, in comparison to other developing regions, Africa has a diverse geography and a low population density. This makes it difficult for companies to install ICT infrastructure and to make a profit from such endeavours. Thus Africa may be losing out on these potential benefits from ICT infrastructure.

Furthermore, in developed countries, markets function efficiently because the prices of goods and services are known or can be accessed cheaply, widely. and readily (Eggleston *et al.*, 2002). In contrast, smallholder farmers in rural Africa are constrained by access to reliable market information due to a combination of lack of cheap, timely, and readily accessible information, poor information delivery mechanisms and infrastructure, and a private sector attitude that typically views them as commercially unattractive (Eggleston *et al.*, 2002; KBDS, 2004). As (Geertz, 1978) powerfully summed it, in many rural areas in the developing world, 'information is poor, scarce, maldistributed, inefficiently communicated, and intensely valued'. The result is inefficiency, characterized by farmers producing the wrong mixture of crops and using inefficient technologies, consumers not receiving the goods they want or are willing to pay for, and dispersions of agricultural prices or deviations from the law of one price[1] (Stigler, 1961; Delgado, 1986; Ravallion, 1986; Crowe *et al.*,1997; Fafchamps and Gavian, 1997; Badiane and Shively, 1998; Eggleston *et al.*, 2002). As Bell (1980) argued: 'In a price and market economy, the condition for efficiency, or optimal use of resources is complete information among buyers and sellers, so that one obtains the "best" price for one's goods or services'.

A functioning or efficient market information system is thus critical to farmer productivity, bargaining power, incomes, and ultimately, poverty reduction and economic development. Eggleston *et al.*, (2002), highlight four important reasons why price information is important to farmers: (a) relative prices allow them to make informed decisions on the crop or mixture of crops to produce, (b) price information facilitates the use of efficient and cost-effective technologies, (c) market information enables producers to know where to sell their produce and (d) what prices to accept.

Increased access to market and other information is having a positive socio-economic impact on the poor in Africa. This is consistent with empirical evidence and historical experiences of other developing countries and some

---

[1] The Law of one price holds that prices for homogenous goods sold at different locations should be equal, net of transportation costs (Eggleston *et al.* 2002, 4-5).

of the now-developed ones. A growing literature indicates that better access to telephony in general and the mobile in particular, by facilitating greater access to market information improves producer prices, by reducing monopoly of power in agricultural markets and farm price dispersion (Hudson, 1995; Bruns *et al.*, 1996; Forestier *et al.*, 2002). Studying the impact of the Grameen Village Pay Phone Project, von Braun *et al.*, 1999) found that the average prices of agricultural commodities were higher in villages with phones than in those without, and that the phones helped farmers to negotiate better rates in foreign exchange transactions, strengthen farmer bargaining power, make better production decisions, and handle medical emergencies and natural calamities better. Garbade and William, (1978); Du Boff, (1980) posited that the telegraph and the trans-Atlantic cable played significant roles in rapidly narrowing inter-market price differentials in the 19th century United States, when a majority of the US labour force was still engaged in agriculture.

Eggleston *et al.*, (2002) go further than this to argue that Information and Communication Technologies did not simply improve prices for producers, but changed the whole nature of economic transactions in the United States, and that they could significantly improve the living standards of the world's rural poor today by enhancing the functioning of relevant markets. *'When markets function well, trade is abundant, and farmers reap the rewards of specialization... additional gains from specialization include developing product-particular skills and knowledge, purchasing inputs in bulk, or reaching economies of scale. A whole region or economy benefits when improved information leads to more integrated markets that widely disseminate new techniques, fertilisers, and other inputs to agricultural production'* (Eggleston *et al.*, 2002).

Thus many extension agencies, governments, NGOs and projects are dabbling in 'e-extension' or 'cyber-extension' to improve their outreach to farmers, and farmers' access to information. However, most of these initiatives are at early pilot stages within Africa, and such use of 'e-extension' is by no means policy, much less practice. But references to ICTs are showing up in government extension strategies, such as the National Agricultural Sector Extension Policy Implementation Framework for Kenya (Agricultural Sector Coordination Unit, 2006).

4. ICT SCENARIO AND USE BY EXTENSION IN SUB-SAHARAN AFRICA: THE FOUR COUNTRY CASES

4.1. ICT Scenario

In spite of its extensive poverty, Africa has been the fastest growing mobile market in the world over the last five years (Coyle, 2005). Some say growth rates in Africa are at about 50 per cent per year (Institute of Development Studies, 2007). Modern telecommunications systems in Africa are largely mobile systems (Scott *et al.*, 2004; Waverman *et al.*, 2005). By 2003, there were 6.1 mobile subscribers for every 100 persons as compared to 3 fixed line subscribers per 100 persons in Africa (Coyle, 2005). In 2005, there were 52 million mobile subscribers (about 7 per cent of the African population) compared to 25 million for fixed lines, with the mobile accounting for at least three quarters of all telephones in 19 African countries (Scott *et al.*, 2004; Coyle, 2005). The Commission for Africa (Scott *et al.*, 2004) estimates that the number of mobile subscribers in Africa will continue to expand at the rate of 35 per cent per year over the next few years. These numbers are conservative, since teledensity figures mask the extent of mobile access because people in Africa, especially the poor, tend to share mobile phones (McKemey *et al.*, 2003).

Not only has mobile phone uptake in SSA been rapid, but the innovative uses of mobile phones have also evolved rapidly. Mobile usage has morphed to serve uniquely African development needs, in addition to performing the conventional telecoms function it has served in the developed world. McKemey *et al.*, (2003) estimate that in typical rural districts of Africa, up to 80 per cent of households make regular use of phones in a number of ways, besides its use as an infrastructure service (see below for details).

The rapid spread of the mobile phone in Africa is partly due to economic and political liberalization of the late 1980s through the 1990s, and partly due to inherent features of mobile telephony itself, which make it relatively more conducive to poorer/marginal conditions/areas and capabilities of poor or low income customers, than either fixed telephony or other forms of ICTs:

1. A shorter payback period on investment compared to fixed telephony. Mobiles involve lower installation costs and faster roll out than fixed lines[2].
2. Complementarities with lower levels of skills than needed for the internet (Coyle, 2005) make it more readily accessible by the poor and the less educated and non-literate.
3. Lower income/social entry barriers than both fixed telephony and the internet due to a combination of lower upfront expenditure and the mobile's suitability to communal or shared usage (Coyle, 2005).

[2] Waverman *et al.* 2005 estimate that a mobile network costs 50 per cent less per connection than fixed lines.

4. Innovative business modelling such as the MTN mobile village Payphones in Uganda, 'air time' credit-swapping services between subscribers such as '*sambasa* me' in Kenya, and lower denominational pre-pay systems.
5. Network effects that generate greater momentum once critical mass is reached (Coyle, 2005).
6. Innovative uses to which the mobile can be put besides its telecoms functions - for example as virtual currency.

However, the rapid growth of ICTs in developing countries in general comes in part from the fact that there is very low initial access to start with (Torero and von Braun, 2005). Also, in spite of patches of success, some claim that only about 30 per cent of agriculture and rural organisations in SSA are connected to the internet.

In Figure 1, Africa is compared to South Asia and to Western Europe with regard to adult literacy rates, number of telephones and mobiles per 1,000 people, and internet users per 1,000 people. While it is clear that Western Europe is far ahead of the African region with regard to literacy and ICT usage, the figure also indicates that Africa is slightly ahead of South Asia in use of ICTs per 1,000 people.

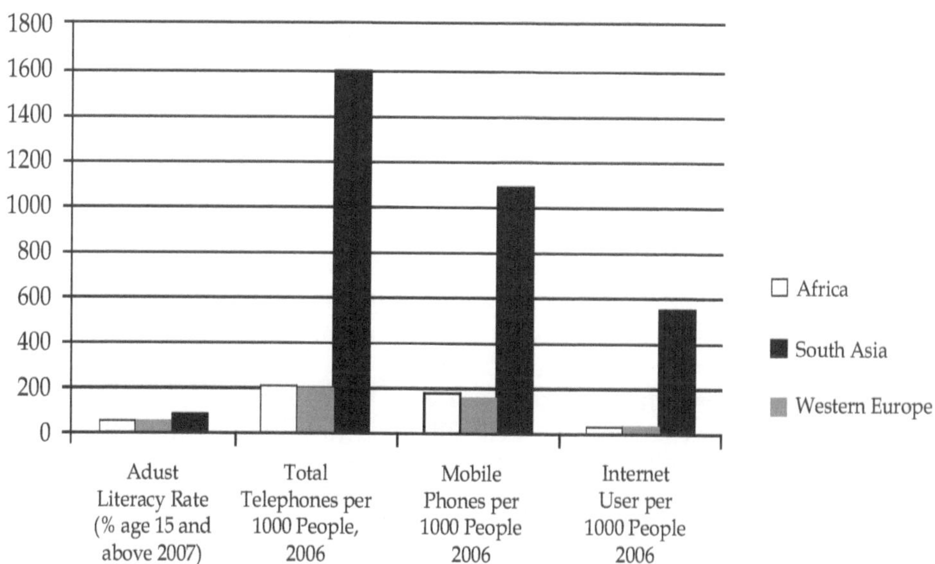

Figure 1 : Use of Adult Literacy, Telephone, Internet, and Mobiles in Various Regions (per 1,000 people)

(Source: KAM, 2008)

However, while on the whole use of ICTs has been growing quickly in Africa, there are big differences between countries (Table 1). While Ethiopia has by far the largest population, it has far less use of cell phones and the internet than the other three countries. Malawi, with just about one-sixth of Ethiopia's population, has just as many cell phone users.

Table 1 : Comparison of ICT users in the four countries

	Ethiopia	Ghana	Kenya	Malawi
Population	82,000,000	23,000,000	38,000,000	14,000,000
Cell phone users	1,000,000	7,000,000	11,000,000	1,000,000
Internet users	291,000	650,000	3,000,000	140,000

(Source: World Bank, 2008b)

Figure 2 breaks down the use of ICTs per 100 people to show a clearer comparison between countries. Ghana has the highest use of telephones, at 1.43 per 100 people. The three other countries, with vastly different populations, each have about 0.80 per 100 people. However, it is important to note that within countries, there are important differences. For instance, in Ghana, the greater Accra region accounts for about 93 per cent of fixed telephone lines (Bertolini, 2006b). Therefore these figures could be misleading with regard to use of ICTs by extension agents and people in the rural areas. Use of the internet in the four countries has an even greater range. Kenya has the highest number at 3.12 per 100 people, almost double the number for Ghana, which has 1.78. Malawi's low rate of 0.40 is still almost double Ethiopia's use of 0.22 per 100 people.

Finally, it is evident that while Kenya and Ghana are very close in terms of per capita use of mobile phones, Malawi is lagging behind, and in Ethiopia it is almost nil, with mobile phone use mainly occurring in cities and other urban areas. This may be due issues of relative levels of wealth, of geography, and of policy. For instance, Ethiopia is a very big country with high levels of poverty. It has many mountains and semi-arid (low population) areas, and has no competition within the mobile phone market.

It is also useful to compare ICT use versus the per capita income of the country. Figure 3 shows the gross domestic product per capita income of each country plotted on the use of telephone mainlines per 100 people. It shows that while Malawi and Ethiopia are very similar with regard to their per capita income and the number of phone lines per person, Kenya has the highest per capita income, and Ghana, with a lower per capita income then Kenya, has a much higher use of telephones per person.

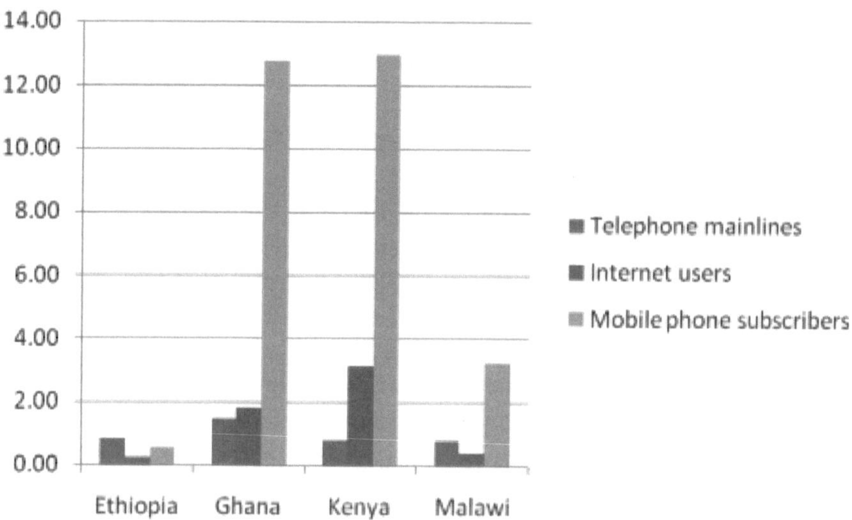

Figure 2 : Use of Telephone, Internet, and Mobiles (per 100 people)
(Source: World Bank, 2008b)

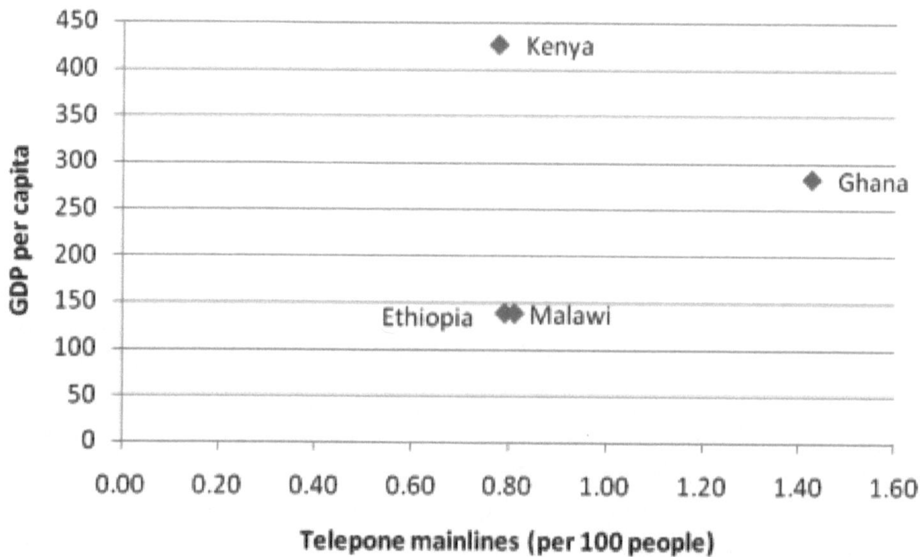

Figure 3 : Numbers of fixed telephone lines per 100 people and gross domestic product per capita, 2005

Note: GDP per capita is gross domestic product divided by midyear population. GDP is the sum of gross value added by all resident producers in the economy plus any product taxes and minus any subsidies not included in the value of the products. It is calculated without making deductions for depreciation of fabricated assets or for depletion and degradation of natural resources. Data are in constant U.S. dollars (Source: World Bank, 2008b).

The Knowledge Assessment Methodology (KAM), developed by the World Bank, allows for comparisons of countries with regard to their knowledge economies. It measures performance on the four knowledge economy (KE) pillars: (1) Economic Incentive and Institutional Regime, (2) Education, (3) Innovation, and (4) Information and Communications Technologies. The Knowledge Index (KI) is the average of the normalized country scores on the key variables in three pillars - education, innovation, and ICT. The Knowledge Economy Index (KEI) measures performance on all four pillars. Variables are normalized on a scale of 0 to 10 relative to other countries in the comparison group. See Figure 4 for a comparison of the four countries with regard to the knowledge economy index, knowledge index, and ICTs. It becomes immediately apparent that Ghana and Kenya rate much higher than Malawi and Ethiopia with regard to the Knowledge Economy Index, the Knowledge Index, and use of ICTs.

It is also evident that different sources and kinds of data are useful for different types of information that extension may be dealing with (Figure 5). Using data from the Ethiopia Rural Household Survey from 2005, it is apparent that in Ethiopia, the radio is much more important than either extension or friends/relatives for topics such as weather forecasts, policy changes, and new market opportunities. It is somewhat important for information on food aid and market prices at the *regional* level. However, for *local* prices on products, fertilizer, and seed, the radio is much less useful. Extension agents are quite important for seed and price information. This shows that the radio is a useful tool at the regional or national level, while friends/relative and extension agents are more useful at the local level.

In East Africa, in a farmer field school project area, both farmer field school and non-farmer field school farmers stated their main sources of marketing information. Local radio/TV and mobile phones are compared to extension workers in Figure 6. Mobile phone use for market information was lowest in Kenya, while local radio/TV was about the same in all the three countries. Tanzania relied most heavily on extension for market information.

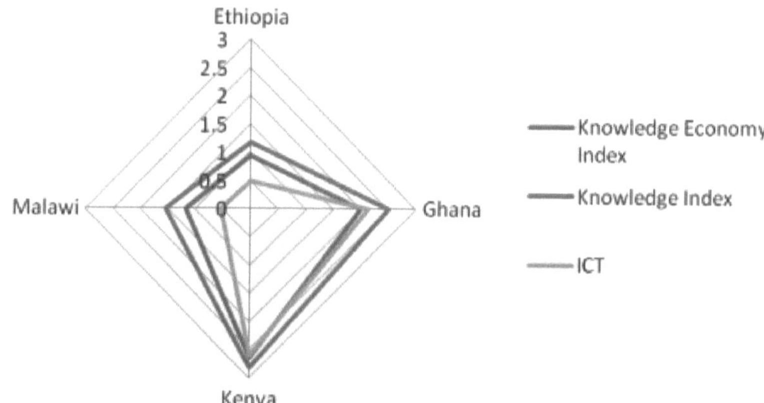

Figure 4 : Comparison of Knowledge Economy Index, Knowledge Index, and ICT
(Source: KAM, 2008)

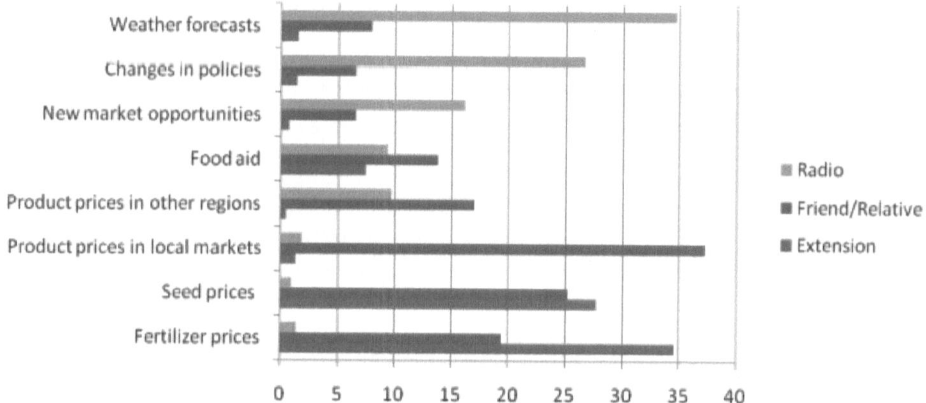

Figure 5 : Sources of Information for Ethiopian Farmers (percent)
(Source: Ethiopia Strategy Support Project, 2005)

4.2. Use of ICTs in Extension in SSA, Focusing on Ghana

In spite of the data described above, information is limited with regard to the use of ICTs for extension purposes in SSA. However, the Forum for Agricultural Research in Africa (FARA) recently developed an inventory of innovative use of ICTs in extension (advisory services) in Africa. This includes voice information delivery services, dial-up and regular radio broadcasts, mobile phones and database monitoring, and e-learning materials in extension (Gakura *et al.*, 2008).

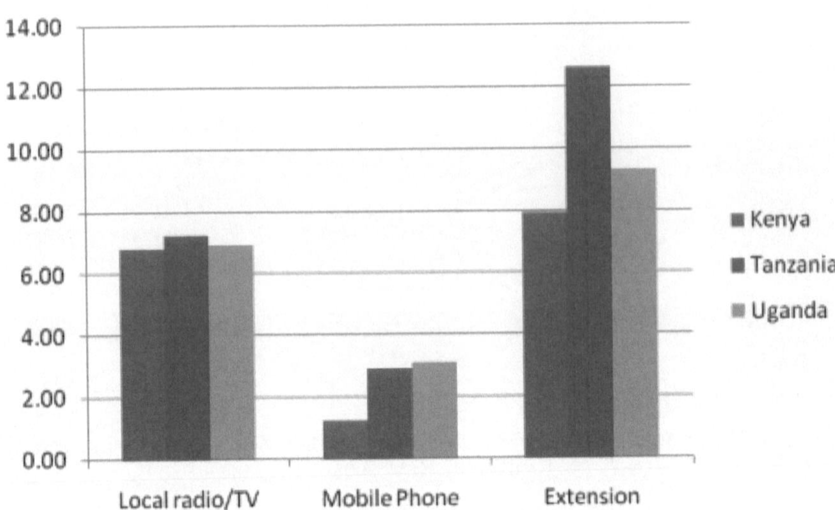

Figure 6 : Comparison of Sources of Marketing Information in an East African Farmer Field School Project

(Source: Author's data)

Perhaps the most information on the use of ICTs for extension in the SSA case study countries can be found from Ghana. In Ghana's telecommunications sector, the media used for information delivery and exchange (extension) among farmers, researchers, and extension officers in Ghana's agriculture include storytelling, radio, newspapers, and extension agent-farmer communication (Addom, 2006).

Radio still remain Ghana's most popular communication medium, although it is being challenged by increased access to television. The State-run Ghana Broadcasting Corporation (GBC) runs radio Radio 1, national broadcaster, broadcasting in English and Ghanaian vernaculars, Radio 2, commercial service, broadcasting in English, and an FM station, Radio GAR. There are officially over 130 radio stations in Ghana granted licenses to operate, and of this number about 96 are currently on air, mostly on FM frequencies (Kafewo, 2006). According to a study conducted by Gadzekpo, (2005) and InterMedia, (2005), FM is the most-used reception platform, with 82 per cent of the population receiving radio *via* FM.

Currently, there is an increasing use of rural radio and FM stations to broadcast agricultural programs to farmers. Most of these programs are in the local languages so that illiterate farmers can be reached. Some presenters allow 'phone-in' by the farmers so that farmers can interact with resource persons. This however is limited only to areas where there are telephones. The TV stations are also broadcasting a few agricultural programs; but the constraint here is that in many rural areas, where the majority of the farmers live, do not have electricity.

The telephone system of Ghana may be described as an outdated and unreliable fixed-line infrastructure, heavily concentrated in the capital, Accra, and a few other large cities. This situation is being overlooked due to the current surge in mobile communication. Competition among multiple mobile-cellular providers has spurred growth with subscribership about 35 per 100 persons and rising. As of 2007, the total number of mobile cellular telephone subscribers was estimated as 7.6 million. This number is expected to rise drastically in the next few years due to the recent sale of 70 per cent shares in Ghana Telecom to Vodafone in August 2008. However, it is unclear how much telephones (fixed lines and mobile phones) are used by extension or by farmers to receive or transmit information.

Ghana's Television Broadcast Stations has also increased to seven, after decades of the state-run Ghana Broadcasting Corporation station - GTV. This includes the Ghana TV, Metro TV (free-to-air TV channel), TV3 (private, free-to-air TV channel), Multichoice (services through satellite), V-Net TV (cable TV channel), Fantazia TV (cable TV channel), and TV AGORO (TVA). Again, information is unavailable as to the use of television by extension.

An estimated number of 24,018 internet hosts (number of computers connected directly to the internet) and 650,000 internet users (the number of users within the country that access the internet) exist in Ghana (CIA, 2007; CIA, 2008). The advent of the internet coupled with the on-going phenomenon of globalisation has led to more efficient means of communication among several actors. ICTs are being used to improve the existing 'old' technologies. The development of ICT in Ghana has been seen in areas of awareness creation, connectivity, national policies, hardware and software improvement, and human resource development. As a result, there have been dramatic moves in Ghana's information communication system for some years now, which are reflected in the use of Business Communication Centres (BCCs), Community Learning Centres (CLCs), Internet Cafes or Cyber Cafes, Community Telecentre (CTC) and the current explosion of wireless technologies, mobile phones and Community Information Centres (CICs).

ICTs are regarded as a tool to improve and enhance two-way information flows and there is substantial evidence that without two-way information flows, development efforts fail (Zijp, 1994; Peters, 2003) pointed out that when used properly, ICTs become key weapon in the war against world poverty, offering tremendous potential to empower people in the developing countries to overcome development obstacles and strengthen democratic institutions in communities. In terms of ICTs, the Ghana National Agricultural Research System (NARS) and researchers have come a long way, but still have a long way to go (Sraku-Lartey and Sam, n.d.). The authors argue that the situation changed from a time when computers and electronic media were unheard of in agricultural research (in the 1970s), to the current state where the NARS and its researchers are using computers quite widely and to a lesser extent internet-based ICT to communicate with each other and to carry out some important aspects of their research activities.

It is estimated that currently less than 30 per cent of research scientists in Ghana have computers on their desks and use them to some extent. Out of nine agricultural-based institutes of the Council for Scientific and Industrial Research (CSIR), only three have developed local area networks (LANs) with Internet access. Even in these institutes, not all the research scientists have personal computers to work with, and many of them have to share machines with their colleagues and subordinates in a common pool. The situation is such that even the Agriculture, Forestry and Fisheries Sector Directorate (AFFSD), which has the responsibility of coordinating agricultural research nationally, has no reliable e-mail or internet facility. Communicating among research scientists in the Ghanaian NARS is therefore a problem at the moment, in spite of a research network created in 1992 under the National Agricultural Research Project (NARP). Scientists who have access to computers use them for a variety of work such as typing reports, analyzing data, searching for literature, and to some extent, staying in touch with colleagues through e-mail. There is room to increase the usage of computers by researchers. Before this can be done, however, the number of computers in the system will have to be increased, and the skills of researchers in the use of computers will have to be updated (Sraku-Lartey and Sam, n.d.).

Several studies have been conducted on the use of ICTs in extension and higher education in Ghana. One of these was an examination of the challenges and prospects of using ICT for extension, conducted in 2004-2005 (Annor-Frempong *et al.*, 2005). The authors found that extension agents had a high demand for the use of ICTs for extension. On the other hand, extension agents required training, infrastructure, financial resources, and technical backup to fully utilise the potential of ICTs.

Another study focused on the state and use of ICTs in some of the Ghanaian universities (Addom, 2004). Information was collected on human capacity, infrastructure, programs, policies, and posture related to ICTs in seven universities in Ghana. The study found that there was good ICT posture in the universities examined; faculty acknowledged the usefulness of ICTs and their desire to take advantage of them. However, ICT facilities available to faculty were not perceived as satisfactory, the human capacity was not available to offer training at the universities, and there was no financial or human capital available to use ICTs in outreach (Addom, 2004).

5. NATIONAL POLICY ON ICT IN AGRICULTURAL EXTENSION: THE FOUR COUNTRY CASES

In spite of the success stories and other advances in ICTs in Africa, many of these efforts have occurred in spite of, and not because of government policy. Instead, it is the adaptation by individuals and private firms that have allowed ICTs to spread and be used in the region. Governments and other actors are now trying to keep up with the boom, which may account for the fact that many of the cases of successful ICT in rural areas use remain isolated and scattered. The question of how to upscale and outscale successes for greater impact remains. This section will examine the ICT policies that do exist on the continent.

As mentioned earlier, ICTs have been shown to bring about positive impacts in rural areas of the developing world. However, to fully utilise ICTs, there is a need for supportive policies that allow for the development of ICT infrastructure, as well as an enabling policy environment (*e.g.* low taxes on mobile phone usage).

ICT policy in Africa is in part affected by the New Partnership for African Development (NEPAD). NEPAD is an organization that develops a vision and strategic framework for Africa's renewal. Each year NEPAD holds a council on ICT issues in Africa. This has to a certain extent affected ICT policy in African countries.

The United Nations, under the System-wide Special Initiative on Africa, has a collaborative effort entitled Harnessing Information Technology for Development (HITD) (Rose, 1999). This involves the United Nations Educational, Scientific, and Cultural Organization (UNESCO), the United Nations Economic Commission for Africa (UNECA), the International Development Research Centre, the International Telecommunication Union, and the World Bank. HITD is to support Africa's Information Society Initiative (AISI), which is a long-term framework for regional telematics (Rose, 1999).

UNECA has been leading another process on the continent called 'ICT for Development' (Dzidonu, 2003). Similarly the United Nations Development

Program (UNDP) led to the Digital Opportunity Initiative (DOI), which developed a strategic framework to guide African countries in developing ICT strategies. According to the DOI, five critical areas to consider in developing these strategies are infrastructure, human capacity, policy, enterprise, content and applications (Dzidonu, 2003).

Another worldwide effort that affects ICT in Africa was initiated by the World Trade Organisation (WTO), which instituted the Basic Telecommunications Agreement in 1998. The Basic Telecommunications Agreement requires member states to open up telecommunications markets to global competition to stimulate greater quality and efficiency (Shanmugavelan and Warnock, 2004). Ghana, Kenya and Malawi are members of the WTO and thus are expected to abide by this agreement. Ethiopia, being an observer government, is expected to start accession negotiations within five years of becoming an observer. Ethiopia's application was received in 2003 (www.wto.org).

Historically, many sectors such as telecommunications were state-owned in Africa. However, national policies with regard to privatization of these systems, together with the Basic Telecommunications Agreement, have led to impact on the quality of telecommunications infrastructure (Torero and von Braun, 2006). Those countries that have undergone reforms (privatization, regulation, competition, *etc.*,) have increased telephone coverage. However, the costs and the low population density in SSA are prohibitive to the use of ICTs, especially in rural areas.

There are specific cases of how regulations and laws affect ICT use in Africa; for instance, in Ghana and Kenya, internet service providers (ISPs) require a license. In Ghana at least, this process can take one month to an entire year (http://www.comminit.com/en/node/242060). Many technical limitations also exist that are not addressed by African policies, including the oversubscription of ISPs on dial-up connections, making the internet exceedingly slow. Due to high demand, however, SSA is taking steps to improve internet connectivity and speed; in Kenya, an undersea fibre optic internet connection from TEAMS (a Kenyan Government project) and SEACOM will land in Mombasa by March and be commissioned in June 2009. Ethiopia is in the process of adding a second fibre optic route out of the country in Djibouti, in addition to the route through Sudan ('Is our bandwidth sufficient to do networked science?' (ILRINet Discussion Forum http://ilrinet.ilri.cgiar.org/archiveDetail.php?artID=2911, March 31, 2009).

Below, the four countries' ICT policies are discussed.

Ethiopia developed The National ICT for Development (ICT4D) Five Years Action Plan for Ethiopia [2006 - 2010] (Dzidonu, 2006). This plan is mostly focused on education and capacity strengthening for federal and regional governments, and in the education and agriculture sectors.

In 1994, Ghana started a program to increase the quality and quantity of telecommunications services. In 1996, the country partially privatized fixed telephone lines. The mobile phone market was opened in 1995 and the internet in 1996 (Torero and von Braun, 2006). Due to privatization, all 10 regional capitals have mobile phone coverage.

The **Ghana** Government through the Ministry of Communications formulated an ICT policy in 2003 dubbed 'Information and Communication Technology for Accelerated Development' (ICT4AD). This policy can be characterised as the foundation upon which Ghana's vision for the information age was built. The policy document specifically refers to the use of ICTs in the agricultural sector.

According to the policy statement, ICT will be utilized to modernise the agricultural sector to substantially improve agricultural value-added and yield and develop a dynamic and vibrant export-oriented agro-business industry. This will be achieved through the following:

1. Deploying and exploiting ICTs to support the various activities of the agriculture sector, including the commercialisation of key sub-sectors and the improvement of current agricultural practises;
2. Developing Geographical Information System (GIS) applications to monitor and support sustainable usage of natural resources;
3. Developing food insecurity and vulnerability information to document who, where and why people are being affected by this issue;
4. Creating ICT awareness for all types of farmers at all levels nationwide, including empowering farm extension workers with relevant ICT skills;
5. Establishing an agriculture information system to provide support for the planning, production, storage, and distribution of natural resources; encouraging market research through the use of ICTs to improve farmers' decision making abilities to align supply with market demands and to access new and foreign markets;
6. Linking farmers and farmers' groups and associations through ICTs to resources and services needed to improve agricultural livelihoods;

7. Establishing linkages between agricultural education, research and development, farming, agro-industry and marketing;
8. Improving research competency and promoting the application and transfer of new technologies, such as biotechnology, to develop a modernized and globally competitive agriculture sector;
9. Promoting the creation of agricultural export production villages;
10. Improving rural infrastructure development and encouraging irrigation farming; and
11. Supporting the private sector to add value to traditional crops while strengthening the production of non-traditional export commodities (Ghana ICT4AD Policy Document, 2003).

Kenya also developed a national ICT strategy in 2006 under the Ministry of Information and Communications. This policy was based on the Common Market for Eastern and Southern Africa (COMESA) model adopted by the COMESA Council of Ministers in March 2003 (de Heer-Menlah, 2002). The vision of the strategy is 'A prosperous ICT-driven Kenyan society', while the mission is 'To improve the livelihoods of Kenyans by ensuring the availability of accessible, efficient, reliable and affordable ICT services' (Kenya ICT4D National Policy, 2006). The Ministry has included issues of gender and youth in the strategy.

Kenya's strategy focuses on issues such as market structure, implementation strategies, access, and the institutional framework of information technology and telecommunications. The objectives of the policy are to (a) improve social welfare of the population, (b) improve economic welfare of the population, and (c) improve the efficiency and quality of public service delivery.

However, other policy issues remain to be dealt with in Kenya. While the government allows competition in the mobile and internet markets, the state-owned fixed line operator has not been privatized (Matambalya and Wolf, 2006).

Malawi has an integrated socio-economic and ICT policy and plan development framework (Dzidonu, 2003). This is based on a methodology developed by the United Nations Economic Commission for African, which was in turn based on a framework from the African Information Society Initiative (AISI).

Malawi conducted a national dialog to ascertain the role of ICTs in supporting socio-economic development in the country (Dzidonu, 2003). This process indicated the importance of ICTs in the country, especially in light of its limited resources and the existing 'digital divide' between Malawi and the outside world, as well as the need to support ICT development further through policies and strategies. While they have as yet no ICT policy, Malawi has a communications policy that aims for: universal access, rural connectivity, liberalisation of the communications sector, and private sector participation in service provision and delivery as some key objectives (Dzidonu, 2003). While extension was mentioned in the document, it is not explicitly linked to ICTs.

Malawi's vision for ICT-led socio-economic development is 'to contribute to the attainment of the aspirations of the Vision 2020 by improving the quality of life of the people of Malawi by enriching their social, economic and cultural well-being through the modernization of the economy and society using Information and Communication Technologies (ICTs) as an engine for: accelerated sustainable development and economic growth; social and cultural development; national prosperity; and global competitiveness' (Dzidonu, 2003). Their mission is 'to, by the Year 2020, transform the Malawian economy into a predominantly information and knowledge-based middle-income economy by modernizing its key sectors using Information and Communication Technologies' (Dzidonu, 2003).

6. CASE STUDIES IN ICTs FOR AGRICULTURAL EXTENSION IN ETHIOPIA, GHANA, KENYA, AND MALAWI

This paper has reviewed the agriculture, extension, and ICT situation in Sub-Saharan Africa in general and in particular the four case study countries of Ethiopia, Ghana, Kenya, and Malawi. This section will move on to discuss specific cases of innovative uses of ICTs in extension, including government, private, NGO, and public-private partnership models.

6.1. Ethiopia

As was seen in section 3, Ethiopia is far behind other Sub-Saharan Africa with regard to access to ICTs. Nevertheless, the country is attempting to improve the lives of small-scale farmers with some innovative use of ICTs in extension.

Nearly half of Ethiopia's rural households are net buyers of food. Traditionally, only one quarter of Ethiopia's grain is brought to market. The Ethiopian Commodity Exchange (EXC) was developed to provide a marketplace where buyers and sellers can come together to trade and be assured of quality, delivery, and payment. The EXC (www.exc.com.et) plans to increase the sale of grain in markets in Ethiopia to one half of the total production. To do this, the exchange provides warehousing, electronic receipts, a reliable payment system, real-time market information, and quality control. The EXC has moved beyond just grains and pulses and now also deals with coffee, a major cash crop in Ethiopia.

The Ministry of Capacity Building is attempting to improve information in rural villages in Ethiopia through *woreda*-net, school-net, agri-net, and *kebele*-net. School-net is designed to be wide-area network linking all schools in the country and serving as an educational development and delivery platform supporting the teaching and learning process (Dzidonu, 2006). *Woreda*-net and kebele-net, on the other hand, promote electronic government and target support the civil services to enhance the operational and functional efficiency and effectiveness of government administration and service delivery (Dzidonu, 2006). Agri-net attempts to network the federal, regional, and local bureaus of agriculture.

Another ICT project in Ethiopia is a partnership between the regional and federal governments and the Japanese International Development Agency (JICA), called the 'Marketing of agricultural products in the Rift Valley'. This marketing project is experimenting with improving information on markets through the use of mobile phones. The collaborative project works with 'farmer research groups' (FRGs), groups of about 20 farmers who meet regularly to learn how to experiment for themselves. The groups are trained in marketing and record keeping. They elect officials and set group norms, including those on the use of mobile phones, which are leased to the groups. These phones use prepaid cards that are bought through group members' monthly contributions. Members can also use the phones for private calls and pay for the calls (Gutu, 2008). One participant mentioned that he received an additional 80 birr (about $8 US) by using the mobile phone to time the sale of his goats. Another farmer learned of a price change *via* the mobile phone and was able to make $100 through selling an ox.

6.2. Ghana

There are many examples of innovative use of ICTs in extension and agriculture in Ghana, of which five are highlighted. These include the Ghana Agricultural Information Network System (GAINS), Community Information Centres, Market Information Systems and Traders' Organizations of West Africa (MISTOWA) a community telecentre, and a special business development unit.

6.2.1. The Ghana Agricultural Information Network System (GAINS)

With support from the International Institute for Communication and Development (IICD), an Agricultural Information Centre was established at the Institute for Scientific and Technological Information (INSTI) in Accra with 10 computers. This Centre acts as the referral centre for information requests on agriculture. It also links electronically with 18 libraries and information centres of agriculture-related research institutes. Its main objective is to provide effective and efficient information delivery to 150 researchers and extension workers by giving them the opportunity to access, share and publish agricultural information on a web platform. The centre also coordinates activities of libraries and information centres of agriculture-related research institutes and faculties of agriculture of the universities in Ghana. Some of the results from the project are seen in the provision of infrastructure (computers, network, and connectivity), a website (http://www.gains.org.gh/), an on-line database, and training of researchers and staff. The project should improve the access to agricultural research information for local researchers. This published information could then be re-packaged by agricultural extension workers to reach the farmers to improve their productivity and quality of produce (http://www.iicd.org/projects/ghana-gains/).

6.2.2. Community Information Centres (CICs)

One strategy the Ghanaian Government adopted to bridge the digital divide between urban and rural dwellers is the establishment of Community Information Centres (CICs) in all 138 districts in the country. The ultimate goal is to establish the CICs in all the 230 constituencies. Equipped with television sets, videos, computers, internet facilities, photocopiers, fax, and other communication gadgets, the CICs will conveniently be termed as an upgrade of the traditional information centres, which were proven to be effective in the fight against rural poverty.

According to the then Minister of Communications, 40 of the CICs have so far been established to open the rural population to the outside world. 'As an effective tool for development, the government views and values the CICs as the most appropriate and modern multi-media technology to link the rural people to the ICT superhighway'. At the inauguration of one of the CICs at Bimbilla in Nanumba North District of the Northern Region, the Minister announced that the government used GH¢ 70,000 (USD 50,000) to build and equip the Centre, while the Bimbilla District Assembly used 50 million cedis (USD 36 million) to furnish it. The program, which falls within the government's development framework for ICT for accelerated development, aims at spreading the use of ICTs to improve the country's economic base and accelerate growth by transforming Ghana into information enabled society.

The initiative is being used to improve upon information provision to farmers by the Ministry of Food and Agriculture (MoFA). The establishment of agricultural information centres in the districts will act as a one-stop shop to provide up to date information on new trends in agriculture. Ultimately, they will be connected to the Agricultural Network (AGRINET) through a Wide Area Network (WAN). These centres are to be managed by extension officers who would try to link up with the research institutes so that they can easily transfer new technologies developed to the farmers. In the future, farmers are expected to learn how to use databases that will be created by MoFA staff to meet their information needs. Most of the centres are to be situated near markets so that on market days MoFA staff will be able to provide information to as many farmers as possible. Some of the research institutes are also linking up directly with the farmers through training courses in the various areas and at open days, fairs and exhibitions (Ghana News Agency (GNA), 16/06/08; SD, Tony Kofi Ankrah, 24/10/08).

6.2.3. Market Information Systems and Traders' Organizations of West Africa (MISTOWA)

'Before we came into partnership with MISTOWA and TradeNet, we were using word of mouth for market information, but after we were introduced to the new software we were able to make our business more profitably than before'(Dorothy Quaye, tomato trader and member of Ghana Agricultural Producers' and Traders' Organization).

This project is a network of regional market information system in Economic Community of West African States (ECOWAS). Its aim is to remove key obstacles to trade in West Africa including: a) lack of access to timely information on prices and market opportunities; b) inadequate business skills of producers and traders to respond to production and market opportunities; and c) unfavourable trading environment, including tariff and non-tariff barriers (*e.g.*, harassments at national borders). MISTOWA has developed a unique and exemplary partnership with the private sector software company, BusyLab, to develop a platform www.tradenet.biz and test new business models for effectively meeting information needs of users. The aim is to improve producer and trader skills so that they can access timely and reliable market information to discover and explore trade opportunities in the regional market. Tradenet is a customizable web platform offering online tools for the exchange of market information, designed to be used by developing-country associations of farmers and commodity traders. The platform aggregates uploaded data on market prices, buy and sell offers, and trader contact information for easier interpretation; offers online space for producer and trader organizations to create websites that can feature the above information; and can be accessed online or *via* SMS (mobile phone text message). MISTOWA complements BusyLab by strengthening capacities of farmers, traders, processors, transporters and other users, through general and targeted training to access real time information on-line or through their cellular phones (http://www.mistowa.org/en/pages/theproject.php). For more information, see Box 1.

Box 1. News Reports on MISTOWA

Buy, Sell, Hold – A plan is afoot to create a pan African market based on cellular phones

The technology revolution may be coming to poor countries via the mobile phone, not the personal computer, as it did in rich ones. And just as the internet encouraged an entrepreneurial ethos, and with it the creation of a few too many dotcom firms, Africa's surge in mobile-phone use may unleash the same sort of business energy, but tailored to local needs. One such initiative is about to begin. TradeNet, a software company based in Accra, Ghana, will unveil a simple sort of eBay for agricultural products across a dozen countries in west Africa. It lets buyers and sellers indicate what they are after and their contact information, which is sent to all relevant subscribers as an SMS text message in one of four languages. Interested parties can then reach others directly to do a deal.

The Economist, (2007), January 25 2007, http://www.economist.com/world/africa/displaystory.cfm?story_id=8597377

Text Messages Could Mean Money for African Farmers

New technologies are starting to bring market information to African farmers straight to their cell phones, and these services are improving. But challenges remain in terms of illiteracy, market reach and expanding cell phone coverage. This type of technology is being fine-tuned by a staff of West Africans and non-Africans working for a company called BusyLab in the Ghanaian capital. According to a British entrepreneur Mark Davies, these information highways in Africa have traditionally been truck routes, transport routes, camel trains through the desert, sort of single corridors of transport, and information sort of just fall off these. This changes when a user gets information on his mobile phone. "You can for the very first time in history get essentially real time information to help a farm know today what are the prices of the commodities in the market today," he added. The idea is to get that information and make it available through text messages when requested (VOA, 2007) (Nico Colombant, 10 May 2007), (http://www.voanews. com/english/archive/2007-05/2007-05-10-voa16.cfm?CFID= 150533003 & CFTOKEN =40006055).

6.2.4. Special Business Development Unit (SBDU)

One of the most frequently asked questions regarding ICTs relevance for developing countries is 'What has a farmer got to do with a computer?' This project from Ghana shows how ICTs can support small and medium farmers in increasing their revenues and improving their farming practices by enabling them to access information on regional market developments and international agricultural know-how (SBDU Project).

The main objective of this project is to efficiently promote Ghana's non-traditional export products in the regional and global market by improving contact between the farmers and their markets. The Ministry of Food and Agriculture (MoFA) has been promoting the non-traditional agricultural exports. However, at the same time, the farmers of these goods encounter major problems in marketing their produce. The farmers lack the necessary information on market developments and therefore have to sell their produce at prices far below their production cost or risk having all their products going to waste. With sponsorship from International Institute for Communication and Development (IICD), it is expected that international information on markets and prices will be made available to the local farmers, thereby allowing them to react to the market changes and also to negotiate as equals with the traders of their produce. Through this project transactions will be done electronically, eliminating excessive travels and also fraud in the industry through the elimination of middlemen.

The project will achieve this objective by providing simple to use, faster and less expensive contacts to external markets using ICTs. ICT services include: a) a website service for non-traditional export crops farmers, exporters, and their customers; b) on-line services on market and crop information, such as products, prices, and new technological developments; c) a database on production, export organizations and their products; and d) the introduction of an e-commerce application supported by training. Eventually an easy, fast and secure on-line ordering and payment system will be added to the services furnished by the Project. Technical equipment required to support these services include: Windows NT server, MS Office 97, Oracle 8.0 work group, customized e-commerce software, a file server, six PCs, and two printers. The successful implementation of this project will allow the Ministry to replicate it in other regions and among farmers. Two centres at Techiman in the Brong Ahafo region and Ga district in the Greater Accra region have been selected. The active participation of farmers was encouraged with the selection and training of farmers in the use of ICTs (http://www.iconnect-online.org/Stories/Story.import5013).

6.2.5. Eastern Corridor Agro-Market Information Centre (ECAMIC) Project

Without a good means of dissemination, the most qualitative agricultural market information is useless since it cannot get to the desired users. ICTs and Market Access are good bedfellows when the right balance is found (Shafiu Shaibu).

Farmers in isolated rural areas in Northern Ghana are often found to be unaware of the value of their crops in main markets. They find themselves in a poor negotiating position with middlemen, who routinely under-represent the final selling price and overstate transaction costs. The same holds for the price of seed, fertilizer, and other inputs, with farmers paying an inflated price for inputs. Farmers may also be unaware of strategic opportunities within their own region: which crops and commodities are fetching higher prices in surrounding towns; which products are in high demand, and so on. Getting accurate and up- to-date price information to farmers, therefore, can have a dramatic impact on their negotiating position in the agricultural economy and on their cash income.

The Eastern Corridor Agro-Market Information Centre (ECAMIC) project focuses on supporting 24 cooperative farmer communities with around 15,000 members in the Eastern Corridor of Ghana. Price information collected at the local district markets is combined with relevant other agriculture information at the ECAMIC office and distributed to district offices through e-mail. Cooperative Information Officers in these district offices use motorcycles to distribute the information to the farmer communities. The officers write the price information on a notice board and explain the other information in face-to-face meetings with community leaders. But more and more farmers are now using mobile phones to receive SMS alerts of district market prices, or they request by SMS prices of their produce from various market centres. The ECAMIC project started in 2003 and is run by the Social Enterprise Foundation of West Africa (SEND).

Currently, market information centres are established at the headquarters in Tamale, and Kpandai, Salaga, and Chamba districts; capacity of cooperative representatives leaders in basic ICT skills has been built; facilitation of procurement and distribution of 200 cellular phones to farm-families; training of 300 farm-family members in the efficient use of cellular phones to transact agro businesses; partner with Tradenet for information dissemination to farmer cooperatives through their mobile phones (http://www.iicd.org/projects/ghana-ecamic/?searchterm=None).

6.3. Kenya

Like Ghana, Kenya has several different innovative projects using ICT for extension purposes. Some of the ones that will be discussed here include the Regional Agricultural Trade Intelligence Network, some radio programs, DrumNet and 'SMS Sokoni'.

The Regional Agricultural Trade Intelligence Network (RATIN) is a regional trading board that links buyers and sellers specifically for grains and beans (see Figure 7). The main activity of RATIN is regional trade analysis of maize, beans and rice using information from various sources as shown in Figure 8 below. Once a person has graded their maize, they can post it on the web, check where there are deficits, and sell the maize online. There is a database of traders online. Transporters and their associations are also linked in. RATIN does not track figures or numbers, but it gives market prices for different countries.

RATIN is connected to the East African Grain Council (EAGC), which is a membership-based non-profit organization that which prepares, disseminates, and promotes the exchange of information on matters affecting the regional grain industry (http://www.eagc.org/).

RATIN is a collaborative effort of several United States Agency for International Development (USAID) projects. The Famine Early Warning System Network Project (FEWS NET) focuses on bringing in the crop production and trade information, and the Regional Agricultural Trade Enhancement Support Program (RATES) deals with changing trade policy to enhance regional trade in maize.

Also related to RATIN is the African Trade Commodity Link, found at www.tradeafrica.biz. Supported by USAID, it is linked to RATIN. It is an online market place for buying and selling agricultural commodities in the region. It also includes information on weather and upcoming relevant events.

Radio *Mali Shambani* brings information to farmers with tips on post-harvest handling, market linkages and animal husbandry. Similarly, the National Radio Channel KBC and the Kenyan Ministry of Agriculture and Rural Development's Agricultural information dissemination unit run a program for farmers called *Sikio la Mkulima,* or the 'farmer's ear' (Summary of the First e-Agriculture.org English Language Forum: Responding to Demand- the Focus of e-Agriculture. 2007).

DrumNet was started in 2002 to provide business services to small-scale farmers in Kenya. DrumNet is a virtual information and service network designed to link clients to markets, information and services through information kiosks and information -brokers. It is run by the non-profit micro-finance organization PRIDE AFRICA, and can be found online at http://www.drumnet.org/home.

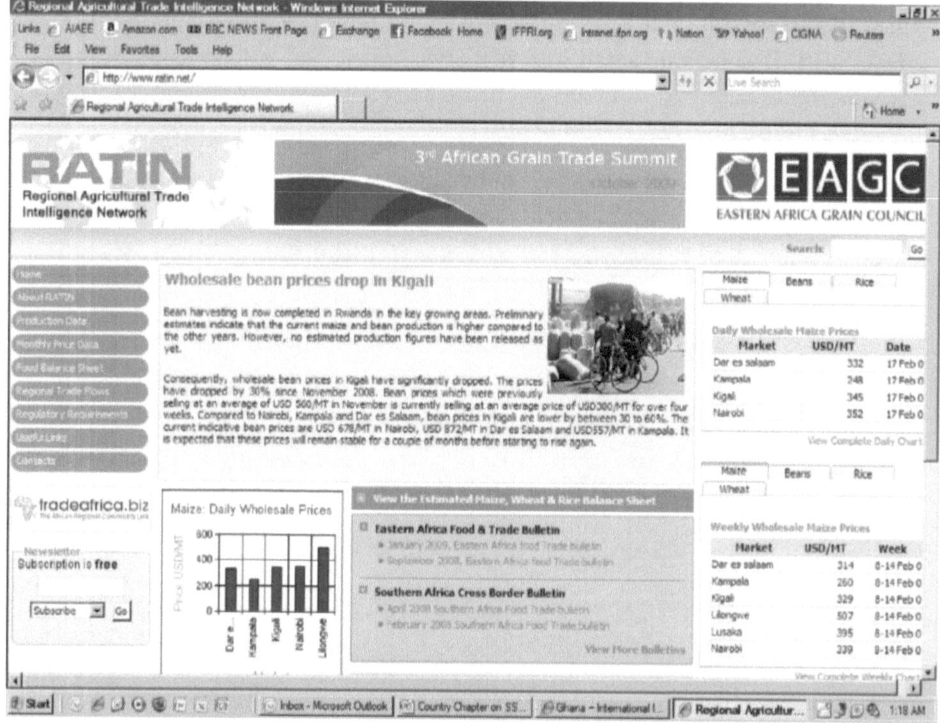

Figure 7 : The RATIN web portal, www.ratin.net

In Kenya, in order to strengthen farmers' bargaining power and increase transparency in commodity value chains, the Kenya Business Development Services Programme (KBDS) contracted the Kenya Agricultural Commodity Exchange (KACE) in 2003 through a partnership with the mobile operator Safaricom, to develop a service known as 'SMS sokoni' that gave smallholders access to daily agricultural commodity prices, extension messages and opportunities to sell or bid through text messages (KBDS, 2004). Farmers, especially the illiterate or those with problems with text messages, can also access market information through voicemail by calling a hotline telephone number. KACE has also developed a number of rural-based market information points (MIPs) which are essentially rural based trading floors linked through an electronic information system that allows farmers to link with buyers in different urban centres (KBDS, 2004, 1). The KBDS records that between May and October 2004 over 24 000 farmers had accessed commodity information on the MIPs in three districts alone. In Mozambique, (Jenson *et al.*, 2004) have shown that farmers with access to market information have seen improvements in farm prices and have increased the value of their sales. Mozambique Agricultural Marketing Service (SIMA) collects and disseminates nation-wide and provincial data on market prices, product processing and availability through a variety of media including text messages, email, internet, national and rural radios, television and newspapers.

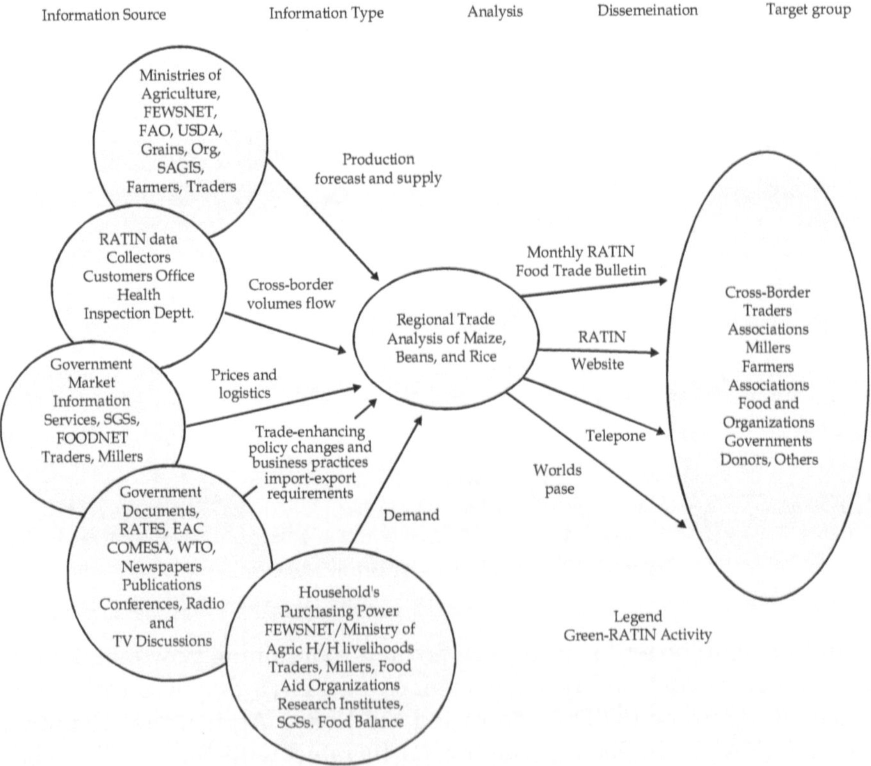

Figure 8 : Activities of the Regional Agricultural Trade Intelligence Network (RATIN)
Source: RATIN web site, www.ratin.net, accessed 6 November 2008

Box 2. Mobile Telephony for Delivery of Animal Health Services

FARM-Africa, an NGO working in Kenya in conjunction with the Kenyan government and other stakeholders, has developed a decentralized animal health care system in their Kenya Dairy Goat and Capacity Building Project (KDGCBP). To link key participants in the system, the project approached the Safaricom Corporation, the Corporate Social Responsibility arm of the mobile phone company Safaricom. The KDGCBP system works with a community animal health worker (CAHW), who purchases a veterinary drug kit and mobile phone at a subsidized price. Animal health assistants and vets working with the project also receive mobile phones. Community phones are also installed in the project at vet shops, which have solar panels and batteries where there is no electricity. The owner of the community phone is responsible for repairs and can make a profit from charging for its use; it is a means of diversifying income for private vets. The phone system allows the animal health care providers to update one another, share information, and conduct referrals. This system has reduced transactions costs and increased efficiency of animal health care in the area.

(Kithuka, *et al.* 2007)

6.4. Malawi

Finally, Malawi also has a couple of innovative programs helping farmers through ICT use in extension. One of these is the Agricultural Marketing and Information System for Malawi (MIS-Malawi). This is owned by the Malawi Agricultural Commodity Exchange, and supported by Telekom Networks Malawi, Rockefeller Foundation, and Bunda College (Gakura *et al.*, 2008). MIS-Malawi has a central hub in Lilongwe; and marketing information centres in Limbe and Mzuzu; marketing information points in four market towns; and six market resource centres. Using short messaging service (SMS), participants can get price information and buy and sell products. Some 30,000 people were using the service in 2007 (Gakura *et al.*, 2008). Buying and selling can also take place *via* a live radio programme called 'Supermarket on the Air'.

In addition, Malawi also uses radio broadcasting; the African Farm Radio Research Initiative gathers, implements, evaluates, and shares best practices with listeners (Gakura *et al.*, 2008). The Initiative partners with the World University Service of Canada, Farm Radio International, and the Bill and Melinda Gates Foundation. It is active in several countries in eastern, west, and southern Africa.

Malawi is also starting to implement 'e-governance', where 70 per cent of the government ministries will be linked electronically. The Malawian government is implementing this plan in conjunction with the World Bank. ICT communications methods will now include email, voice over internet protocol, and video conferencing ('Malawi gears up for e-governance', 31 March, 2009, http://www.digitalopportunity.org/article/view/129772/1).

7. IMPORTANT ISSUES WITH REGARD TO ICTs APPLICATION FOR AGRICULTURAL EXTENSION

There is much hype and excitement about ICTs, and some authors are rightly raising a questioning voice (Koutsouris, n.d.; Michiels and Crowder, 2001) question whether many ICT project have priorities that are external, or whether they are coming from the communities that they are meant to be helping. Koutsouris, (n.d.) questions the 'techno-utopian thesis' that social development will undoubtedly benefit from ICTs. He notes that first access issues must be addressed, as stated by Guise and Eichen, (2002) in Koutsouris, (n.d.):

- *Literacy*: Does the technology require a certain level of literacy?
- *Content*: Is the content demanded and useful?
- *Economics*: Can people afford the services in the long term?

7.1. Haves and Have Nots: The Digital Divide

An important component for using ICTs is access. This goes beyond simply having the correct infrastructure, and also includes issues such as literacy of rural people, the content of the ICT messages (is it useful?) and affordability (Koutsouris, n.d.).

According to an e-agriculture survey, respondents perceived barriers to e-agriculture, in order of importance, to be (a) restricted access, (b) cost, (c) lack of digital content in mother tongue, (d) lack of equipment, (e) lack of infrastructure, (f) unreliable technologies, and (g) lack of skills (Anonymous, 2007).

These barriers had the highest level of reporting in Africa. At a research centre with stations in Kenya and Ethiopia, for instance, the ICT manager stated: 'For an asymmetrical satellite connection of 4Mbps incoming and 1Mbps outgoing in Nairobi, [the centre] spends an extortionate USD 180,000 per year. In Europe you can have shared 10Mbps in your house for USD20 per month!' (ILRINet Discussion Forum http://ilrinet.ilri.cgiar.org/archiveDetail.php?artID=2911, March 31, 2009). This shows the high costs of internet access for an international research centre; for community-based organisations or small-scale farmer, these costs are unattainable and they must rely on slow, dial-up connections for the most part.

According to one expert, the digital divide 'has more to do with the inability of the country to deploy, harness and exploit the developmental opportunities of the emerging digital information and technological revolution to advance her socio-economic development process' (Dzidonu, 2003: 52).

7.2. Women and ICTs

A greater percentage of women than men depend on agriculture as the only source of livelihood. It is well known that in many parts of Africa, women play the leading role in the cultivation of food crops. However, as seen in section 2, women have much lower literacy rates than men. Women also make up the majority of the rural poor.

The United Nations has placed ICT as the third most important issue facing women globally (Hafkin and Taggart, 2001). Smallholders can, in certain cases, take advantage of ICTs to access information. But most rural areas lack the necessary communications infrastructure that could alleviate some of the problems of distance and access, such as mobile phone centres, internet connectivity, and market information services. Therefore, some smallholders benefit while others are further marginalized by the 'digital divide'. Other issues in ICT include cost, literacy, education, language, time, and skills.

Less than one per cent of the population (male or female) have access to the internet in many developing countries. Women make up 22 per cent of all internet users in Asia, 38 per cent in Latin America, and six per cent in the Middle East. Although no regional figures are available for Africa, in Ethiopia, Senegal, South Africa, and Uganda, women make up 14, 12, 51, and 32 per cent of users, respectively (Hafkin and Taggart, 2001).

There are many questions that are unanswered with regard to the use of ICTs by poor farmers, especially women. These include:

- What factors facilitate or constrain the uptake and utilization of ICTs by women?
- How can we best scale and out scale successes for greater impact?
- Why do farmers facing similar resource constraints have different ICT adoption rates?

- Does ICT increase access to resources or does it widen the digital divide? What factors allow the former?
- What policy options can promote networking, ICT use, and learning opportunities for women?

7.3. Capacity and Skills

Agricultural education and training programs, including vocational training, colleges and universities, that educate extension agents, researchers, and other rural development change agents do not always provide graduates with the necessary or sufficient skills to effectively work to bring about sustainable rural development. This includes the use of ICTs in instruction and as part of the curriculum.

At the community level, farmers, stockists, and other rural people are also lacking in skills for utilising ICTs. Simple courses are needed to bring people up to speed on simple ICTs that can be used in rural areas. Literacy is another important aspect of human capital that needs to be further developed.

8. IMPACT OF ICTs IN AGRICULTURE AND THE RURAL ECONOMY

Impact of ICTs on poverty, agricultural growth, and other targets of extension is difficult to show. Very little has been done to rigorously examine the effects of ICTs on rural communities, or even to document the use of ICTs by extension and in rural communities. The recent inventory of innovative farmer advisory services using ICTs by the Forum for Agricultural Research in Africa (FARA) is a welcome addition to this sparse topic. However, much more information needs to be known. While there is much excitement about how ICTs can help the poor, there are almost no studies in SSA showing clearly the link between ICTs and improvements in rural livelihoods, or between ICT growth or access and economic growth? However, at least one study has demonstrated a positive link between telecommunications infrastructure and income. The same study showed a link with telecommunications infrastructure and gross domestic product (GDP) (Torero and von Braun, 2005).

Another study in Africa looked at whether the use of ICT improved the productivity of small and medium enterprises in Kenya and Tanzania (Matambalya and Wolf, 2006). The data showed that ICT had a positive impact on total factor productivity (the total outputs that not caused by inputs). No significant relationship between ICT investment and productivity was found; however, this may reflect the effects of time-lag.

The International Food Policy Research Institute conducted a study on the potential of ICT for development and poverty worldwide, including the determinants of use of ICTs (Torero and von Braun, 2006). One of the countries examined in SSA was Ghana, where they found that age was not correlated to telephone use. However, being female was negatively correlated with telephone use. Other significant determinants of telephone use in Ghana included education, income, and distance to telephone services (Torero and von Braun, 2006). They also found that the education threshold for telephone use was quite low, since many people with no education or primary education were users of the technology.

In the Ghana case study, the data confirmed certain characteristics that increased the likelihood of use of rural household telecommunications and intensity. These included households with male heads, with at least a primary school education, who were more likely to use public telephones. Household income also had a strong positive effect. However, income source, participation in community-based organizations, household location, and size of household did not have the expected effects on telecommunications use or intensity in rural areas (Bertolini, 2006b).

While these determinants of use of ICTs give us clues into who uses ICTs and who can most benefit, these findings still fall short of demonstrating the effects of ICT use on poverty. More studies are urgently needed to fill this gap.

9. LESSONS AND EXPERIENCES FROM ICT USE IN SUB-SAHARAN AFRICA

This chapter has shown that while there is exciting potential for ICTs to help extension in Sub-Saharan Africa to better do their job, there are many constraints to its full use. These include issues such as lack of policy and governmental support, infrastructure, capacity and cost.

While ICTs offer many opportunities for extension, small-scale farmers, and other rural people to improve their lives, they are not necessarily a panacea (Torero and von Braun, 2005). For benefits to occur, countries need to have an enabling policy environment, which includes deregulation, competition among providers, targeted subsidies to address the access gap (Torero and von Braun, 2005).

As noted by (Shanmugavelan and Warnock, 2004), while competition within telecommunications worldwide *via* the WTO's Basic Telecommunications Agreement has improved services, this benefit goes almost entirely to people in urban areas, especially in Sub-Saharan Africa. While in principle African governments are committed to 'universal access' (Shanmugavelan and Warnock, 2004), in practice there is little being done in rural areas to ensure this. There are many optimistic articles about innovative means of reaching rural people through multipurpose community centres, information kiosks, text messaging, and the like, as section 5 has demonstrated. However, these small case studies and anecdotal evidence of the usefulness of ICTs for alleviating poverty need to be validated and scaled up and out. Governments should provide incentives to the private sector and to develop public-private partnerships that can provide ICT services in remote, low-income areas.

At the same time, extension organizations need to be proactive about using all means available to meet, discuss, and learn together with their clientele. Just as extension encourages farmers to diversity, extension organisations should also diversify in terms of how they communicate, and that includes the use of ICT. Extension should be pushing the use of ICTs within their own organization, among staff, and for their clientele. They can also make demands upon the government and ministries of agriculture to include ICT in policies and budgets. Extension organizations can also start to include ICT training in their curricula for staff and as refresher courses. They should also provide ICT training to their clientele where feasible.

While still focusing on tried and true methods of outreach, extension must be willing to take advantage of opportunities before them. At the same time, extension has the chance to be one of the main voices calling for attention to the plight of rural people in Africa, especially women, who cope with long distances, limited services, low income, and many other constraints.

10. THE WAY FORWARD AND RECOMMENDATIONS

For extension, policymakers, and other stakeholders concerned to improve services of rural smallholders in SSA, these parties must seek how to capture the benefits of ICTs for extension clientele: men and women, boys and girls. To do so, there are several steps that agricultural development stakeholders in SSA should take.

First of all, much more evidence is needed that shows the impact, if any, of ICTs on rural household. Not many evaluations have been conducted, especially in SSA (Michiels and Crowder, 2001). These data must take into consideration the policy environment, as well as factors such as age, gender and ethnicity. Rigorous, cross-country impact assessments are highly needed. Policymakers can push for this kind of evidence, and donors can fund studies that measure this impact.

Most ICT projects are at a young stage. This would be a good time to collect baseline data, and to consider how other countries might learn from the lessons of the ongoing projects. Policymakers and donors can require that projects conduct baseline studies, or before and after pictures of the communities, for new projects so that the impact is clearly shown.

Stakeholders concerned must be proactive about including ICT components throughout the agricultural system: from the ministers who make decisions, to managers, to field agents, to community organisations and farmers. Much capacity to fully utilise ICTs is needed at all of these levels, and again, policymakers and other planners can ensure that training on ICTs are included in curricula and in projects.

Finally, there is a need to move beyond the hype of ICTs, and to see what works where well and why. To effectively do this, in addition to the impact assessments mentioned above, there is a need for participatory consultations to understand what is useful to rural people and what is not. There are already many exciting innovations going on in rural areas with regard to ICT use, regardless of government policy and at times in spite of it. This energy in rural communities, and sound local ideas must be taken advantage of and built up to benefit more people.

The following are the concluding recommendations of this paper for agricultural development stakeholders, especially policymakers and extension actors, based on the literature review and experience of the authors:

1. Collect gender-disaggregated evidence on ease of access to ICTs in rural areas of SSA.
2. Provide evidence on how affordable ICT access is to different groups, and how it can best meet the needs of men, women, youth, and pastoralists.
3. Find out if poor people really use ICTs. If so, what are some of the 'best practices' to enable this to happen across the continent.
4. Show clearly what the real impact of ICTs and ICT programs is on people's lives.
5. Diversify in communication methods; use traditional communication channels as well as ICTs.
6. Provide capacity strengthening on ICT use at all levels of the agricultural system.
7. Learn more from the experiences of other countries, of NGOs, and community organisations on the use of ICTs for rural development and how they can benefit rural people.

8. Make clear policies on ICT in all sectors, especially agriculture.
9. Invest much more in ICTs in Sub-Saharan Africa, paying attention to infrastructure, supportive policies, access and capacity.

Acknowledgements

Special thanks to Dr. Cosmas Milton Obote Ochieng for his intellectual contributions, and to Daniel Ayalew for his help with the data.

References

Addom, B. K., 2004. e-Readiness Assessment of Seven Higher Education Institutions in Ghana. *Unpublished Master's Thesis,* Cornell University, Ithaca, New York.

Addom, B. K., 2006. Universities as Alternate Incubators of Information Communication Centres for Ghana's Development, A Book Chapter in: *Meeting the Information Challenge: The Experience of Africa.* Newcastle, UK: Cambridge Scholars Press.

Agricultural Sector Coordination Unit, 2006. National Agricultural Sector Extension Policy Implementation Framework. Nairobi, Kenya. Ministry of Agriculture, Ministry of Livestock and Fisheries Development, and Ministry of Cooperative Development and Marketing.

Amezah, K. and Dorman, E., 2004. Towards Effective Agricultural Extension: Experiences from Ghana. In: *Proceedings of the First Networking Symposium on Innovations in Agricultural Advisory Services in Sub-Saharan Africa,.* 11-14 October 2004, Kampala, Uganda.

Annor-Frempong, F., Kwarteng, J., Agunga, R., and Zinnah, M. M., 2005. Challenges and Prospects for Infusing Information Communication Technologies (ICTs) in Extension for Agriculture and Rural Development in Ghana. Association for International Agricultural and Extension Education (AIAEE) Conference *Proceedings from the 22nd AIAEE Annual Conference,* Clearwater Beach, Florida, Pp: 36-46.

Anonymous, 2007. Analysis of Global e-agriculture Survey. World Summit on the Information Society e-Agriculture Working Group. Geneva: International Telecommunication Union.

Badiane, O. and Shively, G., 1998. Spatial Integration, Transport Costs, and the Response on Local Prices to Policy Changes in Ghana. *Journal of Development Economics,* Vol. 56. Pp: 411-431.

Bertolini, R., 2006a. Institutional Trends and Infrastructural Developments in Telecommunications in Sub-Saharan Africa: The Case of Ghana. In: Torero, M. and von Braun, J., *Information and Communication Technologies for Development and Poverty Reduction: The Potential of Telecommunications.* Baltimore, Maryland: The Johns Hopkins University Press.

Bertolini, R., 2006b. The Effects of Public Telephone Services: The Case of Ghana. In: Torero, M. and von Braun, J., *Information and Communication Technologies for*

Development and Poverty Reduction: The Potential of Telecommunications. Baltimore, Maryland: The Johns Hopkins University Press.

Bruns, B. *et al.*, 1996. Village Telephones: Socio- economic Impact and Implications for Rural Futures. Paper Presented at 6th International Conference on Thai Studies, October.

CIA, 2007. The World Factbook - Ghana. Central Intelligence Agency, Retrieved March 2009, from https://www.cia.gov/library/publications/the-world-factbook/geos/gh.html.

CIA, 2008. The World Factbook - Ghana. Central Intelligence Agency. Retrieved March 2009, from https://www.cia.gov/library/publications/the-world-factbook/geos/gh.html.

Chinsinga, B. and Cabral, L., 2008. Malawi's Agriculture Ministry: Fit for Purpose?' Future Agricultures Briefing.

Coyle, D., 2005. Overview, In Vodafone. 'Africa: The Impact of Mobile Phones' Vodafone Policy Paper Series No. 2.

Crowe, T., Harris-White, B., and Palaskas, T. B., 1997. The Evolution of Local Market Commodity Price Behavior in South India 1972-92. *Journal of International Development,* 9(1) Pp: 101-115.

De Heer- Menlah, 2002. Internet access for African Countries. Ubiquity: A Web-based publication of the ACM, Volume 3, Number 41, Week of November 25, 2002Accessed 7 November 2008, from http://www.acm.org/ubiquity/views/f_kofi_1.html.

Delgado, C., 1986, 'A Variance Components Approach to Food Grain Market Integration in Northern Nigeria', *American Journal of Agricultural Economics,* 68 (4). Pp: 970- 979.

Du Boff, R., 1980. Business Demand and the Development of the Telegraph in the United States, 1844 - 1860. *Business History Review,* 54. Pp: 459-479.

Dzidonu, C., 2002. Empirical study of the development, deployment and the exploitation of ICTs in Ghana. IDRC Final Report on Scan - ICT - Phase II.

Dzidonu, C., 2003. An Integrated Socio-Economic and ICT Policy and Plan Development Framework for Malawi. United Nations Economic Commission for Africa (UNECA) and United Nations Development Programme (UNDP).

Dzidonu, C., 2006. The National ICT for Development (ICT4D) Five Years Action Plan for Ethiopia [2006 – 2010]. Addis Ababa: The Federal Democratic Republic of Ethiopia, Ministry of Capacity Building.

Eggleston, K., Jensen, R., and Zeckhauser, R., 2002. Information and Communication Technologies, Markets and Economic Development. *Working Paper*, Tufts University, Department of Economics.

Ethiopia Strategy Support Project, 2005. Ethiopian Rural Smallholder Survey', Addis Ababa: Ethiopia Strategy Support Project, Ethiopian Development Research Institute, and Central Statistical Agency.

Fafchamps, M., and Gavian,, 1997. The Determinants of Livestock Prices in Niger. *Journal of African Economies*, 6(2) Pp: 255-295.

Forestier, E., Jeremy, G., and Charles, K., 2002. Can Information and Communication Technologies be Pro-poor?' *Telecommunications Policy,* 26 Pp: 623-646.

Gadzekpo, A., 2005. Ghana draft audit report. Accra: World Association of Community Radio Broadcasters.

Gakura, M., Winters, K., and Stepman, F., 2008. Inventory of Innovative Farmer Advisory Services Using ICTs. Accra, Ghana: The Forum for Agricultural Research in Africa.

Garbade, K. and William, S., 1978. Technology, Communication and the Performance of Financial Markets: 1840-1975. *Journal of Finance,* 33(3) Pp: 819 – 831.

Geertz, C., 1978. The Bazaar Economy: Information and Search in Peasant Marketing. *American Economic Review,* 68(2) Pp: 28-32.

Ghana ICT4AD Policy Document, 2003. Ministry of Communications, Government of Ghana.

Guise, J., and Eichen, K., 2002. Information Technology for Development: From Charity to Sustainability. *Development*, 45(4) Pp: 29 – 34.

Gutu, T., 2008. Mobile phones are just more than phones in the hands of FRG farmers. *FRG Update*, 8 May 2008.

Hafkin, N. and Taggart, N., 2001. *Gender, information technology, and developing countries: An analytical study*. Washington, D.C.: Academy for Educational Development and the Office of Women in Development, Bureau for Global Programs, Field Support and Research, United States Agency for International Development.

Hudson, H., 1995. The Economic and Social Benefits of Rural Telecommunications. Unpublished report to the World Bank.

Institute of Development Studies, 2007. Mobile Phones and Development: The Future in New Hands?' *id21 insights 69*, Brighton, UK, Institute of Development Studies.

InterMedia, 2005. *Ghana: Media and Opinion Survey Data for Developing Countries.* Washington DC: InterMedia.

Jenson, M., Myers, M., and Southwood, R., 2004. The Impact of ICT in Africa. Paper Prepared for the Commission for Africa.

Kafewo, S., 2006. African Media Development Initiative Research: Ghana Case Study. African Media Development Initiative Research Project, BBC World Service Trust. Online: www.bbcworldservicetrust.org/amdi.

KAM (Knowledge Assessment Methodology), 2008. The World Bank Knowledge for Development (K4D) website: Knowledge Assessment Methodology. http://go.worldbank.org/JGAO5XE940, accessed 24 March 2008.

KBDS, 2004. *Monthly Newsletter*. November, Kenyan Business Development Service Program.

Kenya ICT4D National Policy, 2006. Kenya ICT4D National Policy. 1st January 2006, Online: http://www.comminit.com/en/node/148320.

Kithuka, J., Mutemi, J., and Mohamed, A. H., 2007. Keeping up with Technology: The use of mobile telephony in delivering community-based decentralised animal health services in Mwingi and Kitui Districts, Kenya'. FARM-Africa Working Paper 10. London: FARM-Africa.

Koutsouris,A., (n.d.), 'ICTs and Rural Development: Beyond the Hype', Unpublished manuscript, Agricultural University of Athens, Athens, Greece.

Matemba, S. and Charman, A. (n.d.). Agriculture Service Provision by the Private Sector in Malawi: The Case of the Kasungu Smallholder Virginia Project. GTZ Agriculture Extension Support Project.

Matambalya, F. and Wolf, S., 2006. Does the Use of ICT Improve the Productivity of SMEs in East Africa? The Case of Kenya and Tanzania. In: Torero, M. and von Braun, J., *Information and Communication Technologies for Development and Poverty Reduction: The Potential of Telecommunications*. Baltimore, Maryland: The Johns Hopkins University Press.

McKemey *et al.*, 2003. Innovative Demand Models for Telecommunications Services', DFID. www.teleafrica.org.

Michiels, S. and L. V. Crowder., 2001. *Discovering the 'Magic Box': Local appropriation of information and communication technologies (ICTs)*. Food and Agriculture Organization (FAO) of the United Nations SD Dimensions. Rome: FAO.

Ministry of Agriculture and Rural Development, 2007. *RCBP (Rural Capacity Building Project). Draft Project Implementation Manual*. Addis Ababa: The Ministry of Agriculture and Rural Development.

Munthali, B., 2004. Agricultural advisory services: Lessons from Malawi. In: *Proceedings of the First Networking Symposium on Innovations in Agricultural Advisory Services in Sub-Saharan Africa*. 11-14 October 2004, Kampala, Uganda.

Ng'eno, J. K., 2004. Agricultural advisory services in Kenya. In: *Proceedings of the First Networking Symposium on Innovations in Agricultural Advisory Services in Sub-Saharan Africa*. 11-14 October 2004, Kampala, Uganda.

Peters, T., 2003. Bridging the Digital Divide. US Department of State. The Evolving Internet. Global Issues, Online: http://usinfo.state.gov/journals/itgic/1103/ijge/gj08.htm.

Ravallion, M., 1986. Testing Market Integration. *American Journal of Agricultural Economics*, 68 (1): 102-109.

Rose, J. B., 1999. Multipurpose Community Telecentres in Support of People-centred Development. In: *Information Technology and Globalisation: Implications for People-centred Development*, London: SFI Publishing.

Scott, Nigel *et al.*, 2004. The Impact of Mobile Phones in Africa', Paper Prepared for the Commission for Africa.

Shanmugavelan, M, and Warnock, K., 2004. *Completing the Revolution: The Challenge of Rural Telephony in Africa.* London: The Panos Institute.

Sraku-Lartey, M. and Sam, J., (n.d), iNARS: A Framework for Developing and Implementing Effective Information Systems for Agricultural Research and Development Forestry Research Institute of Ghana.

Stigler, G. J., 1961. The Economics of Information. *Journal of Political Economy*. 69(3). Pp: 213-225.

Torero, M. and Chowdhury. S., 2005. Increasing access to infrastructure for Africa's rural poor. *2020 Africa Conference Brief 16*. Washington, DC: International Food Policy Research Institute.

Torero, M. and von Braun, J., 2005. ICTs: Information and Communication Technologies for the Poor. *International Food Policy Research Institute (IFPRI) Brief.* Washington, DC: International Food Policy Research Institute.

Torero, M. and von Braun, J., 2006. Institutional and Public Policy Aspects of ICT Infrastructure Provision. In: Torero, M. and von Braun, J., *Information and Communication Technologies for Development and Poverty Reduction: The Potential of Telecommunications*. Baltimore, Maryland: The Johns Hopkins University Press.

UNESCO, 2007. *Correspondence on Adult and Youth Literacy Rates*. May. Montreal: (United Nations Educational, Scientific and Cultural Organization) Institute for Statistics (UNESCO).

VOA, 2007. Text Messages Could Mean Money for African Farmers, by Nico Colombant, 10 May 2007. Available: http://www.voanews.com/english/archive/2007-05/2007-05-10-voa16.cfm?CFID=150533003&CFTOKEN=40006055.

Von Braun, J., Akhter, A. and, Abdul, B., 1999. *Village Pay Phones and Poverty Reduction: Insights from a Grameen Bank Initiative in Bangladesh.* Berlin: ZEF.

Waverman, L., Meschi, M., and Fuss, M., 2005. The impact of Telecoms on Economic Growth in Developing Countries. In: Vodafone, 'Africa: The Impact of Mobile Phones' Vodafone Policy Paper Series No. 2.

World Bank, 2008a. ICT at a Glance. Washington, DC: The World Bank.

World Bank, 2008b. *World Development Indicators 2008*. Washington, DC: The World Bank.

Zijp, W., 1994. Improving the Transfer and Use of Agricultural Information. A guide to Information Technology. *World Bank Discussion Papers 247*. Washington DC: The World Bank.

Authors Information

Dr. Kristin Elizabeth Davis

Knowledge, Capacity, and Innovation Division
International Food Policy Research Institute
PO Box 5689, Addis Ababa, ETHIOPIA
Phone No. (O): 251 11 617 2506
e-mail: k.davis@cgiar.org
URL: www.ifpri.org
Personal Home page URL: http://www.ifpri.org/srstaff/DavisK.asp

Dr. Kristin Elizabeth Davis is a Research Fellow with the International Food Policy Research Institute (IFPRI), based in Ethiopia. Since receiving a Ph.D. in international agricultural extension five years ago, she has conducted research on agricultural extension with a focus on Sub-Saharan Africa. Dr. Davis has published numerous peer-reviewed articles on her research, including two "journal article of the year" articles in different journals. She is the bronze prize winner for the article 'ICTs as appropriate technologies for African development' for the International Finance Corporation-Financial Times Annual Essay Competition with co-author Cosmas Milton Obote Ochieng.

Mr. Benjamin K Addom

School of Information Studies (iSchool)
Syracuse University
126 Remington Avenue, Apt. B, Syracuse NY 13210, USA
Phone No. (R): 315-299-3575
Mobile Phone No.: 315-729-5721
e-mail: bkaddom@gmail.com, tordeben@yahoo.com
URL: http://ischool.syr.edu/
Personal Home page URL: http://ischool.syr.edu/facstaff/member.aspx?id=638

Mr. Benjamin K Addom is a Ph.D. candidate at the Syracuse University School of Information Studies (iSchool) and currently exploring the role of ICTs in the utilization of farmers' local knowledge in agricultural research. With key interest in ICTs and development, Benjamin has participated in a number of international conferences in the past five years including the Oxford Internet Institute. He also has a book chapter, a peer reviewed journal article, and a number of conference papers.

❏❏❏

Chapter 14
SUDAN

Rafaa Ashamallah Ghobrial

1. INTRODUCTION

Sudan lays close to the east-centre of Africa between Latitudes 4° and 2° and longitude 22° and 28° east. It is the largest country in African continent (8.3 per cent) and Arab world and the ninth largest in the world (1.75) covering 2.59 million square kilometers of twenty six states (El Mustafa, 2005). It shares borders with Egypt, Libya, Chad, Central Africa, Zaire, Uganda, Kenya and Ethiopia. It is a criterion for absolute continent in terms of the extent and variety of vast agricultural lands, vast agricultural and pastoral country with rich cultural diversity. It homes about forty million people, seventy five per cent of which inhabit rural areas, speaking over a hundred and thirty indigenous languages and dialects. The country is a gentle sloping land with the exception of Jebel Marra Massif, Red Sea hills, Ingessena hills and the Nuba Mountains (Ahmed *et al.,* 2004).

The Nile and its tributaries traverse it for about 900 km. The rainfall ranges from less than 50 mm in the northern desert to more than 1500 mm in the tropical rain forest in the south.

In general, the soils include aridsols (sand dunes and sandy soils), aridisols and vertisols (clay pockets in the valleys) in the North, vertisols in the central and southern clay plains and oxisols in the south. Variations in rainfall, vegetation and soils resulted in the formation of seven main ecological zones classified as follows: desert (0-75 mm, 28.1 per cent,), semi-desert (75-300 mm, 19 per cent), low rainfall savanna on sand (300-400 mm, 26.6 per cent), low rainfall savanna on clay (400-800 mm, 13.3 per cent), high rainfall savanna (800-1300 mm, 9.5 per cent), Montana vegetation region (800-1000 mm, 0.3 per cent) and flood region (800-1000 mm, 3.2 per cent) which embedded in twenty six states.

The Sudan is affected by natural or induced land degradation processes leading to desertification: degradation of vegetation cover, water erosion, wind erosion and reduction in soil organic matter, soil crusting and compaction. The first four are determinative and the remaining two are subordinate. The area affected by desertification lies between latitude of 10° and 18° N and traverses the country from the eastern to the western borders. It also includes a narrow strip of land along the river Nile stretching northward to the Egyptian border between Longitude 30° and 32° E. Thirteen States are affected by desertification, namely, Northern, River Nile, Red Sea, Northern Jordan, Northern Darfur, Western Kordfan, Western Darfur, Kassala, Gedarif, Khartoum, White Nile, Gezira, and Sinnar. Desertification ranges from very severe in the northern fringe of the semi-desert ecological zone to moderate in the southern fringe of the low rainfall savanna (Ahmed *et al.,* 2004).

Although the estimated arable land is about 200 million feddan (mnf) (1 feddan = 0.42 ha), the land actually under crops is much less. The total land under rainfed cultivation is about 22.5 mnf, of which 6.1 are mechanized and 16.4 are traditional. Crop production in this sub-sector is mainly constrained by limited and irregular rainfall and recurrent drought spells. The potentially irrigable lands are about 11.8 mnf. However, the area of the present irrigated schemes is about 4.3 mnf. Among several government and private irrigated schemes, the Gezira scheme is the largest (2.1 mnf). Most of the agricultural activity lies in the arid and semi-arid zones. The total output of cereals averaged about 3.6 million tons. The contribution of the irrigated, mechanized and traditional rain-fed sectors amounted to 33.8, 44.9 and 21.2 per cent, respectively. It is evident that two thirds of this output is affected by desertification, low and erratic rainfall.

It is estimated that the annual potential available water is 143 km^3, of which 120 km^3 are contributed by rainfall in regions receiving 300 mm or more and 15, 1 and 7 km^3 are contributed by the Nile, seasonal valleys (wadis) and underground waters, respectively. The water actually used amounts to 42.2 km^3 only. The Nile, precipitation, valleys and underground waters contribute 15, 26, 0.2 and 1 km^3, respectively.

Forest resources are found in the savanna and montana ecological zones. They cover about 224 mnf of which 89 mnf are in the arid and semiarid lands and 135 mnf in the high rainfall savanna and Montana zones. The direct contribution of the forest resources to GDP is about 10 per cent. Furthermore, Sudan is the main producer of gum Arabic (80 per cent of world production), which provides about 11.9 per cent of the foreign exchange earnings (http://www.nationsencyclopedia.com/Africa/Sudan-TOPOGRAPHY.html).

The livestock wealth of the country is predominantly owned by the traditional nomadic sub-sector, which represents about 10 per cent of the total population, and livestock contributes about 8 per cent of GDP. The estimated numbers of cattle, sheep, goats, and camels are 112 mn, respectively, with a high growth rate ranging between 2.8 and 6.1 per cent. At present the stocking rate exceeds the carrying capacity of the available seasonal grazing land.

In general, Sudan's vastness and its varied ecological zones resulted in biotic diversity unsurpassed in many African countries. It was estimated that some 226, 106, 938 and 1000 species of mammals, fish, birds and insects respectively, exist in Sudan (http://www.cbd.int/doc/world/sd/sd-nbsap-01-p1-en.pdf).

In the arid and semi-arid lands, 80 per cent of the population depends on biomass as a major source of energy. The environmentally friendly source of energy is expensive and not always available. The following conditions were considered in the formulation of the national plans for combating desertification. The arid and semi-arid lands including the desert, cover an area of 1.78 millions km^2, *i.e.*, approximately 72 per cent of the total area of the country. All this area is subjected to various degrees to one or more of the desertification processes. This fact prompted the elaboration of national plan State-wise *i.e.* a plan for each affected State. Sudan was hit by severe drought spells during the last decades which may recur and accentuate desertification. It is a severe constraint in crop production in the rain-fed sub-sector, and requires an early warning system and preparedness. Each spell is accompanied by mass displacement of people from the affected areas (Abdelkarim *et al.*, 2004).

Most of the population of the affected States is poor and relies heavily on the natural resources for subsistence (cultivation of marginal lands, tree cutting for fuel and construction of huts, overgrazing *etc.*). Poverty leads to subsistence livelihood, which enhances land degradation, and hence desertification, which aggravates poverty. This vicious circle may only be broken if programs for combating desertification are integrated with strategies for poverty alleviation. The conflicts at some parts of the Sudan are a heavy burden on the nation. It created very dramatic socio-economic conditions and caused internal and regional migration and displacement of people from the war and conflict affected areas. They have limited poverty alleviation programs, thus accentuating desertification. Sudan has a relatively weak technological base and insufficient scientific, technical and educational capacity leading to substantial capacity building requirements. Sudan implemented several development projects since it was envisaged that use should be made of the existing international interventions in different sectors.

The ultimate aim of the agricultural strategy is the conservation of environment and natural resources for the benefit of the present and future generations. On this base, the appropriate activities for the various sectors were outlined, namely soil conservation, agriculture, animal resources, forestry, and range and pasture. These strategies are sound taking into consideration the holistic, integrative, multi-sectoral and participative approach.

Sudan carries its own unique resources and problems such as the limited infrastructure and the low level of economic activity. All these, individually and in combination, have a great influence on knowledge needs of people involved in diverse ways of development activities. It can constrain or support knowledge generation, capture, processing and dissemination as well as the development of the related knowledge infrastructure and policies. But there is now also a substantial body of general policy support on accelerating the use of Information and Communication Technologies (ICTs). As has been observed in many studies, for lofty dreams to be actualized there is a need for planning. There is also the need to invest a significant proportion of national wealth to the generation, management and utilization of quality information for development. Since the past centuries we perceive information as mankind's accumulated knowledge, derived from all subjects that could help its users to reduce their levels of uncertainty. Sudan has had a catalogue of long and short-term development plans, the failure of which is largely attributable to lack of proper knowledge management and utilization. This chapter attempts to describe which currently exists in agricultural information sector, identify its main determinants, give short account on the limitations or constraints which determine the agricultural dissemination and extension services and clarify what is necessity to make strategic plan as reality and benefits of affordable ICT opportunities in the country.

2. INFORMATION AND KNOWLEDGE SYSTEMS

An Information system comprises a combination of various components including facilities for collection of data, its storage and retrieval and dissemination to users and their functions which require institutional infrastructure and arrangements for collecting materials from national and international sources, storing, processing, retrieval and preparation of information products and provision of services to meet users of the system. The Sudanese information Systems show clearly that the existing organizations and their resources are inadequate for the country's needs of research and development, education and extension services. Moreover, information systems exist in isolation. This isolation has certain weaknesses, such as neglect of certain areas of knowledge, fragmentation of resources, inadequacy of the services, and costly duplication of material. We noticed that the uncoordinated state of the information systems, which exist today, are not linked with any kind of administrative structure; and examples of wasteful duplication of effect and overlapping responsibilities are numerous. What seems to be lacking is the timely availability of a vital input not only to good planning and decision-making but to the development of a positive response and change and, indeed, to research and practical know-how. Without such perception it is not surprising that information systems generally receive low priority. The assessment of the current situation of information and knowledge systems in the country do not have able viable programmes in the absence of the effective collaborative structures with diverse means and number of shared barriers.

2.1. Agricultural Extension Systems

The broad concept of agricultural extension as an informal way of teaching agriculture is not new; in ancient times, farmers learned from their own experiences and the experience of the previous generations. They passed on their farming knowledge to the offspring. It was an informal way of education, it was not organized, and it had very little resemblance to what now has become conventionally recognized as agricultural extension service. The history of agricultural extension and agricultural technology transfer started in Sudan at the time of establishing the research centers in 1902, with the establishment of experimental cotton farms in Shendi (120 km north of Khartoum) on the main Nile and in Alkamleen (60 km south of Khartoum) on the Blue Nile. In 1904 the department of agriculture established Shambat Research Farm where some botanic and agronomic work was carried out (Edris, 1975). Agricultural extension systems have been developed as part of the irrigation and rain-fed schemes in the country to cover all agricultural extension services which is currently under supervision of Agricultural Technology Transfer and Extension Administration (ATTEA) of Ministry of Agriculture and Forestry; also supported by individual, private, charity and non governmental institutions whose main functions are to develop extension packages in agricultural works, to improve crop productivity and varieties, and to increase the income of the farmers through best means of utilization. Increasing productivity through improving farmers' knowledge by different kinds of media are basic goals of agricultural extension in Sudan (Anas *et al.*, 1991). A number of strategies which strengthened the relation between agricultural extension and research institutions at the national and international research levels have been adopted to extend the opportunities, and disseminate the results of the agriculture researches to be used in the agriculture extension programs. There are necessary channels for the technical information between the agriculture research centers and farmers through the extension agents, and feedback from farmers to the researchers to find suitable solutions. Moreover, all the facilities are devoted for the preparation of the administrators and technicians involved in the technology transfer from the research centers to the farmers with the methods that the farmers accept and understand. Agricultural extension activities have given to Sudan government benefits from the International community, non-governments or foreign countries, through the creation of favorable conditions in relation with the African and Arabs Agricultural extension institutions. Recently, it manages to introduce electronic information systems and networks such as VERCON to link technology transfer sources and different categories of users at different states of the Sudan. It proposes database of agricultural technology transfer and extension during the last years, which are formulated to establish local network area to link different ATTEA departments and be ready in collection, process and allocation of information on internet web page with assistance of local experts and offer proper trainings; establish Intranet at Institutional and State Levels; it attempts to introduce databases in addition to the application of expert systems by evaluation , reporting and encouraging intensive training. The various important issues related to extension services continuously stimulate discussion and raise the awareness of readers / audience *via* different mass media sectors (Source: Agricultural Technology Transfer and Extension Administration of Ministry of Agriculture and Forestry, Khartoum North, Sudan, September 2008).

2.2. Geo-Information Systems

The introduction of Geo-information (remote sensing and geographical information) in Sudan is backdated to early 1980s. Geo-information is accessible in form Landsat data, rainfall estimates (Meteosat), Sudan land cover database, vegetation index, drainage network, basins and water points. There are currently more than 45 institutions from government and private sector working in the field of geo-information with different capabilities. Sustainability development is a major goal in Sudan government strategy. In order to achieve such a goal, a program for human and institutional capacity development was structured to focus on remote sensing and geographic information as the best tools that can handle geo-information which are mainly invested and produced by UN-organizations and non-governmental organizations. In addition to that, Sudan has adopted advanced technologies, which use to enhance the availability of environmental data (acquisition, analysis and output) to run a number of national projects with remarkable wide spectrum training in geo-informatics (Source: Dr. Amana Hamid, Remote Sensing Authority Director, 2007). The National Information Centre "NIC" of Ministry of Council of Ministers has initiated Technical committee for Geographical Information which is composed of key representatives of institutions and professionals and the clients. The context of this Committee is the motivation and coordination among these institutions to be effective in achieving aims and have all benefited from their involvement and outputs as part of the achievements of NIC reported in Arabic, http:// www.nicsudan.gov.sd.

2.3. Documentation and Library Systems

These information institutions are embodied mainly in governmental bodies, non-governmental international agencies working in Sudan and private institutions. Government Libraries have long and proud history in the country as information profession. Each institution manages its own Library Information System (LIS) focusing on the subject area and information needs of the parent body, but there is no extensive co-operation between library and information services in areas of job opportunities, career development, training and co-operative procurement. The university/ research libraries and documentation information centres are in the forefront regarding the use of ICT for knowledge management. These institutions also experience shortage of qualified personnel; and they are run by junior staff. Academic libraries are the central organ of any academic institution and educational instrument. They are not mere storehouse of books attached to a reading room as well as a dynamic instrument of education. They act as intellectual and cultural houses for student, teaching staff, of university and outsiders. They are the local repositories and the principal gateways to current information and the scholarly records for the current and future students and faculty of the University, while also serving other Sudanese community sectors (Ghobrial and Hamad, 2007). They are mainly located in governmental universities (university library /Faculty) and private universities and colleges, *i.e.* single library per university/college. Some of them have started to acquire electronic resources and introduce technology as single efforts but the Ministry of Higher Education and Scientific Research approaches virtual library to provide academic staff and students the necessary information for teaching, learning and scientific research. The draft library Sudanese universities default SUVL part of a project of Sudanese Universities Information Network (SUIN), aims to support educational activities and research and promotion methods, to gain knowledge and the development of management systems and modernization of the university by providing books, journals and electronic teaching materials, research and management information (direct and indirect) by using tools based on information and communication technologies (http://www.suvl.edu.sd).

The Central Library of the Agricultural Research Corporation (ARC) provides specialized agricultural information services to its users and these include access to resources. The library has access to the ARC Intranet and provides access to in-house bibliographic databases developed using FAO-WebAGRIS tools and methodologies and other databases in WINISIS (Arabic version) through this infrastructure. The library is also working towards digitizing the entire ARC research programs output and will have an institutional repository in place (Musa, 2005). The ARC has adopted a system whereby new ARC publications are deposited at the Central Library in both hardcopy and electronic formats. Meanwhile, back issues of the documents are being scanned and converted to electronic formats, mainly to HTML and PDF. The library works with other partners to disseminate agricultural information produced in the Sudan at national, regional and international levels. It is actively working with members of regional and international networks and collaborates with organizations.

Although agricultural libraries and public awareness centres have access to the Internet, on the whole, the use of Web-based facilities to provide access to agricultural information resources is extremely low. Most websites that have been developed only contain basic information (brochure type) and do not link or provide access to the information resources found in the libraries/ information centres or digital collections that have been developed by some libraries. The ARC website (http://www.arcsudan.sd), developed by the Head librarian, Central Library, under the supervision of the Director, Documentation and Information, with the special support of ARC Director General, provides information about the ARC activities and research programs. Plans are underway to provide access to various value- added information services and resources through the website.

Documentation and Information Centre -National Centre for Research (http://www.ncr.sd) is a national Documentation Centre for Science and Technology and is designated as one of the institutions that serve as national depository of scientific and technological knowledge which can play important role in disseminating information and organization activities and creating user awareness and understanding important national concerns based on modern lines. It runs the following activities:

- Collecting, processing and disseminating information on published research findings, dissertations, reports, studies and conference papers as well as on current/foreign located sources relevant to Sudan, documents which exist in single copies available in other institutions, are microfiched and added to Sudan collection.
- Registry of Current Research among the various functions expected of a National Documentation or Information system, the collection of data on Research and Development (R & D) activities and maintenance of a Current Research Registry are paramount. A register of current research - effectively indexed, is not only an important tool with which to locate individual expertise in specific fields, but also to provide information on the trend of investment in Research and Development and other data

needed for the formulation of general scientific policy. Maintenance of such a register of current research, whether in governmental, academic or other institutions, also provides an efficient information service on the equipment and technical facilities available in the country. Initially the register is built manually and the data is converted to machine–readable form as envisaged by the United Nations International Scientific Information System (UNISIST) guidelines on the subject. Another database gives information on National Centre for Research (NCR), research projects in all subject fields.

- The Centre publishes Sudan Science Abstracts "English and Arabic" which is produced from the national bibliographic database, national in its scope; it is then created and maintained, hosted over twenty eleven thousand records related to science and technology. It comes out biannual, of which thirty- two volumes. It abstracts and indexes reports, theses, dissertations, conference papers, published research findings and journal articles relating to Sudan whether generated in or outside Sudan.
- Databases relating to union lists of serials are designed and maintained
- Linkage with international and regional information networks.
- Information Services offered from the library resources, are the existed national databases and international databases on CD-ROM, referral services to other sources that is outside Sudan. The Centre receives bibliographic database as well as access internet services and portals as with the support from outside the country.
- Contributing towards development of a national information system, activities in this regard include carrying out surveys of specialized libraries, information Centres and databases, holding conferences, conducting training courses.

3. AGRICULTURAL CLIENTELE AND STAKEHOLDERS

Adedibu and Adio (1997) stated that studies of information needs and information seeking have provided the clients with basic knowledge for various factors that operate when people need information. There is a bulk of documents that were produced as the result of project related activities. These include work plans, terms of references, office reports, progress reports, project proposals, concept papers, inception reports and or descriptions, final reports, and similar documents involved with projects, project activities, and daily operation. Other documents contained information about institutions in Sudan related to particular topic(s) or sector(s). There were several on procedures for data analysis specific to special software and a few on the application of statistical procedures and data gathering guidelines. Some are basically outdated in no small part because of the very rapid advancement of technology and information sector. They do represent the types and kinds of papers that need to be completed by Sudanese professionals on a regular basis and serve as useful examples. One characteristic of much of the content of the furnished documents is that, in almost, all is more or less uptodate information. Time is an important element in all early warning data, much of the marketing information, some of the environmental information and for others noted as priorities by institution. The useful time span of such data can be measured in hours or days and are, by their very nature, quickly relegated to the archive as a snapshot in time, useful primarily for historical prospective and analysis.

Historical analysis is quite important for determining approaches that work or don't work, what trends are, and how to predict, mitigate, or even avoid serious problems and situations in the future. One would expect that their national role involved in such analysis would increase understanding and awareness of national problems. If this is not a Government function then there is little need for this type of information within the governmental coordinating body.

Others contain useful information on the nature, state of, and characteristics of institutions and organizations in the Sudan. However, organizations may change, become integrated into others, are reorganized or may disappear, budgets are modified affecting available resource, people leave, or advance to new positions, or are replaced *etc*. The usefulness of the information contained in documents of this type decreases with age, sometimes quickly and any existing linkages have to be continually reviewed and updated. A current organizational chart of an institution or its inventories and profiles are almost always more useful in this respect than project documents that are a few years old. At present, the information both for contacts and project activities needed by the decision makers and other stakeholders in the country can most efficiently come directly from the national institutions and organizations at the time it is needed to ensure that it is current.

It is therefore suggested that each of the professionals should develop and maintain national database files for their particular areas of obligation and be responsible for updating these on at least quarterly basis. The necessity of contacting organizations and institutions during updating would also serve to increase the visibility of the coordinating body. There are a number of suitable database programs available for the PC environment that can easily facilitate this. Some database software trainings might be required, but the most important is that the professionals take the initiative to learn how to better utilize the coordinating body's recent extensive upgraded hardware, software and connectivity system.

The effective use of information is to focus on essential information only. In situations where financial resources are scarce, this is, inevitably, the information required to set and achieve immediate policy and management goals. However, some solutions are complex. For example, with environmental issues it is not easy to determine the exact nature of essential information. One approach is to ask decision-makers to articulate their needs directly, but this may not succeed if they have only a hazy idea of their requirements. The price of not pursuing this challenge is heavy, without the right information incorrect decisions can be made and scarce resources used unwisely. With such chaos a realization of the need for organized information within an organization gives birth to the information centre /system/network. (Abdel Karim *et al.*, 2004)

Recognizing that the development of Agricultural Sector is considered as a multi-stakeholder, multi-disciplinary activity, an efficient mechanism for co-coordinating stakeholder contributions is vital. The stakeholders and beneficiaries are classified into:

- Supporters of agricultural information development
- Producers of Information
- Consumer of Information

In Sudan the stakeholders are identified as politicians, policy-makers, resource management agencies, local government administrators, non-government organizations, community based groups, the charity and private sector, industry, scientists, educators, general public, media, international community working in the country, multi-level international, national and local charged with forming and managing technical teams to deliver objectives. This two-tier (horizontal and vertical linkages) structure breaks down barriers to co-operation by facilitating communication at both organizational and technical levels.

Stakeholders may be unable to fulfill their agreed roles due to constraints in human or technological resources, funding, or co-ordination with other stakeholders. This is true especially when stakeholders are collaborating on information production/dissemination and joint operational procedures have yet to be established. In such circumstances, data can be mobilized by addressing key concerns such as data access, data standards, and data quality and data flexibility. The judicious application of information technology (access to the Internet) can also be effective. Attention to these topics is to establish reliable working relationships between stakeholders leading to enhanced productivity and co-operation.

4. AGRICULTURE, LIBRARY AND INFORMATION, COMPUTER SCIENCE, INFORMATION AND COMMUNICATIONS TECHNOLOGIES' EDUCATION

In a knowledge society, the basic duty of education is to familiarize our students with methods of coping with the rapidly growing and changing amount of knowledge. These methods are becoming increasingly important, although this should not make us shy away from communicating certain amounts of factual knowledge. The transfer of encyclopedic or factual knowledge has become pedagogically suspect. This is an understandable evolution, but not necessarily a good one. The basic requirement, however, is to design education in such a way that the focus is more on learning than on teaching. Teaching is a finite activity, which ends as soon as the teacher is no longer around. Learning, however, is infinite, and can take place anywhere, anytime. Learning how to learn should be the main target. Universities have to provide their students with the methods to learn actively, for the rest of their lives. This means, among other things, that our students should be exposed to research methods and to methods of critical evaluation. In turn, this means that we cannot limit university education to just professional training. This will remain important, but it would be dangerously wrong to limit ourselves to that alone. Professional training needs to be embedded in a wider framework, in which more general skills need to be developed, such as language and communication skills, ethical sensibility, and so on (Oosterlink, 2003).

In times of so-called high-tech developments and in a world, which is supposed to be giving so much importance to material goods, this looks a bit soft? The truth is, simply, that the more technologically advanced a society is, the more its competitive edge is found in non-technological skills. This means that modern universities should not forget the aspect of building in their programme. There is nothing wrong with university graduates being excellent professionals, but that is not enough. Immaterial elements in their education may well prove to be the most important and the most valuable competitive edge. Let me just give one simple example. There are many specialists who can perform a large and sophisticated customer or client survey. There may, however, be only a few people who are able to interpret the data and to take the right decisions.

In our era of a knowledge society and a knowledge economy, it is clear that universities have a major role to play. Capital and labor are no longer the dominant production factors. A modern university is characterized by the co-existence of a series of fundamental elements:

- The most important one, and the basis of everything, is knowledge creation. This is obviously, the world of academic research.
- The second characteristic of a modern university is knowledge dissemination, which means that the knowledge created by research is spread among university students, together with a modernized approach to the classics. This is not only meant in the narrow sense of the word, which is dissemination through regular classes, but also the whole complex of attitude formation, value transfer, skills training *etc.*
- And the third component of a modern university's profile can be found in academic service to society. In a way, the previous components, knowledge creation and knowledge dissemination, are also services to society. But here, something more specific is meant. Scientific service to society usually refers to the process of transferring university knowledge to society at large, including the economic world.

In addition, human resources development can be obtained by increase of knowledge and skills and creative potentials that can be applied to improve the quality of life. In spite of low levels of knowledge, inadequate innovative skills at lower, middle and higher levels have contributed to the continuous failures in developing countries in all spheres. Information and knowledge are interrelated. Well-informed, knowledgeable and innovative citizens are causes for human centered development. Information technology facilitates the flow of knowledge in modern society. The failure to use information technology is becoming as negative as the refusal to attend school. It is a choice between being left out or benefiting from enormous benefits of information technology (http://www.uneca.org). With the explosion of information technology, teaching and training methods have changed, which include computer-based-training, electronic text, software platforms, audio, groupware, interactive television, multimedia, online help, teleconferencing, virtual reality and video, and the distribution methods that can include local area network and or Wide Area Network (LAN/WAN), Web, Internet, extranet, CD-ROM, and or DVD, voice mail, audiotape, telephone, computer disk, satellite television, cable television, tactile gear and or simulation.

The second phase of the World Summit on the Information Society (WSIS) at Tunis, the International Telecommunication Union (ITU), together with the International Network of UNESCO Chairs in Communications (ORBICOM), has published the ICT Opportunity Index. The index shows that particularly developing countries have made phenomenal progress over time and, digital opportunities are unequally distributed. The findings of the index suggest that the gap between those with and those without access to ICTs is far from history.

The 2005-working plan of Information, Communication and Technology for Development in Arab Region [ICTDAR] (http://www.ictdar.org) is structured on the following initiatives:

- The youth empowerment initiative.
- An SME support initiative.
- A women empowerment initiative.
- A capacity building initiative for the visually-impaired.

A number of workshops and projects are currently implemented to support the development of ICT policies and e-strategies. There are number of capacity building initiatives in the country which have been supported by development of training materials and content around various themes and for use across Sudan states, cities and localities. The Gederef Digital City organization has recently launched the first telecentre academy "Sudan National Telecentre Academy (SuNTA)" in Africa and Arab World, in supporting the Global Telecentre Movement (http:/ www.sudantelecentreacademy.org). Currently, PROLINNOVA-Sudan, is a non-governmental-led initiative to build a global learning network on promoting local innovations in ecologically-oriented agriculture and natural resource management. The focus is on recognizing the dynamics of indigenous knowledge and learning how to strengthen the capacities of farmers family to adjust to changing conditions; to develop and adapt their own site - appropriate systems and institutions of resource management in order to gain food security, sustain their livelihoods and safeguard the environment (http://www.prolinnova.net/sudan).

4.1. Agricultural and Technology Transfer University Education

Agricultural University education is available at various levels and with fairly good spread in most of the states of Sudan, which has been resulted from the revolution of Higher Education at the beginning of the nineties. The Ministry of Higher Education and Scientific Research took all necessary actions progressively to develop academic agricultural professional education and allowed colleges which offered three-year diploma (middle), in addition, a five year technical B.Sc. program, one year higher diploma, two years M. Sc.

Programs, and three years Ph.D. programs. According to this, to intake of students has so greatly increased. There is correspondingly an increase in the budget of the Ministry, not taking into account the inflation. Academic agricultural professional education has developed progressively. Some of faculties offer an integrated natural resources program including courses in agriculture, animal production, forestry and environment. These programs are based on the modules of the committee of the Agricultural, Natural Resources and Environment and Veterinary Studies of Ministry of Higher Education and Scientific Research (Extracted from Reports of Specialized Committees of Ministry Higher Education and Scientific Research, 2008) including B.Sc., modules for distance learning which are currently run at Sudan Open University. The basic strategic objectives of higher education equip learners with unique programs and skills based on technical education, but technical education does not receive the attention of the academic agricultural education due to the unavailability of skilled trainers in workshops, fields and laboratories and unavailability of tools and equipment, teaching more theory than practical applied studies, migrating the technical level through completion of the academic modules. The proposed e-learning programmes are still under investigations and experimentation. In the past Agricultural Secondary Education was the type of technical education, which was not popular because both students and their parents look at secondary school education as means for university education. This has been led the Ministry of Education to formulate the disciplines related to Agriculture in the Secondary schools curricula as elective subjects.

4.2. Library and Information Science Education

Libraries and Information Science (LIS) educational programs offered middle diploma, B.Sc. and M.Sc. and Ph.D. programs like some of the Arabic countries. LIS colleges are affiliated to the Faculty of Arts following the Egyptian system of Cairo University excluding Juba University. Most of the curricula taught in all the institutions depend heavily on the traditional system, which are considered as negative inputs of development of information sector. There is no real development to the next level in terms of content, revision and evaluation of LIS programs. Like other disciplines there is not even a specialized committee for LIS education in Higher Education or clearing house for information experience on library and information management education in the Sudan. This has resulted into no accreditation standards. These LIS departments are small lecture rooms in the faculties, in addition to absence of ICT laboratories; and acute shortage of schemes will absolutely be reflected on human resources development as well as the development information processing and tools of the teachings. There is no correlation between national information technology policy and library and Information education, development of information, which is unexploited valuable resource. In Sudan, we still ignore what is recommended by UNSECO in 1998 that this sector should be based on academic, personal, and employee competencies (Ghobrial, 2002).

4.3. Computer Sciences and ICT Education

In the mid seventies, computer science was introduced in Sudan. It is high expansion in the computer science and ICT education in different specializations. They are mainly taught at specialized colleges or faculties or departments in engineering and mathematics colleges. In this decade we notice that there is great tendency of expansion in the computer science and ICT education of Higher Education institutions by 88 per cent and especially at private sector. It is considered now as source of financing or funding. Thus, the national strategies and development plans aim to introduce computer science and ICT curricula at all levels of education as basic subjects and means of teaching. The Ministry of Higher Education establishes special committee for this sector, which designs a special module and revises continuously. It is based on basic subject, good knowledge with new technologies, and good understanding of the relation between human and technology.

The development of this sector depends on the following challenges:

- Cultural Challenges
- Demographic Challenges
- Infrastructure Challenges
- Intellectual and technological Challenges

Educational institutions are considered as a big sector of information consumption in educational systems or in the country more than information generation or industry. They almost lack expertise in sustainable publication and management of agricultural information. Our scientists have difficulty in getting and affording the current information about using modern technologies. It is also weakness in operation and cooperation between these educational systems.

5. SUDAN INFORMATION TECHNOLOGY

By definition, Information Technology (IT) is various technologies, which are used in creation, acquisition, storage, dissemination, retrieval, manipulation and transmission of information. IT (core of technology) includes computers, various telecommunication devices, media, broadcasting, press, audio- visuals, and micrographs, *etc.* whereas, Information and Communication Technologies (ICT) include electronic networks - embodying complex hardware and software - linked by a vast array of technical protocols. ICT are embedded in networks and services that affect the local and global accumulation and flows of public and private knowledge. According to the United Nations Economic Commission for Africa, ICT cover Internet service provision, telecommunications equipment and services, information technology equipment and services, media and broadcasting, libraries and documentation centres, commercial information providers, network-based information services, and other related information and communication activities. It is not uncommon to find definitions of ICT that are synonymous with those of information technology (IT). IT can be defined as 'the group of technologies that is revolutionizing the handling of information' and embodies a convergence of interest between electronics, computing and communication. Therefore, in this review, the terms IT and ICT will be used nearly synonymously and in a somewhat broad sense. The terms designate the information processing interaction between providers and users of information and also the development and application of information-processing systems that may not be regarded as part of the development of telecommunications. It is important to emphasize that these technologies only provide new mechanisms for handling an already existing resource: information. Therefore, to understand ICTs, one must first understand information practices and needs.

Recently, the term "technology transfer (TT)" has fallen out of favor among many who view the term as outmoded or too narrow in scope, and who prefer terms such as technology collaboration, technology deployment, technology utilization, *etc.* These are, for the most part, semantic differences, which do little to increase our understanding of these closely, related phenomena. The term "technology transfer" encompasses such a broad range of activities that a general definition brief enough to be useful is impossible to develop. However, operational definitions of technology transfer are easier to devise in a specific context, and are best constructed in terms of specific mechanisms of transfer. From these definitions we can conclude that Information Technology Transfer (ITT) is an ongoing process that continually depends on information flow on the performance of the existing information technology; and regarding information technology utilization Sudan has not yet reached its potential. The problems are underutilization of IT resources; lack of indigenous capabilities and the fact that resources are mainly foreign.

The information institutions need to be at the vanguard of technology transfer from the developed world to the developing economy of Sudan based on ICT.

5.1. ISIS based System

Microcomputer Technology uses CDS/ISIS based Library system that is advanced non-numerical information storage and retrieval software developed by UNESCO. CDS/ISIS is an integrated and Multilanguage package offering all the basic features demanded for a database management system including: database definition; data entry; indexing; retrieval; printing, some utilities; database management and data exchange. Some of Library institutions of UNSECO members have succeeded in encountering CDS/ISIS databases on the Web. The CDS/ISIS software is distributed free to libraries and information centers all over the world by UNESCO - CDS/ISIS has a wide application as an indexing tool but its flexibility has extended its use to library automation applications like cataloguing and information retrieval. This is because CDS/ISIS comes with a built-in subset of the Pascal language that allows local customization and development of user applications and interfaces. Many such customized applications have been reported in the literature (Ghobrial, 1992). Some good interfaces have also been developed to complement the information retrieval capability in CDS/ISIS. The electronic catalogue created form the resource database that can be used to teach and give hands-on experience on on-line information retrieval. CDS/ISIS for Windows (WinIsis) is a developed platform for developing simple applications of its powerful formatting language. New commands for handling presentations are now available as well as hypertext capabilities. The software by itself does not facilitate publishing the database on the Web. In addition, the biggest obstacle for information dissemination to a user community is that bibliographic databases use heterogeneous technologies that inhibit interoperability. The Open Archives Initiative (OAI) addresses interoperability by using a framework to facilitate the discovery of content stored in distributed archives or bibliographic databases through the use of the OAI Protocol for Metadata Harvesting (OAI-PMH) (Hochstenbach *et al.*, 2003).Though the OAI-PMH is becoming the de facto standard, many of the legacy databases or bibliographic systems, for various reasons, are not OAI-compliant. CDS/ISIS is one such database system. In the static repository approach to make CDS/ISIS databases OAI-compliant, the records from a CDS/ISIS database are exported onto a file and the exported records are then converted to the static repository (SR) XML file format. A number of tools for making CDS/ISIS databases on the Internet are available using Greenstone and MYSQL.

A number of tools for making CDS/ISIS databases on the Internet are available such as AGRIS-web, which provides its clients with advanced remote database management functions such as: browse, query, data entry, export and print. Sudan joined AGRIS and CARIS of FAO since, 1976. The Central library of the Agricultural Research Corporation stands as AGRIS Resource Center of the Sudan and it could be retrieved from the AGRIS bibliographic database on the web or CD-ROMs. Recorded by Sudan AGRIS center is fully supported with abstracts. The center also works to share in other AGRIS activities such as the electronic discussion group that has been led by AGRIS secretariat. Most of the institutions have developed electronic bibliographic databases and catalogues using UNESCO's Micro CDS/ISIS (English and Arabic versions) and a few have adopted WEBISIS "WEBAGRIS". These systems are working offline or intranet and not yet represented in their web sites.

5.2. Greenstone (http://www.greenstone.org)

Greenstone is a suite of software for building and distributing digital library collections that provides a way of organizing information and publishing it on the Internet or on removable media (*e.g.* CD-ROM/DVD). It is produced by the New Zealand Digital Library Project at the University of Waikato, and is distributed as open source, multilingual software in cooperation with UNESCO. The dissemination of educational, scientific and cultural information is central to UNESCO's goals, and appropriate, accessible technology is seen as a vital tool in this context. Greenstone's aim is to empower users, particularly in universities, libraries, and public services/awareness institutions to build their own digital collections. The key points that Greenstone makes it its core business to support include:

- Design and construction of collections.
- Distribution on the web and or removable media.
- Customized structure depending on available metadata.
- End-user collection-building interface for librarians.
- Reader and librarian interfaces in many languages.
- Multiplatform operation.

Easily maintained, collections can be augmented and rebuilt automatically. The Sudan Open Archive has adopted this free digital access to knowledge about Sudan. The Archive is an expanding, full text database of historical and contemporary documents, with a linked analytical guide to internet resources. It is being developed with institutional partners in the north and south of the country. It is available online and in free standing CD and DVD formats.

It is a comprehensive system for constructing and presenting collections of thousands or millions of documents, including text, images, audio and video. A typical digital library built with Greenstone will contain many collections, individually organized—though they bear a strong family resemblance. As example, Sudan Virtual Engineering Library - Sustainability Knowledge Network (SudVEL-SKN) is an initiative of establishing the "Nile Basin Virtual Library" which was initiated by the members of the International Institute of Women in Engineering (IIWE) in Paris who were specialists in water and soil. The objective of this Library was promoting researchers in the field of water and soil by transforming the existing documents into a digital form. Based on this initiative proposal of "Sudan Virtual Engineering Library - Sustainability Knowledge Network" (SudVEL-SKN) had been drafted. This library seeks to illustrate how communication and collaboration between engineering institutions and organizations internationally can improve the transfer of technology and information needed to achieve sustainable development globally. In order to provide access to core materials for students and researchers there is a need to build localized databases in education and research institutions that can be accessed offline. The SudVEL-SKN databases have been developed to support research, education and practice in sustainable development in the Sudan. The library has become a part of the learning environment for the University of Khartoum's Faculty of Engineering and Architecture and Institute of Environmental Studies containing both local sources and a selection of international materials. The library will provide easy and efficient offline local access together with international access to students, researchers and academic staff. With the support of UNESCO the prototype version of SudVEL-SKN databases of bibliographic information and inventories of institutions and experts have been developed in both Arabic and English with full accessibility of text material within Khartoum University, which hosts this library.

5.3. CD-ROM Technology

During the flood of 1988 in Sudan, most of the valuable collections of libraries were damaged or lost. At that time, the National Research Centre-Documentation and Information Centre (NRC-DIC) had started to introduce microcomputers and microfiche technology in its services with assistance of UNESCO and Sudan-American Foundation for Education in collaboration with NRC-DIC organizing the first international conference on the use of CD-ROM in developing countries held in Khartoum, 1990, where the participants of 16 countries illustrated the appropriateness of CD-ROM. The verdict of participants was clear that CD-ROM services offered the single best solution to "information gap" or "Book famine". The Librarians expressed their views it uses solution to access to current information at that time. This technology affords libraries in Sudan a unique opportunity to acquire information materials to meet the needs of their users.

Ghobrial (2002) has extolled the virtues of the CD-ROM and highlighted its potentials for researchers in Sudan. Apart from storage space economy, the CD-ROM provides access to information held by important databases without laying cables. This is very significant considering that lack of good telephone services which is one of the major obstacles to computerization and networking by libraries in Sudan during those decades. The Central Library of the Agricultural Research Corporation provides specialized agricultural information services to its users and these include access to resources The Essential Electronic Agricultural Library (TEEAL) which is available on CD-ROM (Musa, 2005).

5.4. Electronic and Networking Technologies

Technologies enable ultra rapid access to the richest sources, wherever they are located in the world's collections, and the rapid exchange of commentaries in electronic forums or video-conferences.

The video-conference system has been applied in institutions of some Ministries such as National Information Centre - Ministry of Council of Ministers which is connected to fourteen States of Sudan, University of Khartoum of Ministry of Higher Education and Scientific research, and Agricultural Research Corporation of Ministry of Science and Technology.

The new possibilities favor an 'extensive' reading, the comparison of diverse texts and viewpoints, multidisciplinary transversality, a "conversation" between readers. They are beginning to have a considerable impact, as much on the individual mechanism of the appropriation of texts, as on the sociology of reading. True poly-textuality in which diverse types of texts and images, sounds, films, data banks, mail services and interactive networks may mutually resist or interfere with one another. This process of reading generates progressively a new dimension - polymorphic, transversal and dynamic. It can be called 'metareading', which is becoming a new driving force of culture *via* online services availed of on campuses of higher education institutions, in internet café and services as well as e-resources (internal and external) as creation of web blogs and Dgroups.

Electronic technologies are considered as effective instruments to direct public interest back to the original objects in their trust, by providing contextual information enlightened with narratives and visualizations with computer-aided renderings and displays. Their appropriate uses of them increase the interest in the indigenous knowledge and they should not leave this opportunity unused to add value to their holdings. They have numerous advantages and may help relieve the traditional conflict between preservation and access and afford multiple, simultaneous uses from a single original in ways that are simply not possible for materials stored in any other form.

5.4.1. Intranet

An intranet is an internal institutional computer network that uses internet protocols, network connectivity to securely share part of an institution's information or operations with its staff. Sometimes the term refers only to the most visible service, the internal website. The same concepts and technologies of the Internet such as clients and servers running on the internet protocol suite are used to build an intranet. Hypertext Transfer Protocol "HTTP" and other Internet protocols are commonly used as well, such as File Transfer protocol (FTP). There is often an attempt to use Internet technologies to provide new interfaces with corporate 'legacy' data and information systems. The components of the Intranet: search tools, taxonomies, content management systems, document publishing tools and portals. Portals are primarily concerned with the presentation of information to users (Onunga, 1998)

Increasingly, intranets are being used to deliver tools and applications, *e.g.*, collaboration (to facilitate working in groups and teleconferencing). Intranets are also being used as cultural change platforms. For example, large numbers of the staff discussing key issues in online forums could lead to new ideas. Efficient knowledge sharing within and between institution departments can be accomplished with an intranet application, which can be design for information and communication. Knowledge sharing should be understood as an effect of coordination, documentation and trouble shooting in daily work practice, rather than an activity in its own right. The usage of intranet is to access information sources and get effective information services in the colleges of some universities, Ministry of Council of Ministers, and some of governmental departments "National Telecommunication Corporation" and research institutions "Agricultural Research Corporation".

5.4.2. Internet

The Internet has advanced far beyond its initial role as a low-cost user-friendly communications and information delivery tool in the memory institutions and is now assuming more general functionality for all forms of electronic communications. The Internet is considered as intermediary to inter-phased international communications, storage source to information, and provider of multi-services to the different categories of the users.

Africa accounts for 14.3 per cent of the world's population, but Internet users in Africa account for only 3.5 per cent of all world Internet users. Out of an African population of about 955 million, internet users are just about 51 million, accounting for an Internet penetration of 3.5 per cent while the percentage penetration for the rest of the world is 24.7 per cent and 21.9 per cent for the whole world including Africa. In terms of use growth, the world growth usage between 2000 and 2008 has been 305.5 per cent (from about 360 million in 2000 to about 1.46 billion users in 2008); Africa's usage has grown at 1,031; 2 per cent (from about 4.5 million in 2000 to about 51 million users in 2008). Sudan is considered as one of the top 10 users in Africa. The statistics do not show how many of the targeted readers have access to the Internet. Nonetheless, the small numbers show how hard it still is for scholars to access information on the Internet. Internet even at higher institutions of learning is still not fully accessible (http://www.internetworldstats.com/stats.htm).

The internet helps also online access to the world's scientific literature with major initiatives HINARI "Health InterNetwork Access to Research Initiative", AGORA "Access to Global Online Research in Agriculture" and OARE "Online Access to Research in the Environment", there is about fifty academic, governmental and civil society research institutions which have joined this initiatives in Sudan. These programs are closely tied to the United Nations' Millennium Development Goals and will be continued at least until 2015. These initiatives represent a truly global public-private partnership for development, providing essential information for life to those who need it most.

- Access to thousands of peer-reviewed international scientific journals online.
- Users can link to and from abstracting and indexing databases.
- Full-text articles can be downloaded for saving, printing or reading on screen.
- Many articles available online before they are published in print.
- Users can search by keyword, subject, author, or language.

Access to information is an essential element in strengthening local medical, agricultural and natural resource management practices, and empowering local experts to find solutions to local health, environmental, social, economic, and food issues. The Sudanese Universities can access through Khartoum University Information Network to a huge number of international data base systems, which cover almost all academic, cultural and scientific specializations such as the academic search premier; medical reference comprehensive; medline; OCLS- world and international abstracts. It provides electronic book (e-book) services, a big number of electronic books have been added to the different specializations.

5.4.3. Global System for Mobile [GSM] Cellular Network Operator

Much of the initial vision for a 3rd Generation (3G) and 3.5G mobile system was the provision of internet services on mobile phones. The General Packet Radio Service (GPRS), often referred to as 2.5G, has three principle features which make it an important element of 3G mobile systems. It introduces a packet-switched provision into core network of the current 2nd Generation (2G) circuit-switched (CS). Global System for Mobile (GSM) communication network is relatively simple and cost effective. It offers always-on-line capabilities to applications running on GPRS enabled phones thus mitigating the dial-in delays experienced in CS provision. It is a vital element in the evolution of 2G GSM system to 3.5 G Wideband Code Division Multiple Access (W-CDMA) versions of the Universal Mobile Telecommunications System (UMTS) Mobile Telephones in the Sudan proved to be the best penetrating the fastest and cheapest technology to make a real uptake, demolishing all features of inequality in terms of connectivity and access. The technology invaded the rural areas equally good as it invaded urban societies sweeping all signs of technological and social barriers. (www.sudan.net)

Internet hosts, apparently despite of the fact that Internet now is evolving independently without any obstruction of physical location, still remain influential when issues such as the content and language used in communication are dealt with. Technology more often than not is well adapted and related in terms of usage to certain unique actual or potential norms and values that shape utility of technology personal computers. Saudani, MTN "Areeba" and Zain offer GSM services in the country. For example, Zain in Sudan (formerly Mobitel) offers mobile operations back in February, 1997. Today it serves the largest number of mobile customers in the country with 3,991,611 active customers as at 31 March, 2008 (www.zain.net). It is clear accessing phone services is very important, the gap will be more completely addressed when access to a higher level of information technologies becomes widespread. The mobile phone has become the right device to deliver e-Services to rural communities and under-privileged populations, the nature and the support of those e-Services still need to be defined. At present, there are three major ways of developing and deploying mobile e-Services:

1. SMS applications: Whereby people send requests by text-messages to a specific phone number, and get the result *via* text-message.
2. Voice applications: Whereby people call a specific phone number that is connected to a voice platform and *via* voice or keypad inputs, receive relevant information.
3. Web applications: People have a Web browser on their phone and a data-service plan which connect them to the Internet and allow them to browse the Web.

6. SUDAN INFORMATION DISSEMINATION

The growing access to computers and electronic communications worldwide makes it possible to offer and acquire automated, updated information almost in every field or activity. Specifically, the practice of scientists to consult electronic data banks as an aid for their research, and to produce and consult scientific bibliography in machine readable form, is extending considerably not only in industrialized countries, but in many developing countries as well. Many academics already use computer facilities for their daily work and scientific interchange. Thus the rapid development of information and communications technologies provide a useful and powerful tool for the construction of a comprehensive information system that includes in principle the whole of the regular, periodic or serial scientific produced in the region. Such a system can also be used in principle to catalogue and classify these publications according to previously defined quality criteria, and it can serve also as a basis to provide both editors and scientists with an efficient channel for the rapid production, dissemination and retrieval of research material. From the general overview, Sudan Agriculture priorities, goals and visions are based on the national integration strategy. So the Agricultural Culture is considered as main element in our life and for development of our societies and human resources.

6.1. Sudan Television and Broadcasting (STB)

STB has merged agriculture in its vital and integrated educational programs. The Administration of cultural education produces about 30-35 per cent of STB programs. It elaborates its unique programs in collaboration with ministries, institutions, planners, decision makers and pioneers in Agricultural sector for guidance and support in the production of progressive programs (Dr. Dirar Mohamed Fadl Elmoula and Ibtisam Kamal, Sudanese Corporation for Broadcasting and Television, August, 2007).

The disseminated agricultural programs cover the following aspects:

- Priorities of food security, strategic storage and agricultural exports.
- The initiated agricultural programs in Sudan States to fulfill National agricultural strategies.
- Follow up of environment, desertification and forestry programs.
- The applications of agricultural technology transfer.
- Cooperation between farmers investors unions and private sector to solve problems related to the production and design attractive and progressive modules.
- Discussion of the problems related to the markets and export.
- Support of agricultural insurance and reduction of the risks.
- Encouragement of reverse migration from urban areas to rural areas to contribute in agricultural, environmental and pastoral development.
- Settlement of Islamic legalizations in land and natural resources utilization.
- Preservation of agricultural research programs.
- Appreciation of the role of women in traditional agriculture.
- The impact of training on the development of skills in STB agricultural extension and programs in the whole country.

The dissemination of these programs requires;

- Digital cameras and related ICT equipment.
- Travel for long distance through the vast country like Sudan.
- Special processing and techniques.
- Arrange informative and scientific trips.
- Use of proper documentation tools and technologies.
- Revisit and reschedule continuously.

6.2. Information Dissemination Challenges

Problems hampering effective dissemination of information to the stakeholders are as follows:

- Isolation of researchers due to lack of technical support for accessing information resources.
- Lack of adequate modern equipment that facilitate the rapid processing and dissemination of information.
- Absence of functional information networks which promote and encourage the exchange of information and experiences.
- Need for and dependence on outside financial support which does not continue.
- Human shortcoming and lack of experiences in information handling, information and communication technologies and both artificial and expert systems.
- Lack of effective mechanism for re-packing research findings in suitable form use as well as appropriate methods for conveying these finding to the user technological advances lag behind and in many cases technological knowledge never reaches the beneficiaries, owing to poor or ineffective methods.
- The indigenous knowledge based on specific sector produce which supplements information derived from fundamental and adaptive research programs which is not integrated in the various approaches practiced in Sudan due to the non-appreciation of these policies.
- Absence of effective means of data collections, flow and dissemination of information in different formats.

6.3. Sudan Agricultural Knowledge and Extension Services Collaboration

Sudanese agricultural knowledge and extension systems are linked to number of regional, international, nongovernmental and humanitarian, agricultural information systems through various collaborative initiatives and activities. These initiatives aimed at improving the management of access to agricultural information generated in the country. These co-operations and partnerships are needed to perform their activities in planned, organized and manageable forms; to reduce victims of information overflow; save time and money; quick decisions; build or strengthen partnership between different stakeholders in the sector; stimulate participatory innovation development systems, and allocating financing trends in ideologies of agricultural and rural development to cope with driving force of globalization.

7. CONCLUSION AND RECOMMENDATIONS

Information for the development of any nation can be both indigenous and international. To harness information from local and foreign sources, the application of modern information management techniques is a *sine qua non.* It has had to battle with very unfavorable climate of information provision and management, a climate that must improve to give way to progress in this century. As development issues, especially in agricultural knowledge and extension services in Sudan to access and utilize information for development, no matter the format, no matter the source. It is important to set up the suggestions for development of strategies of agricultural knowledge dissemination and extension services which are summarized as follows

a. Setting and the development of the strategic plans and policies of preserving, protecting and promoting of agricultural knowledge tailored to the specific circumstances and priorities.

b. Building multi-purposes national committees or councils to concern with identification of knowledge use and diffusion, representation, diversity of culture, different facets of illiteracy, alleviation of knowledge misuse and under-uses, circulation of rare and precious collections.

c. Reforms in information services and use of knowledge techniques as well as reengineering knowledge institution procedure through the use of the information technologies in the digital environment.

d. Developing an effective system to determine the extent to which total quality is implemented in knowledge and emphasized on the quality of organization.

e. Lay emphasis on professional ethics and adopt quality principles to transmit them to the staff *via* various means.

f. Adopt strategic analysis and planning methods which are crucial in the development and the interaction with digital environment.

g. Emphasize group evaluation in the assessment of the performance of knowledge related agricultural extension.

h. Develop the educational building capacity programs in agricultural information dissemination based on ICT and the evaluation of the automatic modern systems. Building capacities, through education, for local people to be able to develop their own services.

i. Design of unified system to digitize national periodicals/information sources and encourage information providers and publishers to promote electronic publishing and speed up the shift from information management and pay attention to standardization processes.

j. Standardize the scientific norms and concepts builds the national thesauri as well as warehouses.

k. Achieving success in any major policy shift that will have profound impacts on society and the nation at large, and policy planners have invariably always directed their efforts at the education system. This is to ensure that changes start at the schools with children who are more receptive and responsive to change. Also the education system is one sector that can respond faster than any other sector of the society to any major changes. This means a greater participation and a more prominent role for the educational institutions in the country.

l. Encourage agricultural institutions to use a wide array of information tools, connect the Internet.

m. Enabling the Mobile Web on existing and future mobile phones available in emerging markets

n. Building and gathering expertise in developing and deploying usable and useful e-Services for under-privileged populations and rural communities

o. Enabling of the Web 1.0, 2.0 or 3.0 in the next few years.

References

AbdelKarim, Abdel Karim H, Musa, Ahlam I, Osman, Elsaddig, Osman, Isam Hassan, Ghobrial, Rafaa, A., 2004. Sudan Agricultural Information Management: Challenges and Opportunities. In: *The 1st Sudan Agricultural Information Network (SNAIN),* RAIN of ASARECA and SNAIN of ARC , 30-31 August, Khartoum, Sudan.

Adedibu, L.O. and Adio, Gboyega., 1997. Information Needs and Information seeking patterns of Medical Students at Lautech. *In: Aslib Proceedings,* 49 (9). p. *238-242.*

Anas. A. A. Hassan, A Alsheikh and Matian Brpwon., 1991. *The guide of technician and agricultural extensionists, in the production of vegetables and fruits.* Khartoum: University of Sudan for Science and Technology.

Ahmed, M.K., Ahmed, S.A. and Abdel Gadir, E.O., 2004. Seasonal Wadis, Institutions and Water Security. In: *Conference on Water Harvesting and the Future of Development in Sudan,* Khartoum, Sudan: UNESCO Chair in Water Resources.

Edris, H., 1975. *Agriculture Research Capabilities and Scope in the Sudan.* Khartoum : National Centre for Research.

El Mustafa M. A., 2005. National Action Programme for Combating Desertification. In: *Workshop on the implementation of the United Nations Convention for Combating Desertification, National Assembly*, Khartoum, Sudan.

Ghobrial, Rafaa Ashamallah., 1992. Evaluation of Information Products and Services of the Documentation and Information Centre, Khartoum. *M.Sc. In Information Science thesis (Unpublished)*, Addis Ababa University

Ghobrial, Rafaa Ashamallah., 2002. Role of Higher Education, Technology and Employee Planning in Development of LIS in Sudan (in Arabic). In: *Workshop on LIS curriculum in the Arab countries*, IFLA/ ALP/Lebanese University, Beirut, Lebanon.

Ghobrial, Rafaa Ashamallah and Abdel Gadir, Hamad., 2007. The way to Knowledge Economy: a Review of Sudan Knowledge Services. In: *The Fifth Session of the Committee on Development Information (CODI-V)*, UNECA, KLIS subcommittee, Addis Ababa Ethiopia.

Hochstenbach, P., Jerez, H. and Van De Sompel, H., 2003. The OAI-PMH Static Repository and Static Repository Gateway. Joint Conference on Digital Libraries 2003, Houston, Texas. *http://public.lanl.gov/herbertv//papers/jcdl2003-submitted-draft.pdf*, retrieved on 20/7/2007.

Musa, Ahlam Ismail., 2005. AGRIS Resource Centre of the Sudan: recent situation and future prospective. In: *ICARDA/FAO /AOAD Training Workshop*, ICARDA, Aleppo, Syria.

Onunga, John., 1998. *The internet*. Nairobi: Information Systems Academy Ltd.

Oosterlinck, A., 2003. Knowledge Management in Post-Secondary Education: Universities. *http://www.kuleuven.ac.be/knowledgetransfer.htm* retrieved 22/5/2007.

1. Internet Pages which are searched January –August 2008

   http://www.greenstone.org

   http://www.nicsudan.gov

   http://www.nationsencyclopedia.com/Africa/Sudan-TOPOGRAPHY.html

   http://www.suvl.edu.sd

   http://www.arcsudan.sd

   http://www.ncr.sd

   http://www.ictdar.org

   http://www.uneca.org

   http:/ www.sudantelecentreacademy.org

   http://www.internetworldstats.com/stats.htm

   http://www.sudan.net

Author Information

Ms. Rafaa Ashamallah Ghobrial

Information Research Professor Assistant
Head of Information Services and Systems
Documentation and Information Centre
National Centre for Research, Ministry of Science and Technology
P.O. Box 2404, SUDAN
Phone No. (O):+249183770776
Mobile Phone No.: + 24991504104
Fax No.: +249183770701
e-mail:rafaasham@yahoo.com, margousrafaa@yahoo.com
URL: www.ncr.sd

Ms. Rafaa Ashamallah Ghobrial has been supervising activities of the information systems and services and Sudan National Focal Points of regional and international networks since, 1992. Contributing in the acquisition, abstracting, cataloguing, classifying and indexing of documents and editing of the Centre publications. Working as Par time-lecturer in the universities, resource persons in national and regional training courses. Contribution in the advisory and planning of information committees inside and also outside the country. Twenty three papers presented in institutional, regional and international meetings. Carrying out research programmes related to informatics and information technologies.

❑❑❑

Chapter 15

TRINIDAD AND TOBAGO

Edwin Joseph

1. INTRODUCTION

Agricultural Extension in Trinidad and Tobago is searching for a way to effectively use Information Communication Technologies (ICTs) to improve the delivery of services to the farming community. The notion of a demand-driven extension service seems to appeal to the Ministry of Agriculture, and many believe that ICT has the potential to become a primary enabler to improving the extension service.

The country has a population of 1.3 million with one of the highest per capita incomes in Latin American and the Caribbean (US$16,090 in 2007). Oil and gas account for 40 per cent of GDP and 80 per cent of exports. The unemployment rate is 5 per cent, and the labour market is characterized by chronic shortage, high price and low productivity particularly in the agricultural sector (Government of the Republic of Trinidad and Tobago, 2008). One of the government's objectives is to develop a vibrant agriculture sector to reverse the decline of agriculture. The government's operational plan 2007-2010 identifies a model for transformation which includes joint ventures with the private sector, investments to develop farms *via* Government to Government arrangements, and improvements in infrastructure and communication (Republic of Trinidad and Tobago, 2007).

2. ISSUES RELATED TO COMMUNICATION, TRANSPORTATION, AND ICT APPLICATION

Unlike other developing countries, Trinidad and Tobago does not lack a transportation and communication infrastructure, and they are considered adequate by regional standards. A total of 188.52 km of access roads was either constructed or improved to benefit 4,740 farmers between 2002 and 2006. Another 32 km of access roads were constructed in 2007 improving access for 876 additional farmers; with upgrades to another 17 km of access roads for 425 farmers. Overall there are 8,320 kilometers (5,158 miles) of roads, half of them paved with main routes covered by 4-lane highways (Encyclopedia of the Nations, 2008).

Telecommunications are dominated by the government-owned Telecommunications Services of Trinidad and Tobago (TSTT), and independent Internet providers are obliged to use its service lines (Trinidad and Tobago Computer Society, 2007). There are a number of Internet Service Providers (ISPs) in Trinidad and Tobago. They offer browser-based access to e-mail and other miscellaneous services such as web hosting, additional e-mail addresses, *etc*. In areas where broadband connection is not available, there are dialup options.

In early 2004, TSTT introduced internet access *via* modem to anyone having a PC, for 75 TT cents a minute (the cost was added to the phone bill of the line used) without paying a regular internet access bill at an ISP. TSTT also launched a service which allowed GSM cell phone or GPRS-enabled device (*e.g.* a PDA) to access email, browse the World Wide Web, transfer files, and use instant messaging. TSTT also introduced mobile wireless broadband internet service in 2006 with unlimited downloads to customers, Wi-Fi was also included, with hotspots in many areas including a location at the International Airport. Trinidad and Tobago Computer Society maintains a listing of hotspots on a Wiki (Trinidad and Tobago Computer Society, 2007).

3. AGRICULTURE AND INFORMATION COMMUNICATION TECHNOLOGY VISION STATEMENT

The Ministry of Agriculture's Vision 2020 statement (Vision 2020 is an initiative of the Office of the Prime Minister which is a strategic planning process with the objective of steering Trinidad and Tobago to developed nation status by the year 2020 through a combination of economic, social and governance reform. *http://www.opm.gov.tt/initiatives/index.php?pid=3001*) outlined broad development objectives that included the implementation of new technologies to reach small, resource-poor, tenure-insecure farmers. According to Vision 2020, it was important that agricultural development should be market-led, technology-driven, and include efficient communications networks. In short, Vision 2020 stipulated that agriculture be managed smartly using an efficient IT support intelligence and information system (Ministry of Agriculture Lands and Marine Resources, 2007).

One may question why ICT has not taken off sooner in the Trinidad and Tobago extension service given its presence and accessibility, and available resources. But it is important to investigate how much people understand social behavior in context of modern communication technologies, and people's perception of ICT as a tool for information in different sectors of the economy. Understanding the role of the Internet in delivering services and acting as a vehicle for communication are probably lacking, and perhaps the introduction of ICT projects may require adequate awareness developed through case studies, so that users would be aware of its potential. We may also need to be aware of stakeholder's media choices for communication and recognize that internet technologies alone will not improve or promote agricultural development initiatives in Trinidad and Tobago. An integrated multimedia communication strategy may be more appropriate, and retooling of extension will become a priority for improving communication, sharing information, and responding to needs.

4. OVERVIEW OF NATIONAL AGRICULTURE SCENARIO

Trinidad and Tobago has a domestic and an export-oriented agricultural sector. The domestic sector focuses on vegetables, root crops, pulses, fruits, condiments, rice, poultry and livestock. The export sector also produces vegetables and fruits, along with cocoa, coffee, sugar, fish, and some agro-processing. Agro-processing in particular has been providing business opportunities for the export market since the early 1990s (Gosein, 1992). But Vision 2020 describes the performance of the agricultural sector as unimpressive, and gives the following reasons for poor performance:

- Lack of adequate marketing and inadequate market intelligence
- Insufficient market-driven research
- Labor competition from other sectors in the economy
- Price stagnation of agricultural commodities
- Poor infrastructure for agriculture
- Lack of credit and other production incentives
- Inadequate response to the needs of the farming community
- Slow pace of mechanization on farms

Agriculture accounts for 16.7 per cent of the land area and employs 5 per cent of the population. From 1996-2002 agriculture's contribution to GDP was 3.3 per cent per annum with domestic and export agriculture contributing equally (50 per cent). Sugar accounted for 47 per cent of the export crop contribution (Ministry of Agriculture Lands and Marine Resources, 2007). In 2004, agriculture's contribution to GDP was 1.02 per cent (The National Medium Term Priority Framework for FAO Assistance – Trinidad and Tobago Draft document, 2006). According to the 2004 Agricultural Census, the total number of holders was 19,143 (15 per cent female and 85 per cent male) with 94.9 per cent in Trinidad, and 5.1 per cent in Tobago. A total of 72.4 per cent of farmers were engaged in Crop Production. Ninety six per cent of all holdings were less than ten hectares with the majority of farms operated by individual households 45 years or more.

When compared to the total number of holders recorded in the 1982 Agricultural Census, these statistics indicated that there was a decline of approximately 37.4 per cent. Growth in real GDP (per cent) between 2002-2007 showed agriculture's greatest decline in 2003 and 2004 (-15.3 and -34.2 respectively); however, in 2006 and 2007 it was reduced to -0.9 and -5.9 per cent. The sector's decline in 2003 and 2004 was due to several economic and other production problems. These included inefficient marketing systems, weak research, praedial larceny, inadequate area under irrigation, flooding and inadequate extension (Government of Trinidad and Tobago, Ministry of Planning and Development Central Statistical Office, 2005).

The Ministry of Agriculture acknowledges that limited resources and small farm sizes restrict global competitiveness in agriculture. The current strategic plan indicates that agriculture for the domestic and export markets are confronted with increasing competition from global suppliers. According to the Ministry of Agriculture, survival of the agricultural and rural sector in Trinidad and Tobago urgently requires significant and rapid transformation. Goals for the sector therefore include envisioning agriculture in the broader context of agribusiness to include production, agro-processing, food manufacturing, special and unique culinary cuisine, culinary arts, food service and agri-entertainment/agro-tourism.

Through the National Agri-business Development Program (NADP), the Ministry of Agriculture is encouraging small farmers and small entrepreneurs to collaborate to form agro-industries for contract production and marketing, and is also favorable towards large technology-driven farms and other large scale agri-businesses (Government of the Republic of Trinidad and Tobago, 2008a). The NADP reflects collaboration between public and private sectors which would allow all the major stakeholders to coordinate the production and marketing of agricultural products (Trinidad and Tobago Agri-business Association, nd). But only 16 per cent of farmers have any formal training in agriculture. Therefore the existing low level of "formal businesses" in agricultural production, and the low level of adoption of new technology may be a consequence of their educational level, which reflects their attitude about farming (Government of the Republic of Trinidad and Tobago, 2008).

A recent study examining farmer's attitude about technology showed that younger farmers had stronger technology beliefs, and farmers with secondary level education also had stronger technology orientation. The "technology belief" factor (farmers' assessment of the risks versus the benefits of offered technologies) was significant among young farmers with at least a secondary level education. According to the researchers, these were important considerations when planning programs aimed at increasing technology use by farmers (Ganpat, *et al.*, 1999).

The ministry started a Youth Apprenticeship Program in Agriculture (YAPA) in 2002 to reduce the age profile of farmers in Trinidad and Tobago, and to provide access to agriculture lands, financing, technical knowledge, and training in farming. The aim of this program is to encourage youths between the ages of 17-25 to engage in technical, professional and business activities in agriculture. To date, over 7,048 youths have benefitted from the program (Republic of Trinidad and Tobago, 2007).

5. NATIONAL EXTENSION SYSTEM: SCOPE AND HISTORY

Studying the history of agriculture in Trinidad and Tobago is valuable for understanding how effective extension has been in meeting the needs of the farming community. When agricultural extension in Trinidad and Tobago began in the 1860s, the emphasis was on the production of sugarcane. Later, the Imperial Department of Agriculture was established to develop agricultural research and education through diversification because of the sugarcane depression in 1896. Agricultural instructors were then trained and were given transportation (horses) and communication tools (language training) to educate small farmers on management practices for other export crops. The newly formed extension service communicated with farmers in their native *patois* to disseminate new information, and was successful in effecting change (Thomas *et al.*, 1985).

By the 1920s, agricultural departments established residential agricultural schools to train potential farmers and future extension agents. However, by the 1960s, extension was characterized by this top-down technology transfer approach. It was also the era of interpersonal communication, community development, and the green revolution. The 1970s extension focused more on technology transfer, and encouraged farmers to produce more annual food crops. The main feature of the 1970s model was Integrated Rural Development.

In the 1980s, the Training and Visit System (T&V) emerged as a promising model. T &V was intended to reform and improve the effectiveness of the extension by building up a professional extension service that would help farmers increase production, raise their living standard, and provide appropriate support for agricultural development (Benor, and Baxter, 1984). This quickly gave way to other approaches such as participatory planning, ecosystems management, and approaches that focused on women in agriculture (World Bank, 1990).

It was during this period that Trinidad and Tobago reduced public expenditures in extension. As a result farmers received reduced extension services (Sustainable rural development: Progress and challenges, 1995). Even though Trinidad and Tobago farmers depended heavily on the extension service for advice, the government withdrew services, and extension could no longer provide information, ideas, and other services to meet food needs. Some researchers identified generic challenges such as budgetary constraints, which made it difficult to explore alternatives (Gershon Feder, *et al.*, 1999). Others blamed bureaucratic inefficiency leading to poor program design and poor performance within extension (Agricultural Extension in Laos: A Pilot Project. Novartis Foundation for Sustainable Development (NFSD), 2001). None questioned the role ICT could have played in developing new approaches.

In the early1990s, participatory methods, collaborative learning, small farmer development, and Integrated Pest Management (IPM) were adopted as more idealistic approaches that allowed extension to cover wider audiences with emphasis on education, leadership development, skill training, and environmental protection. Extension emphasis clearly shifted from production of food, to complex processes such as farming systems research, and improving the organization and management of extension (Harold, 1987).

The late 1990s saw the Ministry of Agriculture in Trinidad modernizing and decentralizing the extension system by setting up mini-central extension systems at the regional and county level which were structured like the national extension office (Singh, *et al.*, 2005). The Ministry's food policy also facilitated traditional and nontraditional exports that included aquaculture and agro-processing, but the lack of irrigation and drainage, and marketing infrastructure were major challenges for transformation of the sector. However, coordination between central and regional extension were not effective (Swanson and Samy, 2002).

6. RATIONALE FOR ICTs IN TRINIDAD AND TOBAGO AGRICULTURAL EXTENSION: RECOGNIZING THE NEED

The International Telecommunications Union (ITU) plays a fundamental role in building confidence and security in the use of information and communication technologies. One cannot overemphasize the need for Trinidad and Tobago to seek opportunities for developing an ICT agenda since ICT developments are occurring all across the globe. ITU statistics show that over the last 10 years, the digital divide between the developing and the developed countries has been narrowing (NationMaster.com, 2003-2009).

The National Information and Communications Technology Plan is an Initiative of the Prime Minister, Patrick Manning, established on 31 October 2002. Under the Chairmanship of Ministry of Public Administration & Information, the objective of the team is to build a National Information & Communication Technologies (NICT) Strategic Plan. The Government's vision to bring Trinidad and Tobago to Developed Nation Status by 2020 recognizes that developments in the ICT sector are crucial. The NICT Project is linked and integrated with Vision 2020 (Villavicencio, and Ponce, nd.).

The first World Congress on Communication for Development brought together, practitioners, academicians, and policy and decision makers who recognized the important role that communication for development plays. One of the key challenges noted was how to close the knowledge and information gap for the poor and allow them to participate in the rapid expansion of information and communication technologies. The congress noted that ICT and development policies and projects must address social equity that allows people to participate in community life (Food and Agriculture Organization of the United Nations, 2007).

Trinidad and Tobago's National ICT Vision is to provide all citizens with affordable Internet access; develop children and adult skills to ensure a sustainable solution and a vibrant future; promote citizen trust, access, and interaction through good governance; and maximize the potential within all citizens, and accelerate innovation, to develop a knowledge-based society. The National ICT Strategy in Trinidad and Tobago includes a Community Connection Program to provide Community Access Centres throughout the country, and a Youth Computer Corps Program to train young adults to work alongside volunteers to help build new skills in the local population and workforce. A Knowledge Innovation and Development (KID) Program was designed to provide computers and high-speed Internet service for the 636 schools in Trinidad and Tobago by 2008. The KID program was also meant to equip teachers to use ICT for education. In addition, a LibraryNet Program was identified to provide all libraries with computers and high speed Internet access by 2008 (Government of Trinidad and Tobago, 2004).

The country is already an active member of the global information society. The National ICT strategy was developed to increase levels of connectivity in schools and communities. The Government envisioned that by 2008 over 50 per cent of homes would have personal computers and affordable Internet access, and over 50 per cent of the population would consider themselves regular Internet users. The strategy also envisioned that by 2008 over 50 per cent of companies would use the Internet as a business tool (Government of the Republic of Trinidad and Tobago, 2003).

But there are several barriers that inhibit increasing use of ICT. They are, high cost of computers, high cost of Internet access, unacceptable internet speeds, and limited community access to computers and the Internet for those who cannot afford home PCs. The Ministry of Social Development established four multi-purpose community based Telecentres that offer free internet access, and provide training in Microsoft Word, Excel, Power Point and Access. The Telecentres are equipped with conference rooms that are accessible for small businesses and community meetings (Government of the Republic of Trinidad and Tobago, 2008b). These centres do offer valuable services but do not satisfy the demand.

7. INFORMATION DISSEMINATION AND ICT OPPORTUNITIES FOR IMPROVING EXTENSION

A study conducted in 2003 and 2004 in Trinidad and Tobago and Dominica identified Farmer Field School as important for information sharing among farmers through experiential learning and participatory leadership. As farmers developed new confidence to communicate with extension officers, key questions were raised about how ICT can be used to generate new exchanges among clientele within the extension environment (Dolly, 2008) With 32 FM and two AM radio stations on the air, Trinidad and Tobago has the capacity to reach every community. Thirty of these stations are available on the internet (Radio Station World: Your Global Radio Station Directory, 2008). Internet radio is an untapped communication source. The Ministry of Agriculture does not own a regular agricultural program, and engages in sporadic radio broadcasts and infomercials in combination with other media, as in the case of a 1995-1997 Hibiscus Mealybug infestation in 2003. The government usually buys slots on commercial radio and TV channels to present informational materials at or around "prime time" (Seepersad, 2003).

Internet radio is cost effective, and has potential to disseminate information to a broad audience. Extensionists in Trinidad and Tobago must therefore be adept in participatory techniques, be able to draw on a mix of communication methods and technologies to increasing incomes for farmers (Neuchâtel Group, 1999). ICTs can bridge the information and knowledge gap between urban and rural areas by linking with providers of information and other essential services that are vital for agricultural and rural development. Internet-based information and communication technologies provide a potentially powerful tool for improving communication between research, extension and even farmers (Riggs, 2003).

ICT interventions can also include other simple technologies like rural payphones (Report of the 6th Consultative Expert Meeting of CTA's Observatory on ICTs, 2006). In seeking to promote the use of the newer ICTs, the traditional ones such as radio and television should be included. One simple reason is that most people in Trinidad and Tobago have access to radio and television while computer ownership is still relatively low. Training effort should be concentrated on providing ICT equipment, training *etc.*, for frontline extension workers while simultaneously encouraging and supporting interested farmers, community based organizations and non-governmental organizations.

ICT has demonstrated its ability to revamp agricultural production in developing countries (Akpabio, *et al.*, 2007). For example, the rapid development and applications of the Internet and other forms of ICTs in the Kenyan agricultural sector have presented a whole new dimension in the transfer and access of agricultural information (Kiplang'at, and Ocholla, 2005) Kenya and Uganda recognized the potential of agricultural knowledge centres as delivery models of ICTs , and researchers confirmed that rural information and knowledge centres provided an opportunity for bridging the information and knowledge gap between urban and rural communities. Education level had a positive correlation with usage of the centres, and the most commonly used information service in Kenya was market information; and in Uganda agricultural production information (Asaba, *et al.*, 2006).

Similarly, a study examining Trinidad and Tobago's e-Preparedness revealed that the market suffered from a lack of widely available electronic payment processing services which hampered domestic merchants in conducting on-line business. Also, a poll revealed that only 16 per cent of Internet users in Trinidad and Tobago make online purchases, with 50 per cent spending less than TTD $1,000. Many local businesses did not adopt e-Commerce as a means of selling products and services to their customers. Business websites were generally informational, with only financial services, the airline industry, and the entertainment sector showing e-Commerce activity. The limited e-Commerce activities were interpreted as a lack of awareness about the potential benefits of e-Commerce, and a lack of confidence in using the Internet for e-Commerce.

The results from a structured survey on the state of ICT awareness and activity among Micro, Small and Medium Enterprises (MSMEs) in Trinidad and Tobago showed that very few were transacting their own business online. Several MSMEs did not have access to online banking facilities, and reported that the cost of internet service and access to computers were deterrence for internet use (Universalia, 2006). Dissatisfaction with internet service has been an issue since 2001 as evidenced by significant proportion of households (40 per cent) who were dissatisfied with the service of Internet Access Providers. They identified high cost, 'too much time for connection' and 'interruption while working, as deterrents' (Republic of Trinidad and Tobago, 2001).

8. CELL PHONE AS A POTENTIAL MEDIUM FOR EXTENSION AND ICT INTEGRATION

In Trinidad and Tobago, about one in three persons has Internet access, with one in five using the Internet every day. However, people have embraced cellular telephones. Fifty per cent (50 per cent) of the adult population use cell phones as the primary medium for communication, and there has been tremendous growth in cellular subscription from 426 users in 1991 to 800,000 in 2005 (NationMaster.com, 2003-2009).

In a recent survey on the impact of cellular phones in Latin America and the Caribbean, the use of cell phones for banking and government service was practically non-existent—even in the more developed mobile markets like Trinidad and Jamaica. The results further revealed that users were rarely taking full advantage of the services enabled by the mobile platform. Text messaging was the only service beyond voice that was gaining rapid adoption. Applications such as m-banking and m-government that could benefit the poor were hardly used (Galperin and Mariscal, 2007).

In contrast to the Caribbean and Latin America, less than one-fifth of Africans have bank accounts, but mobile banking powered by cell phones, is allowing people in Kenya who could never afford traditional bank accounts to send, receive and save money *via* text messages. However, cell phone companies offer banking tools that allow city dwellers to send money to rural relatives, small businesses to pay their employees, and parents to deposit their children's school fees (Bengali, 2009).

Mobile telephony in the Caribbean is highly valued as a tool for strengthening social ties and for increased personal security in the Caribbean and Latin America, but it is proving to be more useful for enhancing business and employment opportunities in Africa (Galperin and Mariscal, 2007). Again, Africa is ahead with an ICT innovation that has created a Pan-African market based on cell phones, with Accra, Ghana putting out an e-bay type store of agricultural products across countries in West Africa. Some companies are even offering the prices of agricultural products in real time (The Economist, 2007). Trinidad and Tobago should investigate similar innovations, and should encourage the use of mobile telephones for business transactions in agriculture.

9. EXTENSION AND e-AGRICULTURE—BEST PRACTICES AND INNOVATIONS

Defining e-Agriculture is complex. It involves the integration of technology with multimedia, knowledge and culture, to improve communication and learning among stakeholders. The goal of e-Agriculture is to improve communication channels and communication practices. As an emerging field, e-Agriculture focuses on the use of internet and other digital technologies to improve information exchange, communication and learning locally, regionally and worldwide (Seywan Institute, 2009).

The Food and Agriculture Organization (FAO) hosted the first e-Agriculture workshop in June 2006, to engage as wide a range of stakeholders involved in e-Agriculture to analyze stakeholders' familiarity with the term. A resulting survey revealed that only 41 per cent of respondents were familiar with the term e-Agriculture. Stakeholder's perception of e-Agriculture also varied widely (e-Agriculture Working Group, 2007).

Not surprisingly, Trinidad and Tobago's e-Agriculture is in its infancy. Awareness and education are factors that impact its adoption. In fact, e-Commerce and e-business are also in their infancy, and there is also a need to educate consumers about their presence and use (Universalia, 2006). The Ministry of Agriculture is currently participating in a number of the priority e-Government initiatives identified by the Ministry of Public Administration and a number of services are now online and are easily accessible *via* an e-Government Portal. This website is available at http://www.gov.tt/ and provides citizens, business and visitors with a one-stop-shop of government information and services (The Ministry of Agriculture Land and Marine Resources, 2005). The link to Food & Agriculture provides information on training and services, fisheries, food safety, imports and exports, agricultural incentives, and land-related services. The portal however is informational, and does not allow a two-way flow of information.

The website e-agriculture.org describes a global initiative that promises to enhance sustainable agricultural development and food security by improving the use of information, communication, and associated technologies in the agricultural sector. The overall aim of e-agriculture.org is to enable the exchange of opinions, experiences, good practices and resources related to e-Agriculture, and to ensure that the knowledge created is effectively shared and used worldwide. Participants are invited to join the e-Agriculture Community and use features such as the e-Agriculture Blog, virtual consultation forums, and other features such as e-Agriculture on del.icio.us (Delicious Social Bookmarking, n.d). Delicious is a free social bookmarking tool used to store bookmarks online, tag them and shares them with colleagues. People can interact on a web-based platform with subject matter specialists and community members around the globe on topics of your choice. e-agriculture.org encourages the two way exchanges that can empower individuals and communities. This should be a logical step for the Ministry of Agriculture in Trinidad and Tobago.

10. INNOVATIONS FOR THE FUTURE: UNDERSTANDING THE ROLE OF TECHNOLOGY

Technology is the systematic study of techniques for making and doing things. Webster's Dictionary defines technology as «the totality of means employed to provide objects necessary for human sustenance and comfort». Technology is also basic to the survival of culture, and introducing technologies into institutional practice must be based on behavioral knowledge of change (Bennis *et al.*, 1969). Technology therefore embraces a growing range of means, processes, and ideas in addition to tools and machines. It involves knowing the cause and effect relationships embedded in machines and methods (Sproull and Goodman, 1990).

Technology can be disruptive or sustaining. Disruptive technologies are innovative that may not fit with traditional standard practices. Initially they may be viewed with scepticism since the initial benefits are not apparent. Extension by nature prefers sustaining technologies because they maintain the status quo (Coates, 2004). But ICT can be disruptive, and is changing organizational structures in the workplace. Knowing how to implement ICT in Trinidad and Tobago can bring extension gracefully into the next century; but present organizational structures must be adjusted to take advantage of this technology (Margaret, 1989). Case studies show that social alliances can effect rapid improvements in local institutions confirming that bureaucratic reform and improved community social organizations allow for powerful coalitions and rapid, self-sustaining development (Monica *et al.*, 2000).

There are other innovations that can be used in conjunction with ICT to improve the delivery and management of extension. Agricultural Extension in Trinidad and Tobago has faced many challenges over the years but has been constantly trying to adapt to the changing scenario (Seepersad , 2000). Since 1999 the Ministry of Agriculture has been experimenting with Geographic Information Systems (GIS) and Global Positioning Systems (GPS) technology to help address the needs of the farming community. There has been rapid development of GIS worldwide, but little preparation of people and institutions for its implementation. Few researchers have questioned the appropriateness of GIS and GPS technology, and the role they could play in developing new approaches in extension management schemes.

Dr. V. P. Sharma of the National Institute of Agricultural Extension Management (MANAGE), Hyderabad emphasized that access to information and improved communication is crucial for sustainable agricultural development. He observed that modern communication technologies can help to improve communication, increase active participation and further the dissemination of information and sharing of knowledge and skills. He believes that "Cyber Extension" will be the major form of technology dissemination in the near future (Report of the First Consultation on Agricultural Information Management, 2000).

New trends in technology have created multiple interfaces for acquiring, storing, and exchanging geographic data in Trinidad and Tobago. In particular, emerging computer and communication technologies have now reshaped possibilities for rapid data transfer and information exchange directly between farming communities and extension services. The digital environment has opened up innovative avenues for collaboratively managing agricultural data through GIS and GPS networks. In addition, an important function within GIS is the ability to integrate temporal data whether as discrete events or repeated observations with location and attribute data to make predictions.

In extension, such an undertaking requires strong linkages, systematic planning, community empowerment, and a compelling focus on learning how to manipulate emerging technologies to fit within a sustainable social, institutional, and environmental framework. GIS can therefore become an effective tool for decision-making in Trinidad and Tobago. GIS has proven to be a quicker and more efficient approach for conducting cost-effective assessments, research and analysis on spatial phenomena. GIS therefore, has the potential to become the hub or major planning tool of the Ministry of Agriculture Land and Marine Resources. It is ideal for acquiring data on rainfall, soil type, land cover, *etc.*, and merging them with current farm or location datasets for display and analysis. As a management tool, GIS allows government to link disparate pieces of information by location, and so becomes the infrastructure for managing information.

Extension administrators must be willing to support innovations designed by extension agents. In short, would GIS be more likely to be adopted if the institutional locus were among extension officers (Shrum and Beggs, 1997). I emphasize the central role in understanding the social dynamics of the adoption of GIS, and providing a reasonable perspective from which to develop new plans, designs, and structures to deliver better goods and services to the farming community in Trinidad and Tobago.

10.1. Defining GIS

A GIS is a computer-assisted information management system of geographically referenced data. It contains two closely integrated databases: one spatial (locational) and the other attribute (statistical). The spatial database contains information in the form of digital coordinates, usually from maps or from remote sensing. These can be points, lines, or polygons. The attribute database contains information about the characteristics or qualities of the spatial features, for example, crop yields, insect migration, and farm populations. Thus, GIS are tools specially developed for effectively handling spatially referenced data.

Although GIS is similar to conventional database technology, its principal distinction is that it organizes, processes, and reports information, which is of a geographical nature. Organized spatial information is an essential element for good management. A Geographic Information System can therefore be summarized as the organized activity by which a researcher can measure aspects of geographic phenomena and processes; and represent these measurements in a computer database to emphasize spatial themes, entities, and relationships; then operate upon these representations to produce more measurements and to discover new relationships by integrating disparate sources; and transform these representations to conform to other frameworks of entities and relationships.

10.2. Defining GPS

The Global Positioning System was developed by the United States Department of Defense in 1973. It is a navigation and positioning tool made up of a constellation of 24 satellites orbiting the earth about 12,000 miles above us at speeds of about 7,000 miles an hour. We use these satellites as reference points to calculate positions on the earth using GPS receivers. GPS technology is now used in countless other professions. There are many applications beyond determining basic positions or locations of objects in space. These include monitoring the movement of people, pests, diseases *etc.*

11. A MODEL FOR INTEGRATING SPATIAL TECHNOLOGIES WITH ICT IN TRINIDAD AND TOBAGO

The diagram below demonstrates how GIS could be integrated successfully and sustainably in an Extension Service. First, GIS and GPS training will help develop cognition through reasoning, computer literacy, and analysis. The intersecting rings represent information processing, problem solving, and applications in a collaborative environment. As extension officers learn about the components of GIS, they develop skills to use GIS and GPS to solve real-world problems, and then learn to work together in a complex but authentic context to develop better cognition skills for using GIS. These are

12. JUSTIFICATION FOR GIS AND GPS INTEGRATION

All over the world GIS is being used increasingly in agriculture. It is now used to assess land use changes through thematic mapping, database development and dissemination of soil and terrain data for sustainable agriculture and environmental protection, and as a tool for monitoring crop diseases (Hijmans, 1997). In the United States GIS allows extension agents to be involved in higher order planning such as program design and research activities (Stan, 1997). GIS recently provided an opportunity for The Extension Data Centre of the Texas Agricultural Extension Service to grow professionally. Staff acquired new skills and brought about a highly diversified group of individuals with the ability to offer essential GIS services and respond to training requests (Jose and Diann, 2000).

The Cooperative Extension Service at Mississippi State University evaluated the effectiveness of incorporating GIS technology into the issues-programming responsibilities of extension specialists and agents. Sample GIS map products were produced to illustrate some possible applications in each program area. Determination of county staffing needs was streamlined using GIS technology. State leaders used GIS to objectively make decisions; specialists and agents in the major areas of programming found many valuable uses for the technology. Through educational and information sharing sessions GIS specialists and other Extension personnel have developed Extension programs utilizing GIS as a program needs assessment and program monitoring tool (Joselito and James, 1997).

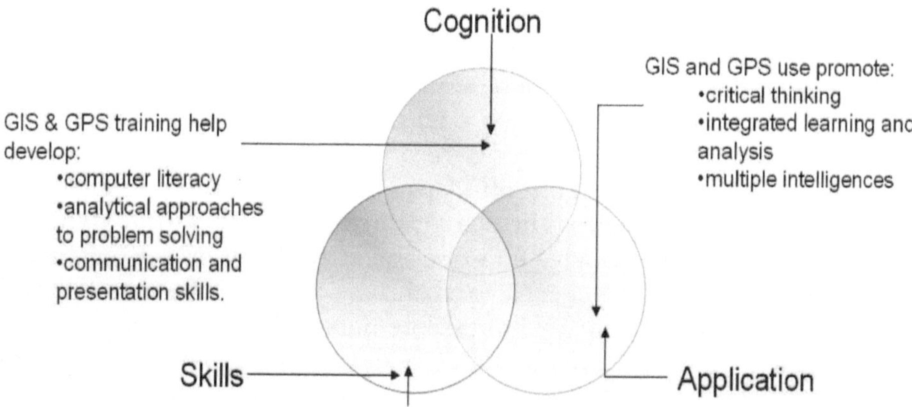

Figure 1 : Core components for sustainable use of GIS and GPS

Recently, Brown County Kansas and Brown County Extension developed an online GIS to improve emergency preparedness in such a way that multiple agencies and the public could access it. The system is accessed using a Web browser. Most exciting are the potential educational impacts of the system. Educational groups, such as schools or 4-H programs, can augment their curricula with local spatial explorations (Brown, 2007). These examples justify at least some experimentation in Trinidad and Tobago to discover a reasonable approach for an accepted model.

13. GIS, GPS, AND MULTIMEDIA INTEGRATION IN TRINIDAD AND TOBAGO

The dynamic and complex nature of agriculture requires an extension system that is responsive and adaptable to changes. GIS makes information more usable, and is flexible enough to integrate with "conventional" databases. Many agree that GIS can also improve an organization's quality of service to its clients/customers by empowering them to speed up their operations. Easier access to information means faster decision-making, and increased confidence of clients and providers. GIS therefore raises the image of an organization and promotes high morale in the workplace. GIS has qualified as a pure public good, and should be appreciated by all because all benefit from its application (Brown, 2007).

GIS, GPS, and multimedia are disruptive innovations that recently gave 15 extension officers in Trinidad and Tobago the ability to acquire, organize, and analyze spatial data (GIS/GPS and Multimedia Training Workshop, 2008). These technologies demonstrated potential for managing extension activities in a way that extension could communicate more effectively with its clientele. Extension officers collected location data on farms using inexpensive GPS units and also used digital cameras to record information. On-line mapping tools and services (Google, Microsoft Maps *etc.,*) were then used to merge the GPS data and digital images. The activity proved to be practical, dynamic, holistic, proactive, and inclusive for extension activities. Working together in groups, extension officers were able to create farm profiles showing important aspects of farms. They were able to explore data and discover patterns and characteristics of farming communities. GIS and GPS technology are therefore indispensible but neglected tools for managing agricultural and natural resources in Trinidad and Tobago. They are effective for program planning and decision-making. Successful integration of GIS, GPS and multimedia technology requires training, peer collaboration, and willingness on the part of participants to learn to use these technologies.

GIS technology was introduced to the Ministry of Agriculture Extension Service in 1999 using a participatory approach (Joseph, 2003). This study focused on the social dynamics of the adoption of GIS, and examined extension officers' responses to the inclusion of GIS into program planning and management (Wellar, 1988 and Onsrud, *et al.*,1992). Participatory design essentially gave extension officers the opportunity to examine issues related to how GIS should fit within the organization to make it meaningful for them as primary users, and become an effective tool for program planning and decision-making.

14. NATIONAL POLICY ON IT IN TRINIDAD AND TOBAGO AGRICULTURALEXTENSION SERVICE

The Government of Trinidad and Tobago is working on a "Fast Forward" agenda to transform the country into a knowledge-based society by 2008 (Government of the Republic of Trinidad and Tobago, 2003). The Ministry of Public Administration and Information (MPAI) developed Fast Forward to provide all citizens with affordable Internet access, and maximize the potential within all citizens (Ministry of Public Administration & Information, 2004). The initiative includes the public and private sectors, and consists of roadmap for an online society and a knowledge-based economy. In short, Trinidad and Tobago's ICT policy has established a goal not only for social, cultural, and economic development, but to develop the country into a knowledge-based society (ICT Policies & Studies, 2003).

The Ministry of Agriculture in Trinidad and Tobago is committed to establishing a distributed computing environment to function in the Internet Age. The Ministry outlined a strategic plan for the period 2003 - 2008 identifying computer training, data security, information sharing, reliable/ cost-effective service, and performance improvements as the five IT-related goals. To meet these goals, the Ministry focused on improving IT infrastructure (hardware, software and networks) and developing standards for intranet/ internet communication. Realizing that transition from analog to digital formats required new operating security measures and procedures for database access and information sharing, the plan proposed a new configuration for IT within the Ministry called the Information and Communication Technology Division.

Today there are established certified Local Area Networks, and a national Agricultural Information System (AGRIS) for recording and delivering goods and services to stakeholders. The GIS workshop was a simple demonstration of practical but effective applications of ICT that can foster a proactive and collaborative environment for immediate application. GIS and GPS technologies are ready for prime time use in the Trinidad and Tobago Extension Services. Training therefore plays a strategic role for extension professionals. It must be continuous, and in many cases change may require employee "retooling" for adaptation to a new political, social, and economic environment.

There are inherent problems with technology. New technologies are often more complex and difficult than earlier technologies. But changing technology isn't the issue in Trinidad and Tobago. Extension's future depends on the ability to interpret trends and use technology to deliver programs and teach problem solving. Knowing how extension professionals will interact with GIS and GPS, as they learn to integrate it on new communication platforms it will help the Ministry to adjust the organizational structure to operate more efficiently. However, Extension cannot do it "overnight" since extension professionals need to be trained to organize and evaluate masses of information generated from a combination of new technologies that can inundate them or overwhelm database and retrieval systems. There is also still a need to maintain and integrate communications from FAX transmissions, conferencing systems, videotapes, telephones, *etc*. The extension agent can't be a specialist in every area anymore, but that agent is now a facilitator for finding the solution to the client's problem using databases and communications systems (Ezell, 1989).

It is therefore critical that policy makers invest and support ICT innovations within the extension service. At the practical level, weak formal research capacity opens opportunities for new or experimental approaches that may prove to enhance older systems. As components are added over time (on the basis of trial and error), the process of testing is never-ending (Janice Jiggins, 1988). This initiative must be accompanied by measures to promote wider use of ICT in the sector (Republic of Trinidad and Tobago, 2007).

15. CURRENT BEST ICT APPLICATION IN TRINIDAD AND TOBAGO: NAMDEVCO

The National Agricultural Marketing and Development Corporation (NAMDEVCO) of Trinidad and Tobago is poised to be the most effective user and promoter of ICT in agriculture for the region, especially in the area of marketing. Their website is still largely informational for the public, but data acquisition is in near real time. In short, NAMDEVCO has been able to deliver services to the farming and marketing community. The NAMDEVCO website offers services, marketing opportunities, fact sheets, media releases, and even member logins. Members have access to a price database that holds over 120 commodities. There is also a feedback form and forum for posting current issues.

NAMDEVCO has recently developed a National Agricultural Marketing Information System (NAMIS) to share agricultural information, and improve domestic agricultural produce marketing. Presently, Extension agents use Palm Pilots (Dell Axims) to collect and update farm gate, retail and wholesale market prices, and upload them to a server in near real time. NAMIS also provides farmers with information that helps them to plan cropping programs, and even arrange sales before harvest time. Sometime in the near future, producers will have opportunities to advertise their product, and there is potential for on-line business transactions (Universalia, 2006). In the near future, NAMDEVCO's website will include on-line maps. The company is investigating ways to integrate GIS and GPS interactively for production and marketing information.

The use of PDAs for data collection is not new. For example, medical practitioners in Uganda have used PDAs for epidemiological surveys on malaria. They were able to access and use medical reference tools and texts, collect data, and share information. PDA made data collection and analysis quick, easy and at a fraction of the cost of traditional pen-and-paper survey (Haji, 2005). NAMDEVCO proposes that in time farmers and agricultural marketing entrepreneurs will be able to provide this data *via* farmer-owned PDA devices; but recognizes that computer literacy among the farming population is low.

NAMDEVCO also produces a monthly bulletin, The Greenvine Newsletter that provides information to the agriculture Sector. "The Market Price Watch" is a service that provides weekly information in three main daily newspapers on wholesale and retail prices for select produce. The National Agricultural Market Information System (NAMISTT) Website provides a range of information including agricultural prices and volume of produce at wholesale and retail markets, buyers and sellers forum and traders contact database.

16. VISIONS FOR THE FUTURE

The extension system in Trinidad and Tobago has become relatively obsolete, and needs modernization. The Ministry of Agriculture is moving in the right direction to establish a demand-driven, gender sensitive, participatory, bottom-up, and interactive approach. It is difficult to prescribe any specific application of information technology for the country. However, the Ministry needs to be more aggressive and study the successes in ICT innovations. Efforts to apply ICT innovations should be applied immediately as case studies at selected locations. Extension officers should work collaboratively in teams with at least one specialist in electronic communication technology to identify specific activities in extension where they can apply communication innovations. There should also be structured training for farmers, extension agents, administrators, and IT personnel to understand how to use tools and equipment that generate useful information for managing agricultural information.

The Ministry needs to improve interactive electronic communication by focusing on Internet applications similar to NAMDEVCO. There must be easy data retrieval systems for farmers to investigate and share information independently. Extension must exploit the free communication tools on the Internet to interact with clientele quickly, reliably, and effectively.

GIS and GPS technology offer opportunities for monitoring the environment, but more so, these new technologies allow new ways for communicating and investigating agricultural phenomena. Through training workshops evaluations, GIS and GPS technologies do have the potential to act as catalysts for change within extension in Trinidad, and the knowledge generated through social interaction will become the fundamental driver for change. Successful implementation of ICT within extension will allow other divisions, units, state-owned enterprises, statutory bodies, and affiliated agencies to examine its potential. The integration of GIS, GPS, and multimedia in extension is new, and as we continue to model the real world using GIS, we realize that organized information informs us on how resources and people can be understood and best served as we address environmental issues.

Technology is more than a collection of machinery. It includes acceptable practices, procedures, and expectations. Strong administrative support and funding must be available to provide the necessary infrastructure for ICT. Administration must also recognize the role of the extension officer in program planning and provide the time and resources for their professional development. Implementation of ICT includes repeated cycles of designs, new ways of thinking, with reinvention becoming a significant part of the process. There is a strong need for linkages among various stakeholders, but this will only occur under the banner of strong administrative leadership. Agencies will have to share information; countries will have to share data and resources. One extension officer in Trinidad and Tobago describes in his own words:

We talk about things should not be done in a vacuum. I feel that we need to sit, and people begin to share the information right away. I feel that I know how we [extension officers] can solve our own [problems]. We need to do that. The point is that no system stuff will happen just like that. There will be a lot of problems, a lot of things you need to iron out. See how it works, and people get familiar with that, then later on – incremental (Joseph, 2003).

In short – we need to begin the process now and address change through participatory methods and communication.

References

Agricultural Extension in Laos: A Pilot Project. Novartis Foundation for Sustainable Development (NFSD), 2001. Available on-line http://www.foundation.novartis.com/laos_project.htm. Novartis Foundation for Sustainable Development (NFSD), 2001.

Akpabio, I.A., D.P. Okon and E.B. Inyang., 2007. Constraints affecting ICT utilization by agricultural extension officers in the Niger Delta, Nigeria. *Journal of Agricultural Education and Extension,* 13 (4). 263-272.

Asaba, Jane Frances *et al.,* 2006. *Bridging the Information and Knowledge Gap between Urban and Rural Communities through Rural Knowledge Centres:* Case Studies from Kenya and Uganda. Quarterly Bulletin of the International Association of Agricultural Information Specialists (1019-9926) Vol.51;p.143

Bengali, Shashank., 2009. Cell Phones Power Financial Revolution in Africa. McClatchy Newspapers. Retrieved March 16, 2009 from http://www.mcclatchydc.com/world/story/61909.html

Bennis, Waren G., Denne, Kenneth D., and Chin, Robert., 1969. *The Planning of Change.* Holt, Rinhart and Winston Inc. New York. Pp. 33.

Benor, D. and M. Baxter., 1984. *Training and Visit Extension.* Washington D.C.: The World Bank.

Brown, Stephen C., 2007. The Brown County Online GIS: An Example of a Multi-Agency Collaborative Mapping System. *Journal of Extension,* [On-line] 45(5) Article 5TOT3. Available at: http://www.joe.org/joe/2007october/tt3p.shtml

Coates, Deborah, 2004. Weblogs as a disruptive technology for Extension. *Journal of Extension,* 42(3). Also Available on-line http://www.joe.org.ezproxy.gvsu.edu/joe/2004june/comm1.shtml

Delicious Social Bookmarking n.d., Retrieved on March 20, 2009 from http://delicious.com/

Dolly, David., 2008. Assessing information sharing and changes in knowledge, attitude, and skill among farmer field school participants from the Commonwealth of Dominica and Trinidad and Tobago. In: *Proceedings of the 24th Annual Meeting E.A.R.T.H.* University, Costa Rica. Also Available from http://www.aiaee.org/2008/Papers/147.pdf

e-Agriculture Working Group, 2007. Analysis of Global e-Agriculture Survey. Retrieved February 12, 2009 from http://www.itu.int/wsis/c7/e-agriculture/docs/survey-analysis-2007.pdf

Encyclopedia of the Nations, 2008. Trinidad and Tobago Infrastructure, Power, and Communications. Retrieved on March 6, 2009 from http://www.nationsencyclopedia.com/economies/Americas/Trinidad-and-Tobago-INFRASTRUCTURE-POWER-AND-COMMUNICATIONS.html

Ezell, Margaret, P.,1989. Communication-Age Trends Affecting Extension. *Journal of Extension*, 27 (3).

Food and Agriculture Organization of the United Nations, 2007. World Congress on Communication for Development: lessons, challenges, and the way forward. Retrieved March 29, 2009 from *http://www.fao.org/docrep/010/ai143e/ai143e00.htm.*

Galperin, Hernán and Mariscal, Judith., 2007. Mobile Opportunities: Poverty and Telephony Access in Latin America and the Caribbean. Retrieved on March 11, 2009 from http://www.dirsi.net/files/REGIONAL_FINAL.pdf

Ganpat, Wayne G., and Bholasingh, Deokee., 1999. Attitudes of Farmers Toward Farming in Trinidad. *Journal of International Agricultural and Extension Education,* Fall 1999.

Gershon, Feder., Anthony Willett, Willem Zijp, 1999. *Agricultural Extension: Generic Challenges and Some Ingredients for Solutions*. Available on-line http://www.worldbank.org/html/dec/Publications/Workpapers/wps2000series/wps2129/wps2129.pdf

GIS/GPS and Multimedia Training Workshop, 2008. Demystifying Global Positioning Systems (GPS) Technology in Extension: Integrating Field Applications with Multimedia and Communications Technology. Ministry of Agriculture, St Clair POS, Trinidad & Tobago W.I.

Gosein, Megnath, 1992. Agribusiness Policy in Trinidad and Tobago for Export Marketing and Agro-Processing., in Farm and Business. Retrieved on February 11, 2009 from http://ageconsearch.umn.edu/bitstream/44550/2/AGRIBUSINESS per cent20POLICY per cent20IN per cent20TRINIDAD per cent20AND per cent20TOBAGO per cent20FOR per cent20EXPORT per cent20MARKETING per cent20AND per cent20AGRO-PROCESSING.pdf

Government of the Republic of Trinidad and Tobago, 2003. Fast Forward: Trinidad and Tobago's National ICT Strategy. Retrieved on March 10, 2009 from http://www.fastforward.tt/default.aspx

Government of the Republic of Trinidad and Tobago, 2008a. Transformation Plan for the Agriculture Sector. Retrieved on March 6, 2009 from http://www.agriculture.gov.tt/applicationloader.asp?app=articles&id=1254

Government of the Republic of Trinidad and Tobago, 2008b. Community-based Telecentres. Retrieved on March 15, 2009 from http://www.gov.tt/citizen/service.aspx?id={954B96C4-FA2E-43BC-B83A-89CED2093162}

Government of Trinidad and Tobago, 2004. National Connectivity Report. Ministry of Public Administration (Telecommunications Policy Unit).

Government of Trinidad and Tobago, Ministry of Planning and Development Central Statistical Office, 2005. 2004 Agricultural Census, Preliminary Report Retrieved February 20, 2009 from http://cso.gov.tt/files/cms/Agricultural per cent20Census per cent20Preliminary per cent20Report.pdf

Haji, Mohamed., 2005. The Developing World and the Role of Information and Communication Technologies. Retrieved March 6, 2009 from http://ocw.mit.edu/NR/rdonlyres/Electrical-Engineering-and-Computer-Science/6-901Fall-2005/69B73C22-D1D2-4A59-ABCD-95A76D0D5ED8/0/developing_world.pdf

Harold R. Baker., 1987. Future options and prospects in rural extension and technology transfer. *Occasional Papers in Rural Extension No.1.* Department of Rural Extension Studies, University of Guelph.

Hijmans, R.J., 1997. GIS for late blight research. In: Denisov, N., C. Heberlein, L. Czaran and O. Simonnet. GIS in Agricultural Research: Awareness Package. UNEP/DEI/TR.97-9. (# 34).

ICT Policies & Studies, 2003. Trinidad and Tobago ICT4D National Policy. Retrieved on January 10, 2009 from http://www.comminit.com/en/node/148369/308

Janice Jiggins., 1988. Farmer participatory research and technology development. *Occasional Papers in Rural Extension, No. 3.* Department of Rural Extension Studies, University of Guelph.

Jose, Reynaldo A. Santos and Diann, M. Mitchell, 2000. Maintaining the relevance of an extension data center. *Journal of Extension*, 38 (2).

Joselito K. Estrada and James R. Steil., 1997. Evaluating the effectiveness of adopting a GIS in the extension service: Evidence from the Cooperative extension service at Mississippi State University. *Journal of Extension*, 35 (5).

Joseph, Edwin., 2003. Responses to the design, application, and management of geographic information system (GIS) datasets (generated models) for managing extension programs. *Ph.D. Thesis (unpublished)*, Development Studies, University of Wisconsin—Madison.

Kiplang'at, Joseph and Ocholla, Dennis N., 2005. Diffusion of information and communication technologies in communication of agricultural information among agricultural researchers and extension workers in Kenya. *South Africa Journal of Library and Information Science*, 71 (3).

Margaret, P. Ezell, 1989. Communication-Age trends affecting extension. *Journal of Extension*. 27 (3).

Ministry of Agriculture Lands and Marine Resources, 2007. *Vision 2020 re Agriculture* Available http://www.agriculture.gov.tt/documentlibrary/downloads/49/Agriculture-Vision2020-part per cent20a.pdf

Ministry of Public Administration & Information, 2004. National Policy on Electronics Transactions. Available on-line http://www.agriculture.gov.tt/documentlibrary/downloads/89/ElectronicTransactionsPolicy.pdf

Monica, Das Gupta, Helene Grandvoinnet, and Mattia Romani, 2000. State community synergies in development: Laying the basis for collective action. *World Bank Group Policy Research Working Paper No. 2439.*

NationMaster.com, 2003-2009. Media statistics Mobile phone subscribers Trinidad and Tobago (historical data), World Development Indicators Database. Retrieved on March 11, 2009 from http://www.nationmaster.com/time.php?stat=med_mob_pho_sub-media-mobile-phone-subscribers&country=td-trinidad-and-tobago

Neuchâtel Group, 1999. Common framework on agricultural extension. Retrieved on March 18, 2009 from http://www.neuchatelinitiative.net/images/cf_en.pdf

Onsrud, H.J., Pinto, J.K., and Azad, B., 1992. Case study research methods for geographic information systems. *URISA Journal,* 4(1). pp.32-44.

Radio Station World: Your Global Radio Station Directory, 2008. Retrieved on March 29, 2009 from http://radiostationworld.com/Locations/Trinidad_and_Tobago/default.asp

Report of the 6th Consultative Expert Meeting of CTA's Observatory on ICTs, 2006. ICTs-Transforming Agricultural Extension? *CTA Working Document Number 8034.*

Report of the First Consultation on Agricultural Information Management, 2000. Food and Agriculture Organization of the United Nations Rome. Retrieved on March 15th from http://www.fao.org/docrep/x7936e/x7936e03.htm

Republic of Trinidad and Tobago, 2001. Utilization of Information Technology by Households, 2001. National Institute of Higher Education Research (NIHERST).

Republic of Trinidad and Tobago, 2007. Vision 2020 Operational Plan 2007-2010 Transformation in Progress 2007 Report. Available Online: http://www.vision2020.info.tt/pdf/Vision2020_2007report.pdf

Riggs, Michael, 2003. Agricultural information management: Global and community development. *Quarterly Bulletin of the International Association of Agricultural Information Specialists (1019-9926)*, Vol.48;p.32

Seepersad, Joseph., 2000. Some challenges facing Extension in the 21st century: A Caribbean perspective. *Proceedings of the 10th Annual AIAEE Conference*, Washington, USA

Seepersad, Joseph., 2003. *Case Study on ICTs in Agricultural Extension in Trinidad & Tobago*, Prepared for CTA's ICT Observatory 2003 ICTs—transforming agricultural extension. Wageningen, 23-25 September 2003

Seywan Institute, 2009. What is e-Agriculture. Retrieved on March 20, 2009 from http://eagriculture.org/?m=200901

Shrum, Wesley and Beggs, John J., 1997. Methodology for studying research networks in the developing world: Generating information for science and technology policy, in Knowledge and Policy: *The International Journal of Knowledge Transfer and Utilization*, Winter 1997, 9 (4). pp. 63.

Singh, R.H., Rankine, L.B. and Seepersad, G., 2005. The Caricom regional transformation program for agriculture. A review of agricultural policies: Case study of Trinidad and Tobago. *Report prepared for the CARICOM Secretariat.* Available on-line http://www.caricom.org/jsp/community/agribusiness_forum/agri_policy_trinidad_tobago.pdf

Sproull, L.S. and Goodman, P.S., 1990. Technology and organizations: integration and opportunities. In: Goodman, P.S., Sproull, L.S. and Associates, (Eds). *Technology and Organizations*, pp. 255, San Fransisco: Jossey-Bass.

Stan, Geertman., 1997. Geographical Information Technology as a Planning Tool. In Geographical Information '97. April 16-18, 1997. *Joint European Conference and Exhibition. Austria Center, Vienna. Proceedings,* Volume 1.

Sustainable rural development: Progress and challenges, 1995. Special report is based on the *Fourth Progress Report on the WCARRD Program of Action.* Sustainable Development Department Food and Agriculture Organization of the United Nations FAO.

Swanson, Burton E. and Samy, Mohamed M., 2002. Decentralization of Agricultural Extension Systems: Key Elements for Success. Available on-line http://info.worldbank.org/etools/docs/library/51025/ZipAgExtension1/ag_extension1/Materials/May6Session1/Decentralization-India4-18-03_paper.pdf

The Economist, 2007. The Spread of Mobiles in Africa. Retrieved on March 11, 2009 from http://www.economist.com/world/mideast-africa/displaystory.cfm?story_id=8597377

The Ministry of Agriculture Land and Marine Resources, 2005. *Information Technology Strategic Plan Version 1.1.* Government Document.

The National Medium Term Priority Framework for FAO Assistance - Trinidad and Tobago Draft document, 2006. Available http://www.caricom.org/jsp/community/donor_conference_agriculture/agri_profile_trinidad_and_tobago.jsp

Thomas, H. Henderson and Michael Quinn Patton, 1985. Agricultural Extension For Rural Transformation: The CAEP Model. In: Gomes, P.I., (Ed.). *Rural Development in the Caribbean*. Kingston, Jamaica.

Trinidad and Tobago Agri-business Association, (n.d). Description of the National Agri-business development Program for Trinidad and Tobago. Retrieved from http://www.agribusiness-solutions.org/?q=system/files/V.Stewart_ INNOVATIVE+BUSINESS+SOLUTION+-SUPPLY+VALUE+CHAIN.doc

Trinidad and Tobago Computer Society, 2007. Internet Options in Trinidad and Tobago. Retrieved on March15, 2009 from http://www.ttcsweb.org/articles/isptnt/

Universalia, 2006. Assessment of E-Readiness of Micro, Small and Medium Enterprises (MSMEs) in Trinidad and Tobago. Retrieved on March 12, 2009 from http://

74.125.95.132/search?q=cache:FiM1KPQ5HAwJ:ebusiness.fastforward.tt/business/LinkClick.aspx per cent3Ffileticket per cent3DdOKIWKaVWWM per cent253D per cent26tabid per cent3D168 per cent26mid per cent3D640+Assessment+of+E-Readiness+of+Micro,+Small+and+Medium+Enterprises+in+Trinidad+and+Tobago&cd=1&hl= en&ct=clnk&gl=us

Villavicencio, Daniel & Ponce, Lluvia Ponce, (n.d.). The Republic of Trinidad and Tobago. Retrieved March 11, 2009 from http://portal.unesco.org/education/en/files/55575/11998954195Trinidad_y_Tobago.pdf/Trinidad per cent2By per cent2BTobago.pdf

Wellar, B., 1988. A Framework for Research on Research in the Information Technology - Local Government Field. *Proceedings of the 26th Annual Conference of the Urban and Regional Information Systems Association*, Los Angeles, California (Washington D.C.: Urban & Regional Information Systems Association), Vol. 4, pp. 379-386.

World Bank, 1990. Agricultural Extension: The Next Step. *Policy and Research Series No. 13*. Washington, D.C.

Author Information

Dr. Edwin Joseph

Associate Professor

Geography & Planning

Grand Valley State University

B-4-200 Macinac Hall, 1 Campus Drive, Allendale MI, USA

Postal Code: 49401

Phone No. (O):616-331-2990

Mobile Phone No.: 616-916-1883

Fax No. 616-331-8635

e-mail: josephe@gvsu.edu, eddijay@gmail.com

URL: www.gvsu.edu

Dr. Edwin Joseph is a tenured Associate Professor in the Department of Geography & Planning at Grand Valley State University. He is a GIS Outreach Specialist and an ESRI (Environmental System Research Institute) Authorized Instructor for ArcView GIS. Dr. Joseph was employed as an extension officer with the Ministry of Agriculture in Trinidad for 15 years. He conducts workshops and training in ICT, GIS, GPS, and multimedia integration using action research. Dr. Joseph has a keen interest in agriculture in developing countries.

□□□

Chapter 16

ZIMBABWE

Christopher Tafara Gadzirayi, Eliada Gudza,
Godfrey Sibanda Title and Justin Mupinda

1. INTRODUCTION

1.1. The Agriculture Sector of Zimbabwe

Zimbabwe has a total area of 390 580 square kilometers of which one per cent is made up of rivers and lakes, 38 per cent is arable land and the rest is suited for livestock and wildlife production. The country is bordered by South Africa, Botswana, Namibia, Zambia and Mozambique. It is located between latitudes 15,30 and 22,30 degrees south of the Equator and between longitudes 25 and 33,10 degrees east of the Greenwich Meridian (CSO, 2006).

The Agriculture sector forms the backbone of Zimbabwe's economy and accounts for about 17 per cent of the GDP. Zimbabwe's agricultural sector comprises mostly smallholder farmers who include communal, resettlement and small-scale farmers. However, this sector is besieged by a plethora of issues, namely high stocking rates, sub-optimal climate, and deforestation and land degradation. The Land Reform and Resettlement Programme which commenced in 1980 resulted in hundreds of thousands of Zimbabweans being resettled on former white commercial farms, in either Model A1(communally owned) or A2 schemes (large scale commercial farms). Thus the challenge to policy makers was now to ensure that future agricultural production is not compromised by current practices. The new farming sector largely comprises resource poor farmers, academics, corporate leaders, high-level policy makers and traditional farmers. Most of the new farmers lack requisite Agricultural skills and knowledge needed in modern Agricultural practice. Land reform measures led to the exodus of some white farmers in the beginning of the 21st century which crippled the economy depriving it of technological know-how and causing shortages of basic commodities.

The modernization of agriculture therefore, among other intervention strategies, through the use of ICTs can contribute directly to food security at national and household levels, beneficiation of agricultural produce, effective land management and an increase of national wealth.

1.2. Contribution of Agriculture to the National Economy

Agriculture employs 60-70 per cent of the total population and it still produces 60 per cent of the raw materials needed for industry. However, owing to a decline in production it is contributing only 15-19 per cent of the country's total GDP. Still, even though it accounts for a relatively low percentage of GDP in absolute terms, it is considerably higher than the 2 per cent observed in most developed market economies. This is an indication that the nation has moved to a mainly subsistence agricultural system. There is yet another phenomenon that greatly hinders the emergence of sustainable agriculture and marketable surplus. The country has high death rates due to HIV infections and unavailability of treatment which leaves population growth at a low 0.62 per cent making increase in food production difficult due to the dependency between population and food production. Rural population averages 8,175,494 (70.3 per cent of the total population) while the urban population is about 4,459,183. Female rural population, which forms the bulk of farmers in Zimbabwe, is 36.5 per cent of the total population. This implies that any innovative intervention strategies aimed at improving Agricultural practices should target women as they contribute greatly to Agricultural output. The country is largely youthful, 43 per cent of the population is below 15 years and 4 per cent is above 64 years (CSO, 2006).

In order for Zimbabwe's agriculture sector to effectively contribute to the national economy, there is need for an increase in food production and a healthy labour force, since labour is the only variable factor of production in a subsistence economy.

1.3. Support Systems

Traditionally the Zimbabwean government has funded most farming projects particularly in the small and communal farming sector. However, owing to the volatile economic climate characterized by hyper-inflation of more than 1million percent, government coffers ran dry resulting in farmers getting their farming support largely from NGOs, church organizations and private companies under contract farming schemes. Commercial Banks also assist with finance under the Reserve Bank Farmer support programmes. Agricultural Extension support from government agents has been minimal to nil, as much agricultural advice come from the private sector.

2. AGRICULTURAL EXTENSION SYSTEMS IN ZIMBABWE

2.1. Organizational Background

The agricultural extension system was introduced in 1927 by Emory D. Alvord, who started with nine agricultural demonstration workers. Later, the departments of Conservation and Extension (Conex) and the Agricultural Development (Devag) were established. The former had the institutional mandate to provide advisory services to white commercial farmers, while the latter was meant to service native smallholder communal farmers. At Independence in 1980, the Department of Agricultural, Technical and Extension Services (AGRITEX) was formed as an amalgamation of Conex and Devag. In its early days, AGRITEX incurred a loss of experienced staff between 1981 and 1985 resulting in both institutional memory and technical expertise in dealing with farmers being lost. This was largely so because AGRITEX was a product of two organizations with different philosophies and experiences in dealing with farmers from different socio-economic backgrounds. The Agriculture Extension workers stood the challenges of the time and established itself as a service for all farmers, but especially smallholder farmers (AGRITEX, 1994).

2.2. Agricultural Extension Approaches in Zimbabwe

2.2.1. The Group Development Area Approach (GDA)

The approach is used when a large number of farmers are involved in the learning process. The GDA approach is based on area and project development through community participation. Local people provide labour while government or donors provide the required inputs. The GDA approach allows extensionists to service and penetrate difficult areas and introduce agricultural extension technology. The main weaknesses are that it is usually difficult to direct services to the needy since the focus is on a particular individual or group. The approach can preclude support and services for other deserving clientele. A heavy reliance on donor support makes projects vulnerable in the event of donor fatigue, and the approach assumes that all farmers face similar problems.

2.2.2. Master Farmer Training

The aim of master farmer training is to communicate modern, scientific farming techniques, largely small holder and communal farmers. Master farmer certificates and badges are awarded to farmers who adopt and practice improved farming methods. This approach is premised on the concept of "trickle-down" theory of extension. A few innovative farmers receive new information, which they are expected to pass on to other farmers through farmer-to-farmer interaction. The approach results in high adoption rate of new innovations as witnessed on hybrid maize varieties among communal farmers in Zimbabwe. The programme failed to produce expected yield increases due to differences in agronomic practices between experiment sites and farmer field conditions. The programme tends to benefit only better-off innovative farmers and may have little contact with target groups, thereby actually increasing income differentials among farmers.

2.2.3. The Radio/ TV Group Approach

The radio listening group approach is widely used in communal areas of Zimbabwe without access to electricity. Farmers gather together in groups of up to seven to listen to radio programmes that address needs of various agro-ecological farming regions. Similarly groups of farmers with access to television come together during presentation of farming programmes. Farmers then discuss the agricultural issues raised in the programmes, and help each other to overcome any difficulties of understanding before applying any of the programmes' messages or technologies that are relevant to their farming practice. The approach helps farmers to learn from each other's experiences as well during the post listening or viewing period. In Zimbabwe, radio and TV programmes provide farmers, from across the country, with platforms to chart to one another through telephone and answer sessions.

2.2.4. The Training and Visit System (T&V)

T&V extension approach was developed for the World Bank by Daniel Benor. It focused on upgrading the technical expertise of field extension personnel. In Zimbabwe, confirmed agricultural practices from government and private research centres, are translated into 'packages' of practice recommendations. They are then passed down the extension organization's hierarchy from agricultural specialists to extension officers, who adopt and adapt recommendations to their specific areas before passing them on to farm level extension assistants. Extension assistants then pass the recommendations to communal and smallholder farmers through extension farmer groups. Agricultural specialists act as a link between research and extension, while periodic training and visits are designed to facilitate linkages between extension and farmer groups.

The approach proved an excellent extension system in mechanised projects, which followed strict schedules. The T&V system was stopped after ten years when evaluations found it inappropriate for developing countries with limited resources and ragged physical environments that are impassable. The system tended to be too mechanist in approach resulting in a top-down orientation and was not demand driven. It lacked the flexibility to make it more relevant to the needs and environment of smallholder farmers (Pazvakavambwa, 1994).

2.2.5. Farming Systems Research and Extension (FSRE)

The approach was developed as a direct response to the failure of various prescriptive agricultural extension models that failed to consider the unique socio-economic circumstances of smallholder farmers resulting in recommendation of inappropriate technologies to the agro-ecological region. The FSRE extension approach is problem solving oriented, systems-oriented, interdisciplinary, farmer-oriented and interactive. It places emphasis on problem diagnosis and on-farm trials as a way of facilitating linkages among the farmers, researchers and extension agents.

FSRE has largely been championed by the Farming Systems Research Unit within the Department of Research and Specialist Services (DR&SS) in Zimbabwe. Extension workers identify trial farmers and recommend them to DR&SS for and on-farm trials. Research specialists monitor farmer performance as they work in their trial plots at the research centre. The farming systems perspective involves greater use of local resources and indigenous knowledge systems. Past and current FSRE efforts have tended to emphasize traditional cropping systems, at the exclusion of livestock husbandry systems. It has been criticized for drawing extension considerations from practice than from theory.

2.2.6. Commodity-Based Approach (CBA)

In Zimbabwe the CBA approach largely organized through parastatal or private companies and is very important for cash crops or export crops such as cotton, tobacco, sugar cane or fresh flowers. The tobacco commodity-based approach has been very successful among small scale commercial farmers owing to timely input supply and effective monitoring. For the cotton production sector, the approach has been greatly helped by a crop research programme supported by effective commodity-based extension, which has also established some successful out-grower schemes. However, the system is negatively affected by side-marketing of cotton products. In sugar cane production yields have risen among small holder farmers as a result of the commodity-based approach in which private firms offer extension and processing facilities. In horticulture and flower production, the approach has been widely used to establish out-grower schemes and provide extension and input credit services to participant farmers. Despite an illustrious history and remarkable achievements, the commodity-based extension approach has drawbacks. If unchecked, the CBA often gives monopoly power to the parastatal or private companies which can exploit marginal and resource poor farmers. Also a decline in trade and pricing often result in poor returns to farmers who happen to be price takers.

2.3. Reflections on Zimbabwe Agricultural Extension Services

- There is little information on how many farmers are actually reached and serviced by AGRITEX.
- The majority of commercial farmers rely on support services from private companies as they are far and wide apart for AGRITEX to reach them.

- The adoption and use of supportive technologies, such as fertilizers, pesticides and recommended agronomic management practices, are generally well below 40 percent across the country (Towonezvi, 1995).
- AGRITEX is only able to recommend technologies that were made available a decade or two ago due to limited research taking place owing to financial limitations.
- Most communal farmers, who have not adopted recommended technologies, tend to use either practices from their own indigenous knowledge system experiments.
- AGRITEX has not systematically adopted farmer driven technologies that work nor adapte to local farm situations.
- Extension farmer contact is severely curtailed by lack of operating budget, transport and brain drain among government staff.
- Staff development programmes, which attempt to strengthen the professional image of AGRITEX extension staff is being hampered by lack of funding.
- NGOs and private organizations provide extension and agricultural services to large-scale and smallholder farmers but for targeted enterprises.
- Modern participatory extension approaches, interactive learning approaches, participatory rural appraisals, rapid rural appraisals, participatory technology development, farmer field schools, innovative farmer workshops, and look-and-learn tours are also being used in Zimbabwe.

Worldwide emerging extension approaches, farmer-first, farmer-back-to-farmer, farmer-to-farmer extension and facilitation - extension agents respond to farmers' requests and programmes and timely reactive extension services are being adopted as sustainable extension techniques in Zimbabwe. Such bottom-up approaches enable farmers to take the initiative, make decisions and own the programme, Action research project-based extension approaches in which a group of farmers work on a project at their local farm, such as mushroom production, while learning the production aspects that will allow them to implement that project on their own, are being practiced.

Although Zimbabwe is trying the above innovative agricultural extension approaches, new digital systems such as ICTs that facilitate faster dissemination of technologies seem to be stealing the limelight. A variety of

ICT tools are now being used to communicate, create, disseminate, store, and manage Agricultural information among large and small scale farmers. ICT technologies being used in Zimbabwe include virtual reality, computers, internet, radio and television, land lines and cellular phones, CDs, journals, news letters and tele-centres among other technologies.

3. ICTs IN ZIMBABWE'S AGRICULTURAL EXTENSION SYSTEM

Zimbabwe is located within the coverage areas of two geostationary communication satellites operated by International Marine Satellite (INMARSAT), Atlantic Ocean East (AOE) and the Indian Ocean West (IOW) lying 35 800 km above the earth. The country is also served by Communications Satellite (COMSAT) and Pan American Satellite (PANAMSAT) that provides rented channels and direct satellite to television. For good reception there is need for a large aperture satellite dishes at fixed locations. There is also a low Earth Orbit system for the coverage of rural areas with respect to the Global System of Mobile (GSM) communication but is too dear for communal and small holder farmers (Zimbabwe e-Readiness Report, 2005).

3.1. ICT in Zimbabwe Agriculture Sector

ICTs can be conceptualized as technologies that are used to collect, store, analyze and pass on information in various forms through various media. The widespread diffusion of ICTs in Zimbabwe's agricultural extension sector has been recognized as a formidable force that can accelerate adoption of new farming technologies. ICTs are powerful instruments for increasing national economic growth and development. Through adoption of ICTs, developing and developed countries alike are now able to emerge with sustainable solutions to long standing issues such as poverty reduction, wealth creation, equity, as well as social justice through internet volunteer consultancy (World Bank, 1995).

In view of the above, Zimbabwe developed a harmonized sector-based ICT policy framework. The Zimbabwe ICT policy framework enables the nation to co-ordinate various initiatives in the public and private sectors and of other stakeholders within and outside the country in order to optimize the allocation and utilization of resources in the development and use of ICTs across all sectors of the economy. The national policy framework provides requisite guidance to the formulation and implementation of ICT strategies and programmes in Agricultural Extension systems.

3.2. Zimbabwe ICT Policy and Its Application to Agriculture Extension

3.2.1. Zimbabwe Millennium Development Goals (MDGs)(2005)

These are goals that were adopted by Heads of State and Government at the fifty-fifth session of the United Nations General Assembly in September 2000. The Zimbabwe Millennium Development Goals (MDGs) recognizes the role of ICTs as tools that add value and contribute significantly to increased food production and eradication of hunger by 2015 among other key MDG targets. Zimbabwe ICT policy was therefore conceived within the framework of the MDGs and adapted to suit local situation.

3.2.2. Policy Objectives

The major ICT Policy objectives are to:

- Provide strategic direction and guidance for sustainable national development through the development and applications of ICTs in Zimbabwe.
- Ensure provision and maintenance of infrastructural facilities necessary for ICTs development, such as reliable supply of electricity, communications and transport.
- Promote and support the systematic, relevant and sustainable development of ICTs.
- Embark on extensive educational and training programmes to provide adequate supply of qualified ICTs personnel and knowledge workers in all sectors.
- Establish structures for effective implementation of ICT strategies.
- Establish institutional mechanisms and procedures for determining sectoral application priorities.
- Encourage the development and use of, and ensure equitable access to benefits offered by ICTs across gender, youths, the disabled and the elderly.

3.2.3. ICT Policy Statements

- Develop and improve ICTs infrastructure for all sectors of the economy (communications, electricity and transport).
- Encourage full utilization of existing communications infrastructure to reduce resource wastage.
- Implement an integrated and equitable framework for accelerated ICTs development and uptake.

- Increase bandwidth on the national backbone and international gateway(s) systems to enhance speed and efficiency of operations. Develop supportive and enabling infrastructure to ensure equitable access to ICTs by all citizens including disadvantaged groups and rural communities.
- Promote local production of ICT products to ensure relevance of content and use of appropriate technologies that meet international standards.
- Establish institutional mechanisms to co-ordinate inter-organisational planning, policy-making and implementation of strategies to develop ICTs taking into account the convergence of broadcasting, telecommunications and on-line computer services.
- Implement measures to develop and retain skilled human resources in the ICTs sector.
- Rationalise the ICTs tariff structure to make ICTs more affordable and accessible.
- Introduce and enforce stringent quality of service standards in the provision of ICTs.
- Create a conducive environment for investment in the areas of ICTs through public private partnerships in all sectors.
- Promote local research and development in software and hardware relevant to all sectors of the economy.
- Promote awareness and use of ICTs (GoZ, 2002).

3.2.4. Status of ICTs in Zimbabwe

The Zimbabwe government demonstrated its commitment to ICTs application through:

- Establishment of the Cabinet Committee on Scientific Research, Technology Development and Applications.
- Expanded rural electrification programme.
- Widening area network used among others.
- Deregulation of the telecommunications sector.
- Phenomenal growth in internet users of up to 400 per cent in 2004.
- Establishment of regulatory frameworks for the ICTs sector.
- Massive computerization of government ministries and all institutions of learning.

3.2.5. Fixed Network Subscribers 1998 – 2004

In Zimbabwe, the incumbent operator, Tel-One has experienced steady growth from 1998 where main telephone lines in operation were 236530 and increased to 287854 in 2002. The main telephone lines per 100 inhabitants grew from 2.13 in 1998 to 2.47 in 2002. The number of public payphones increased from 2864 in 1998 to 3234 in 2000.

3.2.6. Mobile Subscribers 1998 – 2004

Currently there are three mobile telephone operators namely Net-One, Econet Wireless and Telecel.

3.2.7. Applying ICTs to the Department of Agricultural Extension Services

The sector offers extension education services and has research stations in all agro ecological zones of Zimbabwe. Most district centres have no access to the Internet because of poor telephone infrastructure. The problem is worsened by constant vandalism done to national property by thieves.

Access to telephony is quite low within the Agricultural Ministry. A total of 655 centres share 273 computers and about 232 telephone lines, which account for about 35 per cent access rating. This shows the poor access to telephony of 30 per cent (Table 2).

Table 1 : Access to ICTs in the AGRITEX division (2005)

Type of center	No. of Centers	No. With at least a computer	Total number of computers	No. with telephone	Internet access	E-mail access
Provincial	8	8	24	8	1	8
District	56	34	34	25	0	15
Research station	32	32	80	32	2	4
Total countrywide	96	74	138	65	3	27

Source: Zimbabwe e-Readiness Report, 2005

Table 2: Accessibility Rating of ICT in the Agriculture Sector (2005)

Communication Modes	per cent Rating
Conventional telephone lines	30
Mobile telephones	10
Satellite communication	0,1
TV Networks	10
Video conferencing	0,1
Regular mail	40
Fax and e-mail	10
Private Wireless Communication	5

(Source: Zimbabwe e-Readiness Report, 2005)

Often, computers are found in offices and agricultural centres located in places with electricity. The computers are mainly used for administrative purposes as opposed to agricultural extension services. Communal and small holder farmers are mostly located in the remote areas with no access to ICTs related services as there is no electricity. Most Internet and Tele-services are found in urban centres, a few agricultural centres and large scale commercial farms (Ministry of Agriculture and Rural Development). At times cattle auctioneers and Horticulturalists use websites for advertising and marketing commercial stock and fresh produce respectively. Computer controlled irrigation systems are also found in large estates where computer based packages are in use as well as in a number of hunting safari where, GPS is used to identify the location of a shoot. Lack of ICT access among remotely deployed Extension Staff, resettled farmers and the farm labor force confirms the digital divide in Zimbabwe's agricultural sector, more so given the fact that agriculture accounts for more than 65 per cent of the country's labor force (Gudza, 2008).

There is a glaring need for ICT accessibility and usability within the Agricultural Extension division if Zimbabwe is to realize its full potential in improving the country's Agro-based economy.

Commendable strides have been made in using television and radio in educating farmers in programmes like Talking Farming/Murimi Wanhasi/ Umlimi Walamlha.

Zimbabwe Farmers Union (ZFU), an umbrella group for smallholder farmers in Zimbabwe , and several NGOs are developing support systems in the implementation of public access to communication and information centres in rural areas through establishment of telecentres. The ZFU has over 160 000 members spread throughout the country. The strength of the organization is the 6,000 Clubs, which form the local level organizational structures in the farming areas. Availing ICTs through telecentres at all levels of farming would help in the speed transfer of agricultural knowledge among the disadvantaged farmers.

3.2.8. ICT Training in Agricultural Extension

Education and Training Sector

The Zimbabwean government's conviction that education is the best form of investment in human capital saw ICT teaching being introduced in the curriculum of Agricultural Education and Extension programmes within colleges and universities throughout the country (Tables 3, 4 and 5). This indicates that Zimbabwe's Agricultural Extension sector has a high potential to be a knowledge society as the majority of its people can read and write. The adoption and deployment of ICTs will further enhance the process of learning, teaching and training (SARUA, 2007). A total of 600 degree holders and 1253 certificate and diploma holders in ICT are produced annually from public institutions. Most Agricultural graduates with ICT related courses get more lucrative jobs outside Extension Education.

Table 3 : Level of access to computers and Internet in Zimbabwe Polytechnic colleges

Criteria	Harare	Bulawayo	Kwekwe	Mutare	Gweru	Masvingo
Total No. of computers	125	30	107	110	60	75
No. of network points	150	48	80	117	100	90
No. of users	3519	4590	3620	2243	2000	2040
Student access	ICT and Engineering students	ICT Students	Limited	All students	ICT students	All students
Bandwidth Cost	64kbps	56kbps	56kbps	56kbps	56kbps	56kbps
Type of Link	Radio	Radio/Leased line	Dial-up	Dial-up	Dial-up	Dial-up
Internet service provider	ZAR Net	ZAR Net	Africa Online	Africa Online	ZAR NET	Com One
Quality of ISP	Good	Average	Good	Poor	Good	Poor

Source: Zimbabwe e-Readiness Report, 2005

Table 4 : Level of access to computers and Internet in Zimbabwe Teacher education training colleges

Criteria	Belvedere	Joshua Mqabuko Nkomo	Hillside	Mkoba	Mutare	United college of Education
Total No. of computers	200	12	47	28	15	14
No. of network points	200	6	39	28	1	8
No. of users	1524	240	1300	660	19	250
Student access	All	limited	limited	All	limited	All
Bandwidth	64	64	128	128	56	--
Required Bandwidth Cost	128Kbps	128Kbps	128Kbps	128Kbps	128Kbps	128Kbps
Type of Link	Radio	Dial up	Dial up	Leased	Dial up	Dial up
Internet service provider	Ecoweb	Com one	ZARNet	ZARNet	ZARNet	ZARNet
Quality of ISP	Good	Average	Good	Good	Poor	Fair

Source: Zimbabwe e-Readiness Report, 2005

4. ZIMBABWE ICTs SCENARIO IN THE LAST TWO DECADES

4.1. Innovative ICT Initiatives for Agricultural Extension in Zimbabwe

4.1.1. e-Hurudza

e-Hurudza /Umtshayi Wenala /Electronic Farm Manager was developed to empower those with relevant agricultural knowledge and those who have not had the privilege to acquire it.

From October 2002 Jawbone Enterprises, Department of Science and Technology Development and the Department of Agricultural Engineering have worked collaboratively to develop an electronic farm management system that would address the challenges faced by farmers in their quest to boost agricultural production (Jawbone Enterprises, 2007).

e-Hurudza Software

The Electronic Farm Manager covers topics on various crops and livestock. e-Hurudza covers crop and livestock cycles, rotations, the relevant input requirements and applications, farm administration including financial reports, marketing and sales.

Language

e-Hurudza/Umtshayi Wenala being a locally developed product, addresses the farmer in Zimbabwe's three main languages, English, Shona and Ndebele.

Characteristics of the Electronic Farm Manager/ e-Hurudza/ Umtshayi Wenala

- There is knowledge base and empowerment tool where researchers can also contribute in providing their model.
- Farmers can be advised through Alarms that serve to alert and warn them on the stage reached and possible loses if they exceed planned targets.
- e-Hurudza / Umtshayi Wenala promotes Inter-dependence by allowing farmers to access models done by other experienced farmers on growing crops and livestock.
- Planning is made easy by the budget tool in this package where input prices are captured and continuously updated to enable the farmer make informed decision when preparing a budget.

- Salary bill and deductions are done automatically by the system when daily labour activities are captured. Report are produced each week, per stage and per project.
- Input funds will be directed to Input suppliers without going through farmers hands hence the elimination of misappropriating agricultural funds.

Knowledge Base and Empowerment

- The first step the inexperienced farmer has to take is to inquire and understand how to grow commercial crops, their varieties, disease control, chemical application *etc.* All this information is available on e-Hurudza/Umtshayi Wenala.
- e-Hurudza/Umtshayi Wenala provides basic and intensive instruction on how and where to grow/rear a particular crop/livestock.
- It advises the farmer the most suitable crop for the region he/she is situated in according to the soil type and rainfall pattern.
- Colleges and universities use the software throughout their training period and make their models and findings known or available through the Internet. On leaving college, graduates will be fully equipped to manage the farms at world-class standards.

Alarms

- The alarm feature in e-Hurudza/Umtshayi Wenala warns the farmer on action to be taken at each stage of crop/livestock life cycle.
- In crop production the farmer is alerted when it is time to plant, apply chemical or fertilizer.
- In financial management the farmer is alerted when production costs begin to eat into the profit margin.
- In this module the amber light flashes when you begin to eat into profit margin. The amber gives way to the red light when it has become unprofitable to continue with the project, which should force the farmer to review the whole project or revise his management plan.
- This inhibits the farmer from undertaking unbudgeted activities and over spending.

Table 5 : Level of access to computers and Internet in Zimbabwe Universities

University name	Africa	Buse	U/Z	Midlands	Masvingo	Chinhoyi	NUST	ZOU	Catholic	Women's university
Total No. of computers	300	300	1500	250	100	250	800	90	30	18
No. of network points	1000	350	3000	500	14	200	4000	151	50	22
No. of users	1600	1000	10000	6200	800	1560	3000	150	100	30
Bandwidth	128 Kbps	64 Kbps	1,5 Mbps	128 Kbps	64 Kbps	128 Kbps	1Mbps	64Kbps56 Kbms	64Kbps	64Kbps
Required Bandwidth	2Mbps	2Mbps	4Mbps	2Mbps	2Mbps	M2bps	2Mbps	2Mbps	128Kbps	256Kbps
Type of Link	Leased line	Leased line	Leased line	Leased line	Dial up line	Leased line	Radio link &dial up	Radio link &dial up	Leased line	Dial up
Internet service provider	Com one	Com one	Com one	Com one	ZARNet	Com one	Com one	Mweb&Ecoweb	ZAR Net	Com one
Quality of ISP	poor	very poor	poor	poor	poor	Fairly good	poor	poor	Fairly good	Very good

Source: Zimbabwe e-Readiness Report, 2005

Inter-dependence

Zimbabweans by nature have a tradition of sharing and e-Hurudza/ Umtshayi Wenala promotes this by allowing farmers access to best practices and models provided by the relevant seed companies, Agricultural Colleges and Agricultural Technical and Extension Institutions. When a farmer accesses these models, and chooses one similar to his plans on the farm, the system will prepare a budget for him and he can make any adjustment as he sees fit.

e-Hurudza/Umtshayi Wenala has a function, which provides a link between the main stakeholders, who depend on each other to ensure a bumper harvest-stakeholders such as, seed, fertilizer and agricultural chemical suppliers, financial institutions, Government, farmers' unions *etc.*, for price updates, loans, stock availability and marketing through the Internet, websites and e-mail.

Harvesting and Marketing

- When the crops and livestock are ready e-Hurudza/ Umtshayi Wenala takes the project to harvest and storage stage.
- The system will lead the farm manager to the sales section, which will record all the transactions done on the project before producing profit and loss together with the balance sheet for this project.
- This therefore means that both government and unions will be able to market the produce to potential buyers internally and externally even before the harvest since the system can give them actual quantities to expect in the nation's basket.

Financial Attributes (Budget tool)

- Input prices are captured and continuously updated to enable the farmer make informed decisions when preparing a budget.
- e-Hurudza/ Umtshayi Wenala allows the farmer to budget all inputs in a categorized manner.
- When approaching financial institutions that give loans, the farmer simply prints out the detailed cash flow projection and presents it to the lender. However, it would be expedient for the bank to have the software so that they can also keep an eye on how the farmer is making use of the borrowed money. If the farmer should approach the bank for a soft loan during the crop life, they can be in a position to view the desirability of giving the loan first hand.
- e-Hurudza/ Umtshayi Wenala captures models done by experienced farmers and when accessed, the system prepares a budget for the farmer saving him the agony of entering all the required inputs - he can work around the model to add his/her personal touch.

Wage Bill

- e-Hurudza/ Umtshayi Wenala also accounts for each man-hour on the farm and prints management reports as required, per stage, per project at least at the closing of each month.
- Salary bill (Tax, NSSA and other deductions) and pay-slips are also generated by the system.

Security

e-Hurudza/ Umtshayi Wenala does have tight locks. There are many hands on the farm and all can access the system as long as they have the right username and password. The Supervisor, clerk and the store men can have access to different sections of the system. Who gets the key to what is left entirely to the farmer's discretion. For instance, the payroll can be accessed by the farm owner only or/and his manager. Meanwhile, the clerk can capture data on all other activities freely without accessing the pay-roll arena on the same computer.

The system records/logs activities against a name, time and date. The owner at any stage can, find out through the system, when and how long the clerk has worked and which sections of the system he has roamed. Since the farmer holds the master key (password) he can monitor activities on his farm from another town or country as long as he has access to his computer *via* the Internet.

Added advantages

To the farmer:

- e-Hurudza/Umtshayi Wenala encourages those aspiring to go into farming as it enables them to see profitability or otherwise of their intended venture before embarking.
- The ability to market the produce before harvest.
- The importance of planning for crop rotation.
- Preparation of projected cash flows and profit & loss accounts for submission to financial institutions where funding is required.

To Government:

- The ability to know gross produce at national level, which will trigger alarms to import or not.
- The ability to advise farmers on national requirement for national consumption as well as international requirements for foreign currency earnings.

4.1.2. Taguta Software

This locally produced Zimbabwean software product, Taguta, enables farmers to create cropping scenarios from a total of 240 possible varieties and computes comprehensive five-year financial forecasts corresponding to a variety of user defined scenarios.

4.1.3. Virtual Reality (The Rural Life Skills Development Project)

In Africa, lack of resources and often, poor literacy pose a huge challenge to learning - both at school as well as in adult-based learning. This renders some interventions ineffective due to inaccessibility of appropriate content. In the case of Zimbabwe, against a background of economic downturn and poor rainfall in most parts of the country and at a time when land was being given to more and more people new to farming practices, this project presented itself as a timely and appropriate intervention.

Field surveying for water conservation–theory to practice follows theory

The project involves two primary partners, World Links Zimbabwe and Naledi3d Factory, who through funding from the W.K. Kellogg Foundation set to address the following goals:

Help rural communities better understand and, therefore, be better empowered to address local issues that impact on rural development and in particular, focusing on agricultural and other life-skills development

- Demonstrate the use of VR-based learning content and especially the use of Interactive3d Learning Objects (i3dlo's) as a new, innovative visually interactive communication / learning medium
- Implement agricultural capacity building through focused community-based training of individuals
- Facilitate the transfer of skills to Zimbabweans to be able to "localise"*Interactive3d learning objects* before they are taken out *"into the field"*.

The project involved creating a range of Virtual Reality based i3dlo's that would help to address the above aims and especially in the field of agricultural and life skills development in rural communities. This was done largely from drawing out of the Swedish Cooperative Centre's (SCC) self study material, developed for the Small Holder Drought Mitigation Programme (SDMP) in rural communities, for example:

- Water and Soil Conservation with drought in mind
- Root Crops production
- Principles of Vegetable Production
- Conservation Farming with drought in mind

Virtual Reality in Africa

According to Dave Lockwood, the leading and driving force at the Naledi3d Factory, a private company based at the Innovation Hub, Pretoria, South Africa, the "interactive3d learning object" (i3dlo) concept began to emerge about five years ago, as a marriage between interactive 3d simulations and the educational concept of the learning object. Conceptually, i3dlos are efficient 3D knowledge modules that can be used in different ways, adapted to suit local needs and even easily translated into local languages. Such VR file localization can be a relatively simple process using freely available web based tools.

The reason why VR is such a powerful tool for communicating information in the learning process is that it fits very well with how the brain works as sited in the World Links Zimbabwe's Team Talk Newsletter (September 2007) article addressing the question, "What is VR"? The article explains that the area of the human cerebral cortex devoted to the processing of the visual inputs is in order of magnitude greater than that devoted to text and speech input. We dream visually. We read visually. Therefore by using virtual reality to show concepts and knowledge in a visual and effective way, we can dramatically enhance the level of understanding of any subject.

Due to its visual nature, Virtual Reality (VR) can overcome the language as well as literacy barriers, as **the learning is shown -not told**; which has always been a challenge in traditional, descriptive text-based extension education. The ability to transfer knowledge and skills to those with lower literacy skills makes VR a very exciting, new communication medium in the African context - one that can lead to social development, reduce poverty at the personal level and help to build economic growth at the national level.

Having seen how VR was being employed in a project on Hygiene in Uganda and also based on demonstrations by Naledi3d Factory, World Links Zimbabwe, and in particular its Director, Eliada Gudza, also began to share this vision and in 2003 the two organisations started to explore ways to work together to take this approach down to grass-roots level. It was with the support of the W.K. Kellogg Foundation (firstly in 2005, through a small intervention aimed at bee-keepers) and then later, through a two year project that this collaboration was able to develop to the practical level.

The above goals were addressed first through the establishment of technically enabled VR Centres within four out of the fifty school and community based World Links Community Telecentre network. These centres were equipped with equipment to facilitate the use of i3dlo's by both students and members of the local community.

VR Centre Equipment

- Ten Pentium III Refurbished Computers.
- Two Pentium IV Dell PCs (New)
- One Laptop
- One Digital Camera
- One Laser Printer
- One Data projector

Each VR centre had a deliberate blend of old and new computers, a LAN and a shared printer. The laptop and data projector were also provided for use within the centre and for local outreach programmes along with a digital camera for capturing local events and local features thereby generating images that would help inform future i3dlo content development activities.

Community awareness meetings were held in order to raise awareness, plan and to set up a VR Committee to coordinate activities at each centre. These meetings involved discussions with the school administration, community leaders, local farmers and youth on the agrarian problems that existed in the community and ascertain whether introducing the VR Project and its approach would be welcome and beneficial to their community.

Blending Technology, Modern and Traditional Ideas

The VR Project was premised on the paradigm that for Zimbabwean farmers to succeed in solving their agricultural problems in their communities they needed to re-adopt the ways of problem-solving used by their fore-fathers such as the *African Culture, the Padare* concept (evening gatherings, often for men, where people say their problems and fellow countrymen suggest solutions), Ubuntu (an African common value system based on the paradigm *I am because we are)* but at the same time adopting modern technology and new forms of thought such as *Spiral Dynamics* (a model that describes and makes enormous complexity of human existence). The use of VR therefore made farmers take ownership of issues of Agricultural development in their communities by using technology that helped them to see how best to use local inputs and principles to solve farming problems. The project brought together adults, youths, men and women of different ages together sharing knowledge and skills in ways commonly not possible using traditional means of learning.

Training Workshops

These were typically run over two days and focused on "training the trainer" in addition to members of the local community.

By way of example of the workshop process, a workshop at Mandedza VR Centre (March 2008) was attended by 23 people (farmers, teachers, AGRITEX officers, school students and administrators and local chiefs). After going through the two plough-related i3dlo's (amongst others), most farmers realized that they were using their own plough (which typically were not supplied with manuals) incorrectly, even removing some parts to make it lighter, not understanding that these parts were not only important, but used for example to adjust plough depth. These two i3dlo's generated some of the richest discussions and showed what could be achieved using i3dlo's, compared to using the real object.

Another example was the example of an Agriculture teacher and local farmer at the Rusike workshop who had read and tried to make the A-Frame but could not complete the calibration following text-based instructions but upon using the i3dlo, he was able to complete the tool with much ease.

Some of the adult participants found so much fulfillment and boost in self-confidence from just having the first experience of using a computer to being able to manipulate the mouse unassisted. Its like their ability and desire to learn, not just within the context of the workshop but in general, was awakened.

By the end of 2008, five learning objects had been translated by World Links Zimbabwe into the Shona language and two of these into the Ndebele language. The combination of the translation process (itself a significant technology transfer component) with community-based implementation is what we refer to as "localisation".

Developed i3dlo's in Summary

Water Conservation – In-field water conservation

The main drawback of contour ridges lies in the fact that, during seasons with lower rainfall, only crops close to the actual contour ridge get enough water whilst crops further in-field suffer. This can be addressed *via* a number of simple yet effective techniques. This i3dlo shows some of these techniques to capture more run-off water during the rainy season, as well as to irrigate at low cost during the dry season. The main outcomes are:

- Understanding the importance of capturing rain water during the rainy season
- Understanding how to irrigate during the dry season
- The importance of pot-holing
- How half-moon ridges can be effective
- Buried bottle watering
- The making and use of clay pipes in vegetable gardens

Note, there are other techniques that can also be applied, including tied ridges, tied furrows; mulching, strip cropping and mixed cropping (not addressed).

Water Conservation – Harvesting water from rooftops

This *i3dlo* shows how to set up a system to harvest water from roofs, which can provide a useful supplementary water source for schools, farms, homes and businesses. The main outcomes are:

- The main components of a rooftop water harvesting system
- Planning for water yield and the impact on storage size
- How to build a water storage tank (using the example of a 20,000 litre tank)
- An example of a more advanced storage system, water pumped to a roof-top storage system.

Soil conservation: Soil erosion – raindrop, sheet and rills

This *i3dlo* was designed to help farmers recognise whether an erosion problem is emerging on their land. It gives a basic understanding of the mechanistic processes involved in the different types of erosion that are commonly experienced. The main outcomes are:

- How to recognize water problems in our fields
- Types of erosion - raindrop, sheet and rill erosion
- The importance of soil cover – mulching
- Impact of slope length, steepness and soil type

Soil conservation - Gullies - formation and degeneration

This small i3dlo shows the user how gullies are formed and how to stop them from spreading further and causing more destruction (see gully control

- stone checks). Understanding how gullies are formed can help the farmer understand the damage they can cause, as well as the importance of taking steps to stop them forming. Once gullies are allowed to form, they are difficult to repair. At best, measures can be taken to stop them from growing further. If they are allowed to continue growing in size, they can cause huge damage and significantly impact on the amount of productive land available to the farmer - leaving the farm land unfit for farming - impacting on the revenue generating potential of the farming unit. The main outcomes are:

- How gullies can develop from small rills
- How gullies permanently impact on productive farming land
- The importance of preventing the development of gullies and
- A trigger for discussion around these issues and how they can be avoided

Soil conservation - Gully control using stone checks

This i3dlo shows the user how to repair gullies and reduce the speed of water flow in the gully by building stone checks. In this way, future growth from erosion is controlled and the damage impact minimized. It is aimed primarily at helping farmers to understand how to build stone checks as a way to stop gullies from further degenerating. It doesn't address other issues, such as the erosion process, or how gullies grow. Note that the vertical scale of the gully shown has been exaggerated for effect. The main outcomes are:

- Understanding how stone checks can help mitigate the further growth of gullies
- How to build a stone check for gully control
- Gully control steps.

Soil conservation - Pegging out contour lines – using an A-frame

This i3dlo shows one simple way to peg out a field using the A-frame - and to accurately define the contours across the field. This i3dlo is aimed therefore at helping farmers to understand the importance of ploughing along contours, as well as how to develop a clear understanding of lines of level contour by surveying the field using an A-frame and by pegging (marking) out the field accordingly. The main outcomes are:

- Why a field should be pegged
- Things to think about before starting to peg
- Making an A-frame
- Pegging out using the A-Frame

Knowing how the contours run across your field is important when ploughing in order to minimise, or control water run-off. Ploughing along the contour (as opposed to down the slope) helps to ensure that as much water as possible remains in the field and soaks into the soil. If water runs down the slope, it will wash away topsoil and nutrients, to the detriment of the field environment and the farmers' ability to generate income.

Soil conservation - Pegging out contour lines – using a Line Level

This i3dlo is aimed primarily at helping farmers to understand the importance of ploughing along contours and how to achieve that aim; as well as surveying the field using a Line Level and by pegging (marking) out the field accordingly.

- Why a field should be pegged
- Things to think about before starting to peg
- How to make a Line Level

- Pegging out using the Line Level
- Understanding how to identify and peg out level contours, and to identify slopes of 1:250 (for Contour Ridges) and 1:100 (Tied Ridges).

Mechanisation - Animal-drawn ploughs – main parts

The plough used in the demonstration is based on the **Mielie Brand** range of ploughs, manufactured by Zimplow Ltd (a Zimbabwean company, one of the largest manufacturers of animal drawn implements in Africa). The setting for the i3dlo is an open field where the single-furrow plough is demonstrated. The plough, ripper and high wing ridger are inter-changeable blades that can be attached to the same handle-beam assembly, making it logical to group these items together.

- Understand the main parts of a standard single-furrow plough and how the parts fit together
- Applicable life skill – how to get things done

Mechanisation - Animal-drawn ploughs –simple settings

Setting a plough is arguably one of the most important aspects of ploughing. If the plough is set incorrectly the operator and the draught animals struggle and may even be injured. Productivity suffers and furrows are not prepared correctly with serious consequences for crop production.

The three important adjustments demonstrated are (1) adjusting the depth of the furrow; (2) the width of the furrow and finally (3) the position of the wheel (and it is also important to emphasise that the wheel must never be used to adjust the height).

The setting for the i3dlo is an open field where the single-furrow (standard) plough is demonstrated.

- Understand the main settings that can be made on a standard single-furrow plough
- Understand how to adjust the plough-beam clearance and shape
- Furrow depth and width on a single-furrow plough

Crop control - Pests, management and control

Africa is host to a wide range of pests that attack and often destroy entire crops in a short space of time. It is vital, therefore, that farmers be provided with the knowledge to recognise the presence of these pests at an early stage so that they can take preventive measures quickly before the damage reaches disastrous levels.

This i3dlo covers major crops and vegetables, *i.e.* maize, sorghum, onions, tomatoes, cabbages, rapes, kales and sweet potatoes. The pest (aphids, bagradra bugs, thrips, armoured crickets. birds, cut- worm, maize beetle, stalk borer, termites, red spider mite, diamond-back moth, sweet potato weevil and rats) is shown together with the damage that it inflicts. The user is then given the opportunity to apply a treatment to eliminate the pest. These treatments are often chemical in nature but an attempt has also been made to, where possible, recommend non-chemical treatments such as inter-planting, crop rotation, *etc.* These non-chemical treatments are also cheaper and also more environmentally friendly.

Cropping - Sorghum - planting for success

As with any crop, sorghum crop yield is dependent on proper planting and growing practice. This i3dlo focuses on planting aspects of sorghum crops, highlighting the typical stages of growth - from seedling to germination, maturity and the harvesting stages. Land preparation involves making furrows 1m apart, adding fertilizer, and planting the seeds in the soil. All of these aspects are shown and the learner is introduced to typical practices. However, local variations in practice are invited to be raised in facilitator-led discussions.

The key areas addressed include:

- Planting (furrowing, fertilizer, seeds per hole and the physical planting)
- Germination stage
- Intermediate stage
- Maturity stage
- Harvesting stage

Cropping - Sorghum - origins and diversity

Sorghum originated in Africa where it was domesticated around 3000 BC. The use of the crop spread from Africa to India and China, and much later to the US through the slave trade. This distribution process is shown in this i3dlo and the learner is introduced to the 5 main races (group of species) found in Africa, namely Caffrorum, bicolor, caudatum, durra and guinea. Races of sorghum were developed by the adaptation of cultivars to address the following issues:

- seedling emergence and early root development
- good tilling
- longer growing cycle to make the best use of infertile soils
- resistance to insects and molds
- tolerance to bird pests and striga

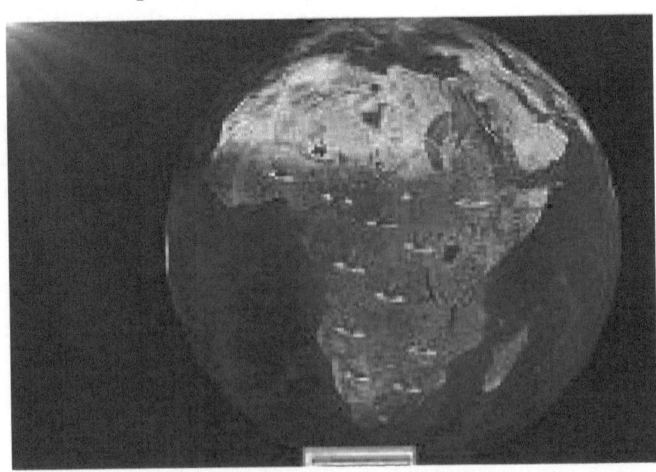

The typical distinguishing features between races include growth habit; panicle characteristics; seed shape and color and adaptation.

Cropping - Sorghum - plant characteristics

It is important for the user to recognize a sorghum plant and in the process also become aware when sorghum plants in the field are experiencing growth or other problems (in most cases, the plant characteristics will be visible affected). This i3dlo looks at the following:

- Seeds
- Root structure
- Height

These characteristics are conveyed in this i3dlo through interactive sequences where the learner has to match the correct seeds to the correct plant (using a number of main crops available in the world today - maize, rice and wheat - as comparisons). The learner has to click on the crop that relates back to the root structure shown and observe the difference in height.

Cropping - Sorghum as a food source

This i3dlo looks at the many applications of sorghum as a food source. The user is introduced to a typical kitchen environment where a number of different foods made from sorghum have to be identified.

These include the following foods and beverages:

- Porridge
- Pancake
- Biscuits

- Bread
- Cake
- Popcorn
- Beer

Cropping – size of land

The "size of land" i3dlo introduces the learner to concepts around the amount of land required to be a viable and economically sustainable sorghum farmer. This i3dlo aims to address the minimum amount of land required to:

- Make a profit
- Support a family of eight (break even) and
- Maximise profit - looking at cost per ton and associated income streams.

This is achieved by allowing the learner to identify plots in each of the following land types (category) from the i3dlo and discuss/ think through the impact of farming with sorghum on those particular plots:

- On a incline
- On a hilltop
- On a flat surface
- Flat surface with erosion
- Next to a river
- Next to a river with erosion

Cropping - yield

In any farming environment, it is important for the farmer to be in a position to understand the optimum conditions required to achieve optimum yields. In this i3dlo we look at four of the main conditions that need to be considered when striving to achieve the best output/ yield for sorghum. These are:

- Rainfall
- Climate
- Soil
- Fertilizer

The learner is introduced to a typical farming environment and by using sliders, the i3dlo shows what yield can be expected relative to the magnitude and type of condition that has been changed in the simulation.

In all, sixteen (16) i3dlo's were developed (others still undergoing localisation) as inputs to the training workshops. The completed ones were also made publicly and freely available *via* the Naledi3d Factory's i3dlo website (in English and where available in Shona). It is hoped that this will lead to the creation of a dedicated web based repository of i3dlo's covering a wide range of topics.

The Localisation Process

This is perhaps a feature that lies at the heart of the design and implementation of i3dlos. The first process chart below describes the overall technical process. The second describes stages involved in the translation, recording and editing of the text and audio files and their incorporation back into the i3dlo. It is this that makes VR and in particular i3dlos so attractive and a versatile tool that promises to become an essential and necessary feature of most learning and training programmes. This project gives us pride as a team of Africans to be using this cutting edge technology in what would be commonly called poor and under-resourced communities.

Farmers' Contribution

Selected farming communities are expected to provide infrastructure, security and sound leadership for the centre. Participant farmers also provide regular feedback that can be used to develop new content or perfect existing ones. More importantly, farmers are expected to make practical use of skills learnt in solving their farming problems.

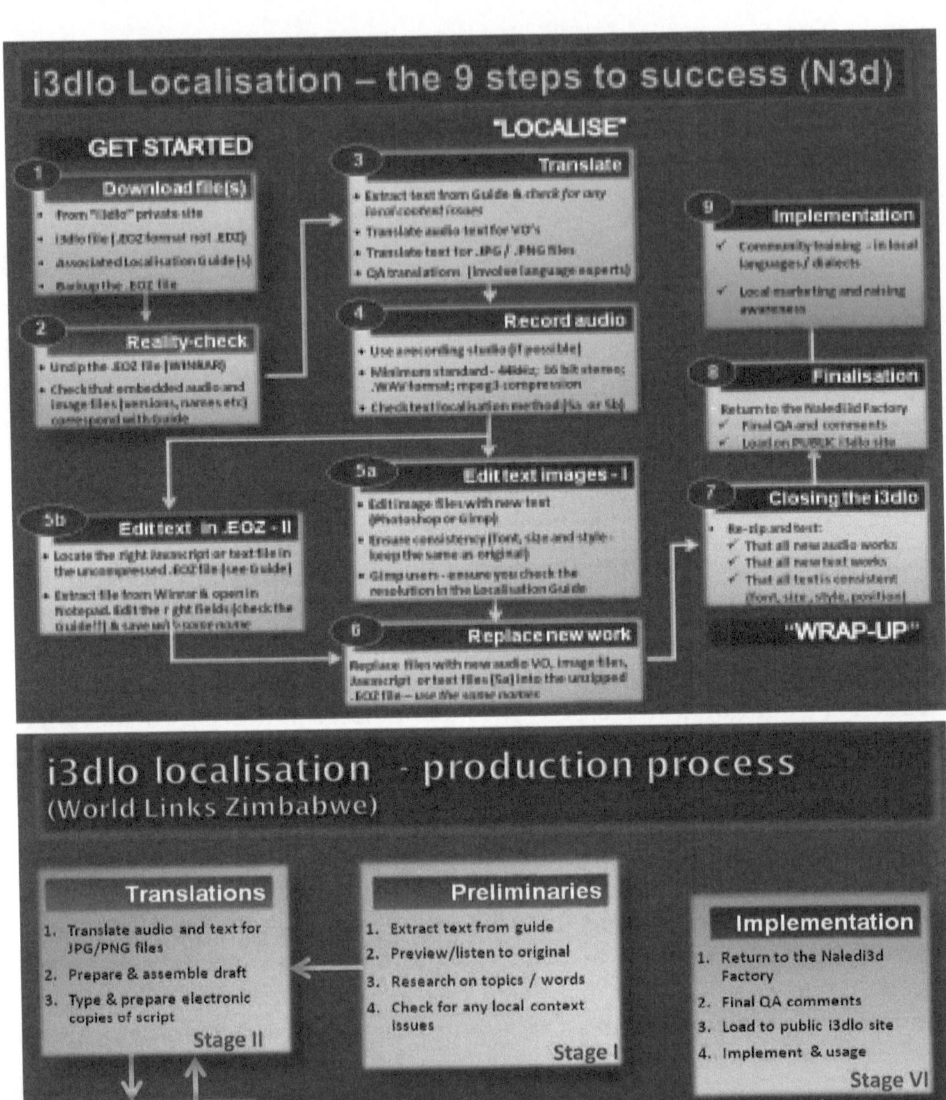
i3dlo Localisation – the 9 steps to success (N3d)
GET STARTED
1 Download file(s)
2 Reality-check
"LOCALISE"
3 Translate
Extract text from Guide & check for any local context issues
Translate audio text for VO's
Translate text for .JPG / .PNG files
QA translations (involve language experts)
4 Record audio
Use a recording studio (if possible)
5a Edit text images - I
Edit image files with new text (Photoshop or Gimp)
Ensure consistency (font, size and style - keep the same as original)
Gimp users - ensure you check the resolution in the Localisation Guide
5b Edit text in .EOZ - II
6 Replace new work
Replace files with new audio VO, image files, Javascript or text files (5a) into the unzipped .EOZ file – use the same names
7 Closing the i3dlo
Re-zip and test:
That all new audio works
That all new text works
That all text is consistent (font, size, style, position)
8 Finalisation
Return to the Naledi3d Factory
Final QA and comments
9 Implementation
Community training - in local languages / dialects
Local marketing and raising awareness
"WRAP-UP"
i3dlo localisation - production process
(World Links Zimbabwe)
Preliminaries
1. Extract text from guide
2. Preview/listen to original
3. Research on topics / words
4. Check for any local context issues
Stage I
Translations
1. Translate audio and text for JPG/PNG files
2. Prepare & assemble draft
3. Type & prepare electronic copies of script
Stage II
Checkpoint 1
WLZ Review translation
Recording audio
1. Studio rehearsal
2. Studio recording
3. Edit recording
4. Check text for localisation method
Stage III
Checkpoint 2
WLZ Review audio
Editing text images
1. Check image files with new text
2. Ensure consistency of font size and style
3. NB. Keep same as original
Stage IV
Test / check
1. That all new audio works
2. That all new text works
3. That text is consistent (font, size, style position)
Stage V
Implementation
1. Return to the Naledi3d Factory
2. Final QA comments
3. Load to public i3dlo site
4. Implement & usage
Stage VI

4.1.4. TV and Radio "Talking Farming" Programme

Radio and TV play a crucial role in disseminating information on agriculture in Zimbabwe. Enjoying greater coverage than TV, radio broadcasts have been produced and transmitted over a longer period than TV programmes. The main aim of broadcasts is to assist farmers by providing them with knowledge on farming and agriculture.

Worth mentioning in the initial stage is a special programme which specifically targeted farmers in the Mt.Darwin Chimhanda farming scheme. Broadcasts started in 1983 and lasted for a decade. When the Educational Channel, Radio 4 was established, agricultural programmes were recorded at the Audio Visual Services studios in Mt. Pleasant Harare and transmitted on Radio 4. When production and transmission of educational programmes for schools ceased in 2003, Agritex officers in Harare and Bulawayo continued with recording and production using Zimbabwe Broadcasting Co-opeation facilities. Currently Shona and Ndebele programmes are produced and transmitted on National FM. The programme "Kurima Nekuchengetedza Zvipfuyo" (farming and looking after livestock) covers a wide spectrum of topics to assist farmers.

Unlike TV, radio has maintained its lead as a medium of communication providing information, education and entertainment. It is cheaper and more affordable. With the advent of the dramatic land redistribution exercise in 2000, television, which now complements radio, became a regular avenue for dissemination of information on agriculture.

"Talking Farming/Murimi Wanhasi/Umlimi Wanamuhla" is a radio weekly and TV programme targeted at educating farmers on good agricultural practices. It focuses on various sectors of agriculture such as livestock production, crop production, horticulture, post harvest technologies and soil fertility Management and irrigation. The programme is presented twice a week in local languages. Agricultural experts start by presenting topical issues related to the agricultural topic under discussion. The listeners are invited to phone and contribute to the discussion as well as present farming problems. Expert agricultural panelists provide solutions *in situ* or refer farmers to their local extension advisers. Funding of the programme is done by the Government through the Ministry of Agriculture.

4.1.5. Telecentres

"A telecentre is a public place where people can access computers, the internet and other technologies that help them gather information and communicate with others at the same time as they develop digital skills. While each telecentre is different, the common focus is to support community and social development – reducing isolation, bridging the digital divide, promoting health issues, creating economic opportunities, reaching out to youths and utilizing appropriate technologies. Telecentres exist in almost every country on the planet, although they sometimes go by different names (e.g. ….rural Information Centre, village knowledge centres, infocentres, community technology centres, community multimedia centres or school based telecentres…" (Source: http://en.wikipedia.org/wiki/Telecentre)

In Zimbabwe telecentres take a variety of forms, such as: school-based telecentres, community-based telecentres, government information centres, project/field based community centres, privately-owned internet café, privately-owned phone shops, mobile phone shops and mobile computer labs.

There are Agricultural Extension centre meant to serve the farming community in Zimbabwe and are mainly located in towns and cities with a few pilot projects in communal areas. Most of the centres are run by NGOs who focus on particular farming systems such as Organic Agriculture. Telecentres provide telecommunication services such as telephone, fax, e-mail, typing and printing services, video based learning, library services with CDs for different crops and animals, mannuals, internet access, web based learning and telecollaborative projects.

While there are well equipped telecentres in Zimbabwe, only the well-off farmers who are computer literate benefit from such centres at the exclusion of the needy poor farmers in communal and small holder farming sectors. Horticulture and mainly cash crop farmers make use of such sites in order to get latest information particularly on markets. There is need to have computer training programmes in the telecentres targeted at small holder farmers.

4.1.6. Newsletters/ magazines

They are produced as and when relevant information is needed and on monthly basis respectively. Newsletter mainly provides facts on animal and crop production cycles. Monthly magazines proffer detailed feature articles on specific agricultural enterprises, chart columns for farmers and question and answer sections in the magazines. These magazines are mostly sold in book shops and distributed to subscribed members.

4.1.7. Research Journals

These are mostly university-based refereed academic journals containing latest Agricultural research articles. They can be accessed through various websites of universities. Information from such articles is mainly accessed by research centres, NGOs and specialized farmers who have access to internet. One of the biggest limitations in Zimbabwe is that there are no research journals written in local languages.

5. e-AGRICULTURE DISTRIBUTION IN ZIMBABWE

According to the Harvard University Guide (HUG) model, Zimbabwe's ICTs in Agriculture are on stage one because of poor access to telecommunications and a low tele-density of less than two people per one hundred people. Majority of small holders and communal farmers have no knowledge of what computers can do for them. The few who herd about computers only managed to do so through their school going children.

6. ICT RESEARCH IN AGRICULTURAL EXTENSION

Assessment of ICT in Agricultural Extension was based on the provision and penetration of infrastructure such as electricity, telephone and the availability of internet in the government departments serving the agricultural industry and the farmers.

6.1. Findings

Access to ICTs in the Agricultural Extension sector and general Agriculture industry is very limited and consequently the contribution of ICTs to rural and economic development has been marginal.

- Poor access to electricity and telephony infrastructure in rural areas are major contributors to the poor penetration of ICTs in this sector.
- The penetration of telephony, the main vehicle for transmission of voice and data in Zimbabwe is low (about 30 per cent) in this sector as is access to internet (less than 5 per cent).
- There is paucity of agricultural information with local relevance and almost complete absence of such information in local languages. This limits the application of internet to the daily lives of residents of rural farming areas and its relevance to their networking and with other communities.
- There is an increase in use of GPS in safari hunting for identifying locations of shoots.
- Local efforts to develop software for farm management and farm operations need nurturing and strengthening.
- The lack of specific and comprehensive Agricultural Extension policy framework on ICTs is a major constraint in development and use of these technologies in the sector.

- The Agricultural Extension sector has well developed sector which could spearhead the penetration of ICT into the farming communities (Zimbabwe e-Readiness Report, 2005).

The limited use of ICTs in the Agricultural Extension sector is due to poor access to supportive infrastructure. The situation is aggravated by the high cost of ICT equipment and software and by limited ICT skills .The expression of interest by farmers to make use of e-Hurudza and GPS demonstrates that farmers can adopt ICT quickly if its results are evident and is affordable and easy to use.

It is in view of the above factors that the Agricultural Extension sector was assessed as being on stage one according to the HUG Model. The model is internationally recognized and mostly suitable for developing countries. The model uses a four stage scale, 1=low rate of readiness and 4=ideal level. Five attributes, namely, network access, networked learning, networked society, networked economy and network policy were looked at.

7. THE FUTURE OF ICTs IN ZIMBABWE'S AGRICULTURAL EXTENSION SECTOR

For sustainable ICTs application in the sector, there is need for:

- Complete electrification of large and small scale farming areas.
- Promotion and development of ICT skills and appreciation of the internet as important additions to the skills of all farmers.
- Development and implementation of a nationwide network and connectivity between all provinces and districts. The network will facilitate national communications and data transfer between farmers.
- Developing and implementation of policies to encourage the use and investments in ICTs in agricultural development, production and marketing.
- Promoting development of local languages and distributing information on the internet which is relevant to the farming community.
- Encouraging use of local software suitable for farming communities.
- Setting up ICT telecentres in support of the agricultural reforms in Zimbabwe.
- Ensuring that ICT services are accessible to women who make the majority of farmers.

- ICTs application in the Agriculture Extension sector should take a cross-sectorial approach in order to maximize resource utilization.

References

AGRITEX, 1994. The Master Farmer Training Scheme, Ministry of Agriculture, Harare, Zimbabwe.

CSO, 2006. *Statistical Bulletin*, Central Statistical Office, Government Printers, Harare, Zimbabwe.

GoZ, 2002. An overview of the Zimbabwean Economy, *Reserve Bank of Zimbabwe Publications*, Government Publication, Harare Zimbabwe.

Gudza, E. 2008. World Links Zimbabwe, A Programme Overview: ICTs in Agriculture Extension Workshop, Bindura University of Science Education, 22nd June, 2007, Bindura, Zimbabwe.

Jawbone Enterprises, 2007. e-Hurudza, Harare, Zimbabwe

Pazvakavambwa, S, 1994, Agricultural Extension. In: Rukuni, R. and Eicher, C.K. (Eds). *Zimbabwe Agricultural Revolution*, University of Zimbabwe Publications Office, Harare, Zimbabwe .

SARUA, 2007. Southern African Regional Universities Assocation.www.sarua.org.

Tawonezvi, H.P.R., 1995. A study of Agricultural Extension Services in Zimbabwe, Agricultural services institutions and smallholder development workshop, Vumba, 27th-28th March 1995, University of Zimbabwe Rockfeller Foundation Land Tenure Studies project.

Zimbabwe e-Readiness Report, 2005. *Zimbabwe e-Readiness Report*, Harare, Zimbabwe, www.ict.org.zw

World Bank, 1995. *Agriculture Technology Notes.* No.11, The World Bank, Washington DC.

Authors Information

Mr. Christopher Tafara Gadzirayi

Senior Lecturer
Agriculture Education and Extension
Bindura University of Science Education
P.Bag 1020 Bindura, ZIMBABWE
Postal Code: 99999
Phone No. (O): +263-4-446049 ; Business+263-71-7531/4
Mobile Phone No.: +263-23-814-924
e-mail: gadzirayichris@yahoo.co.uk, cgadzirayi@buse.ac.zw

Mr. Gadzirayi holds an M.Ed in Agriculture Education and Extension from the University of Zimbabwe, has 25 years work experience in the field of Agriculture Extension Education. He worked at Belvedere Technical Teachers' college for 15 years as an Agriculture lecturer, facilitates sustainable organic farming community projects, is a board member to several organisations practising organic farming and is currently a lecturer in the department of Agriculture Education and Extension at Bindura Unversity of Science Education. He has published more than 25 international articles on indigenous knowledge systems, Extension Education and Organic Farming.

Mr. Eliada Gudza

Executive Director
Southern Africa Regional Office
World Links Zimbabwe
10 Windermere Lodge, Mazowe Street, Fife Avenue, Harare
Pox CY 2968 Cuseway Harare, ZIMBABWE
Postal Code:99999
Phone No. (O):+263-4-253924
(R): +263-4-2935377
Mobile Phone No.: +263-912-239534
Fax No.: +263-4-253924
e-mail: eliada@world-links.org, eligee@gmail.com
URL: www.world-links.org.zw

Mr. Eliada Gudza, a mathematics teacher by profession has been involved in the development of ICTs in education for nearly 30 years. He holds Bachelor's Degree in Mathematics Education from Nottingham University (UK). Having worked in the Ministry of Education for 18 years as teacher and curriculum developer between 1981-98, and served in the Secretariat of the Presidential Commission of Inquiry into Education and Training, he joined the then World Links for Development Programme in 1999 as Education Coordinator and rose through the ranks to be the World Links Regional Coordinator for Southern Africa and Eritrea. Now Executive Director of World inks Zimbabwe, an organization that he co-founded and commands 45 school and community based Telecentres countrywide, Eliada has been on many programs and ICT missions in Botswana, South Africa, Rwanda, Ethiopia, Eritrea and Nigeria. In 2002 Eliada was awarded the prestigious Geoff Fairall Award for his contribution in the development of ICTs access and usage by youths in Zimbabwe. He is also a Board member of the e-Knowledge for Women in Southern Africa (EKOWISA) and Tonga Online.

Mr. Godfrey Sibanda Title
Technical Manager
Research Council of Zimbabwe
Science and Technology Ministry
45 Colin Avenue, Letombo Park, Amby, Harare, ZIMBABWE
Postal Code: 99999
Phone No. (O): 263-4-794112
Mobile Phone No.: +263-11874729
e-mail: gsibanda@gmail.com

Mr. Sibanda holds an MBA and M.Sc. in renewable energy. He worked as a physics and maths teacher for eight years, and four years as a lecturer in Maths and Physics at Belvedere Technical Teachers' College. He was a subject manager with Zimbabwe Schools Examination Council for four years and is currently working as a Technical Manager with the Research Council of Zimbabwe.

Mr. Justin Mupinda
Country Programs Coordinator
World Links Zimbabwe
P.O. Box MP965, Mt Pleasant Harare, ZIMBABWE
Postal Code: 99999
Phone No. (O): +263-4-308286
Mobile Phone No.: +263-912-364027
Fax No.:+263-4-253924
e-mail: justin@world-links-sar.co.zw, worldtpo@africaonline.co.zw
URL: www.world-links.org.zw

Mr. Justin Mupinda is the Country Programs Coordinator for World Links Zimbabwe. A teacher by profession who studied Technical and Vocational Education, majoring in Computer Studies. Has been a facilitator and coordinator for online professional development courses for ICTs in Education. He has co-authored ICDL (International Computer Driving Licence) study guides developed for the Computer Society of Zimbabwe. He is a Committee member of IT institutions or initiatives in Zimbabwe such as the Civil Society Training Service Centre, the Zimbabwe FOSS.

❑❑❑